EPIGENETICS

History, Molecules, and Diseases

EPIGENETICS

History, Molecules, and Diseases

John M. Greally

New York Center for Rare Diseases
Children's Hospital at Montefiore
and
Departments of Genetics and Pediatrics
Albert Einstein College of Medicine

COLD SPRING HARBOR LABORATORY PRESS
Cold Spring Harbor, New York • www.cshlpress.org

Epigenetics: History, Molecules, and Diseases

Publisher	John Inglis
Project Manager	Barbara Acosta
Editorial Assistant	Danett Gil
Permissions Coordinator	Carol Brown
Production Editor	Kathleen Bubbeo
Production Manager	Denise Weiss

Cover description: Dr. Yuelin Song used a Reporter of Genome Methylation (RGM) system (Song et al., *Mol Cell* 75: 905 [2019]) targeting the mir290 locus on the *M. castaneus* (CAST) allele with a green fluorescent protein (GFP) reporter gene, and a red fluorescent protein (RFP) to the same locus in the 129 mouse strain. The images are from F_1 offspring mice, showing blastocysts that express only the CAST (green) or 129 (red) copy, or both (yellow), with the nonexpressed copy silenced by DNA methylation (*image courtesy of Dr. Yuelin Song*).

Chapter-opening icon: Figure title: 5-Methylcytosine. The 3D representations presented in this DOI were automatically generated by NIH's 3D workflows using data extracted from PubChem ID 65040 in PubChem. The image is reprinted from NIH 3D courtesy of a CC-BY license that is applied to the image in its entirety. (NIH 3D. (2017). *5-Methylcytosine* (Version 2.x). NIH 3D. https://doi.org/10.60705/3DPX/5106.2)

Library of Congress Cataloging-in-Publication Data

Names: Greally, John M. author
Title: Epigenetics : history, molecules and diseases / John M. Greally, New York Center for Rare
    Diseases, Albert Einstein College of Medicine and Montefiore Medical Center.
Description: Cold Spring Harbor, New York : Cold Spring Harbor Laboratory Press, [2025]
    | Includes bibliographical references and index. | Summary: "Epigenetics is the study of how
    cells in an organism develop different features without making changes to their DNA sequence
    and pass these on to their progeny. The term has come to imply different things to different
    people. This book reviews the history of the field since the word epigenetics was first coined
    and explains the biological mechanisms underlying the phenomena that have been observed"—
    Provided by publisher.
Identifiers: LCCN 2025022627 (print) | LCCN 2025022628 (ebook) |
    ISBN 9781621825432 hardcover | ISBN 9781621825449 epub
Subjects: LCSH: Epigenetics | Medical genetics
Classification: LCC QH450 .G74 2025 (print) | LCC QH450 (ebook)
LC record available at https://lccn.loc.gov/2025022627
LC ebook record available at https://lccn.loc.gov/2025022628

All World Wide Web addresses are accurate to the best of our knowledge at the time of printing.

For a complete catalog of all Cold Spring Harbor Laboratory Press publications, visit our website at www.cshlpress.org.

This work is dedicated to my parents John F. and Marie Greally. When your father is a Pathologist (and Ireland's first Immunologist) and your mother is a Clinical Geneticist, a book like this by their firstborn is probably inevitable. They have always inspired me to push the limits of scientific understanding while keeping a focus on caring for those with medical problems. Thanks, Dad and Mom. This book is my small way of sending a bit of love and recognition of your inspiring careers.

Contents

Acknowledgments

I would not have started this project if Richard Sever had not persuaded me against my better judgement to take on what ended up being an unexpectedly large amount of work. Thanks for pushing and supporting me, Richard.

I was able to immerse myself in epigenetics, my chosen field of research, because of the generosity of Dr. Ruth and Sandy Gottesman, who took a chance on me early in my career and supported my leading a new Center for Epigenomics. My sincere thanks to these extraordinary philanthropists and supporters of medical research and education.

My crisis of faith in epigenetics prompted me to take a sabbatical from Montefiore Einstein in the Bronx in 2014–2015. I greatly appreciate the support of my institution during this transition period of my career. Thanks to the New York Genome Center for allowing me to be part of their community during that time, Brad Bernstein for hosting me at the Broad Institute, and Edith Heard and Déborah Bourc'his for looking after me at the Institut Curie. To be able to immerse myself in the research environments at these different centers of excellence was inspiring.

I'm in debt to George Davey Smith, Ezra Susser, Brad Bernstein, and Wendy Bickmore for reading and critiquing some draft chapters. You were very generous with your time and expertise; I am fortunate to have you as my colleagues.

There were many people who gave me their direct input when I reached out with questions. I would like to honor, in particular, the memories of Art Riggs, David Allis, and Jim McGrath, who all passed away during my writing of the book, after massive contributions to science. I feel privileged to be able to recount their insights in this book.

Listing them in alphabetical order, I thank the numerous other colleagues who generously gave me their perspectives, guidance, and advice: Steve Baylin, Brad Bernstein, Adrian Bird, Carolyn Brown, Job Dekker, Melanie Ehrlich, Andy Feinberg, Susan Gerbi, Jamie Hackett, Matthew Hall, Oliver Hobert, Bill Holloman, Lily Huschtscha, Peter Jones, Karl Kelsey, Jim McGrath, Kevin Mitchell, Marilyn Monk, John Pugh, Ollie Rando, Jan Sapp, Kunio Shiota, Davor Solter, and Brian Strahl.

I genuinely appreciate how Dan Landau and Franco Izzo went out of their way to generate beautiful new plots of their data for this book, an extraordinary contribution.

I have had so many trainees and early-stage colleagues who have entrusted me with the development as scientists. I hope you see your influences in this book; the many conversations we had over the years were of great value to me as I thought through problems and ideas with all of you. My hope is that this book represents a starting point for you to bring better ideas to our field.

Finally, I thank my wife, Geraldine McGinty. I can't imagine doing any of this without you.

John M. Greally
New York, 2025

CHAPTER 1

Introduction

> ▷ **SEMANTIC**
>
> *Adjective* /sɪˈmæn.tɪk/
>
> Of, relating to, or arising from the meanings of words, the study of meaning

Semantic has become a term of dismissal. *That's just semantics..., I'm not going to enter into a semantic argument with you..., if you want to get semantic about it...*—you use the word today to suggest that the discussion is not worth your while. It's as if semantic is taking on the meaning of pedantic, the excessive, inappropriate, and annoying attention to minor details, in Finland described (impolitely) as *pilkunnussija* or (politely) one who has intimate intercourse with commas.

In telling the story of epigenetics, you must embrace semantics while keeping your inner pedant firmly under control. Today, this rich and fascinating avenue of scientific enquiry has not only captured imaginations, it has also captured multiple meanings. Two people discussing epigenetics may think they are talking about the same concept but have completely different and even opposing views of what it means. Someone seeking to construct a hypothesis based on epigenetics can find scraps of evidence from the multiple definitional buckets, building a structure that to the uncritical eye may look solid but lacks the compatible mortises and tenons for stability.

We can illustrate this house of epigenetic cards with a made-up example of a terrible study[1] idea justifiable by assembling the sprawling definitions of epigenetics.

---

[1] The fact that I had to search to make sure nobody is actually doing this study at present is worrisome by itself.

> ▶ HYPOTHESIS: SUNLIGHT CAUSES OBESITY IN GRANDCHILDREN THROUGH "EPIGENETICS"
>
> - Somebody's epidemiology study showed that today's grandparents were more sun-exposed as children, and today's grandchildren are more obese, so sunlight is a candidate for nongenetic (epigenetic) effects on the health of these kids.
> - We know that sunlight helps to create the active form of vitamin D in human skin, and vitamin D binds to transcription factors to regulate gene expression (epigenetic).
> - Part of a gene regulation response is mediated through changes in methyl groups added to DNA (epigenetic).
> - The vitamin D response to sunlight is therefore changing DNA methylation (epigenetic).
> - DNA methylation can be passed from parent to daughter cells and from generation to generation (epigenetic).
> - The sunlight response can therefore be passed on to the next couple of generations in a non-DNA-mediated way (epigenetic).
> - By eating copious amounts of kale or another currently fashionable superfood, you can reverse the DNA methylation changes and change your health, overcoming the epigenetic curse of your ancestors (epigenetic).

This logic, deliberately exaggerated, is uncomfortably close to that underpinning real studies being funded and published today. Each step of the logic contains a basis in fact but also a weakness. In the final chapter we will break down this series of statements based on the deep dives of the intervening chapters, so that the reader can come away equipped to look more critically at today's less well-supported research while retaining the excitement that should be prompted by our insights into epigenetics today.

This is why semantics must be an early and prominent part of a systematic exploration of the world of epigenetics. There are good reasons that there are multiple uses of this word—rooted in distant and recent history—involving some very human instincts and decisions. This book starts with an archaeological dig that reveals how we got to today's definitional ambiguity and then carries all the definitions through the chapters as we explore their implications for human diseases.

So What Exactly Is Epigenetics?

At present, the common understanding of what is meant by the word epigenetics is that it describes how the DNA sequence of the genome is used, a higher level of regulation that can be influenced by the environment of the organism. Having said that, some people extend the use of the word to mean long-term memory, including from parent to daughter cells after a cell divides. Extending this further, some include the inheritance of a memory of a past event from one generation to another in multicellular organisms. Those who study regulators of gene expression, chromatin states, nuclear architecture, and other biochemical processes influencing the genome often describe their work as the study of epigenetic regulators. Researchers seeking a mechanism for how the environment leads to changes in health see these biochemical genomic processes as strong candidates for mediating these effects. The use of studies to test for changes in these biochemical genomic processes in people with certain diseases or exposures has become a major area of research, the epigenome-wide association study.

The common theme is one of a molecular mechanism for disease that resembles that due to genetic influences but is not mediated by DNA sequence changes. This concept is such an intriguing possibility that it has resurrected the idea that a characteristic acquired by someone during their lifetime can be passed on to their offspring, the idea of Lamarckism. All of these definitions and ideas have entered into the broader discussion of epigenetics, which has led to confusion when multiple definitions collide but has also been a major reason for enthusiasm about the possibilities in this field of biology.

WHY THE EXCITEMENT ABOUT EPIGENETICS?

The idea that there is a layer of information beyond the DNA sequence itself is very attractive, especially for those trying to understand puzzling questions about why certain individuals develop phenotypes and others do not. For example, there are conditions that are highly heritable, indicating that they involve substantial contributions of DNA sequence–mediated susceptibility. However, even for these conditions, identical (monozygotic [MZ]) twins, who share (almost) exactly the same DNA sequences (apart from a small number of de novo and somatic variants), are not always both affected, but have higher rates of concordance than for nonidentical twins or siblings, who only share half their DNA. This higher rate of concordance has been used as evidence revealing the genetic influence on these phenotypes, but the fact that the MZ twins are not 100% concordant for the conditions has been taken to indicate that nongenetic factors are also involved—what has been called "missing heritability" (Kaprio 2012).

Nongenetic factors in disease susceptibility could include numerous potential influences, but mostly end up grouped into the very broad category described as the influences of the environment. For example, it is quite reasonably assumed that someone having excess weight involves not only a genetic predisposition but also the influence of the individual's diet (Walter et al. 2016). Apart from diet, the environmental exposures tested in these "Gene X Environment" interaction studies often include medications, toxins, psychological stress, metabolic factors, infections, and a range of other influences that are frequently studied using epidemiological approaches.

At the same time, there is interest in how genetic nondeterminism can be turned to one's advantage. If genetic susceptibility increases risk but does not uniformly lead to disease, is the opposite possible? Can some sort of intervention tilt someone's balance to make them relatively less likely to develop a condition? If there is an environmental influence that tilts risk toward disease, the logic is that there should exist a counterinfluence that decreases this risk.

A convergence of these interests has occurred in the study of epigenetics. If environmental exposures and stresses exert their effects on cells, there must exist a molecular mediator of these responses. There are multiple biochemical processes involved in expressing a specific set of genes in a defined cell type—if this property indicates a malleability of programming of the same DNA sequence in normal development, could it not also be malleable as a response to the environment and change the properties of a cell in a way that promotes or reduces risk of disease?

Epigenetics as a concept has stimulated interest for other reasons also. The property of malleability has also prompted the idea that therapeutic approaches involving new types of pharmacological agents or other interventions could reverse detrimental changes. The idea has been that, unlike gene therapy to reverse DNA sequence mutations, which is resisted by cellular processes that guard against DNA damage of any kind, the inherent reversibility of epigenetic regulatory mechanisms should make them easier to treat.

Furthermore, there is a definition of epigenetic properties that involves how a cell remembers past perturbations and exposures, retaining an "imprint" of these stresses at the molecular level in the absence of the original stimulus. The potential that epigenetic mechanisms could confer a memory of past events on a cell is possibly their most intriguing property, prompting studies that link molecular processes to events that occurred earlier in the life of an organism, during the period of its development, and even to events that occurred in prior generations that appear to have left residual phenotypic characteristics.

In short, the excitement about epigenetics is prompted by several quite distinct avenues of research, which in turn reflects the breadth of the processes currently encompassed by the word "epigenetics." This broad field of research includes many disciplines now worth our attention.

THE FOUNDATIONS OF MODERN EPIGENETICS

Although this book emphasizes the links between epigenetic processes and phenotypes—in particular, mammalian organisms and disease phenotypes—the roots of the field include many nonmammalian model organisms. In fact, probably the first striking example of what later came to be called an epigenetic process was from studies of the fungus gnat *Sciara coprophila*, which was found to have a mechanism for sex determination involving selective elimination of paternal X chromosomes from the cell nucleus (Crouse 1960). The fact that the embryo could distinguish between maternal and paternal chromosomes was assumed to involve an "imprint," or a molecular memory of the origin of a chromosome, based on the germline in which it had been most recently. Implied was that this could not be determined by DNA sequence differences between the chromosomes, as it was always the paternal chromosome that was lost whatever its genetic characteristics.

This observation combines two elements to which this book will return in more detail—epigenetics as a cellular memory and epigenetics as a molecular process. In the early days, epigenetics was more synonymous with cellular memory, but, perhaps inevitably with the explosion of biochemical and genomics assays and analytical methods, the research began to be focused on the molecular processes involved.

Technology has been a major driver of insights into the human and other genomes and has also driven insights into the molecular regulators of the expression of the genes embedded in these genomes. One early tool that proved extremely useful was the use of enzymes that selectively failed to cut DNA when the sequence at which it acted was modified. Normally we think of the DNA making up the genome as being composed of adenine, cytosine, guanine, and thymine (A, C, G, and T). What became apparent was that cytosines could be selectively modified, adding a small molecule consisting of a carbon and three hydrogens, a methyl group, creating what can be thought of as a fifth nucleotide, methylcytosine. When researchers studied DNA from living organisms, they were intrigued that methylcytosine looked like it varied in content between cell types, in response to environmental exposures, and with gene activity. A pivotal biochemical discovery was that the pattern of DNA methylation on a DNA molecule acted as a template to recreate the same pattern in daughter cells after their division. This finding linked the ideas of epigenetic memory and the biochemical mediators at the molecular level.

The ability to test the methylation of DNA was exciting but was soon accompanied by the realization that not all organisms methylate their genomes, raising the question whether DNA methylation is of major importance in the regulation of memory and gene expression. A more universal property of eukaryotes (from the Greek ευ true, κάρυον, nut, referring to the readily visualized nucleus of everything from yeast to mammals) is the packaging of DNA by histones to form chromatin. By chopping up the chromatin into small pieces and then isolating proteins or modifications of proteins using antibodies, we were able to enrich regions of DNA where these (modified) proteins were located. This chromatin immunoprecipitation (ChIP) technique added layers of insights into the regulation of genes and was applied to understand the memory involved in phenomena like imprinting.

Although initially it was only possible to explore one locus at a time in the genome for DNA methylation or chromatin properties, or sequences that were highly repetitive in the genome, the advent of DNA microarrays representing large numbers of loci and the later development of massively parallel sequencing technologies transformed the insights we could obtain. The ability to look broadly throughout the genome began to be referred to as the study of the "epigenome" and the technologies as "epigenomic." Examples of these technologies will be profiled more comprehensively in Chapter 6.

In addition to the biochemists, epidemiologists became deeply involved in the field of epigenetics. An influential early figure in the field was David Barker (1938–2013) from the University of Southampton in England. His insight was that adult disease risk may not be mediated solely by events during adulthood but may also be influenced by events occurring as early as during our prenatal development in the womb. This was the inspiration behind the field that came to be known as the developmental origins of health and disease (DOHaD). When considering a mechanism for how a temporally remote perturbation could manifest its effects later in life, attention focused on the same epigenetic mechanisms that mediate memories in normally developing cells instead mediating disturbed and disease-predisposing signals following a stress earlier in life.

Other epidemiological observations were also prompting consideration of epigenetic mechanisms. Two separate studies were being performed concurrently that were asking similar questions. One was a study of individuals in Överkalix in Sweden whose grandparents had experienced a limited food supply; the other examined the offspring of Dutch mothers who suffered famine during the winter of 1944–1945. In each case there were associated phenotypes—increased disease risk and mortality in the Swedish cohort and obesity in the Dutch cohort. Once again, in considering a likely mechanism for this presumably nongenetic heritability of these disease risks, epigenetic mechanisms became the focus of attention.

Events occurring that mediate human diseases tend to get the headlines in the popular press, but during this period the mammalian models were much less fruitful than the studies in other organisms. In plants, it was being shown conclusively that information could indeed be transmitted from parent to offspring through mechanisms not involving DNA sequence changes. The phenomenon of paramutation was especially exciting, in which the presence of a silenced allele somehow transmitted that information to an active homolog, which became silenced and maintained that state, even in the next generation of plants in which the silenced allele that started the process was no longer present. The field of epigenetics was therefore drawing together researchers from an unusually broad range of disciplines in the hope that there could be an exchange of insights in this emerging area of biology.

MAJOR PARADIGMS IN EPIGENETICS

If we were to ask ourselves to choose a few examples of replicable or especially intriguing findings that link epigenetic changes with mammalian phenotypes, there are a few that stand out. As will be discussed later in Chapter 7, there is a general problem with interpretability of the large number of epigenetic association studies that have been published to date. This significantly limits the number of studies that can confidently be believed to have the ability to stand the test of time. The four studies listed below have been influential in shaping our opinions about what epigenetics can offer practically when trying to understand processes involved in human disease. Although each will be discussed in more detail later, it is worth including them in overview here to help describe how epigenetics became such an intriguing field.

The Viable Yellow Mouse

The names of mouse strains are now standardized but in the past were descriptive of the trait being manifested by that strain. Of the mice with spontaneous mutations causing yellow fur color, there was a separate strain that did not survive embryogenesis when the mutant allele was present on both parental chromosomes and was described as the lethal yellow mouse strain. In contrast, the viable yellow mice survive even when homozygous for the mutation, prompting their descriptive name.

In the 1990s, the work of two researchers converged to create a compelling story that contributed to the foundation of the field of epigenetics. In Arkansas, George Wolff, at the National Center for Toxicological Research, had been working for decades on the viable yellow strain of mice. His work and these mice will be described later in Chapter 7, but they are worth introducing here. These mice were fascinating—littermates with exactly the same genetic state had a range of coat color phenotypes, from "pseudoagouti," the same brown color as animals lacking the mutation, to being covered entirely with yellow fur. This dissociated genotype from phenotype, a nice example for those seeking to understand nondeterminism of the genetic makeup of the organism. In 1965, Wolff also noted that the more yellow the mouse, the greater their tendency to gain weight (Wolff 1965). In Australia three decades later Emma Whitelaw (University of Sydney) cloned the mutation causing the viable yellow phenotype and found it to be due to an unexpected mechanism. Mammalian and other genomes contain virus-like sequences that are capable of replicating themselves and inserting a copy elsewhere in the genome, collectively called transposable elements. This is what had happened to cause the viable yellow mutation—an endogenous retrovirus type of transposable element called the intracisternal A-particle (IAP) had copied itself and landed upstream from the *nonagouti (a)* gene (Morgan et al. 1999). The function of the *nonagouti* gene includes the addition of dark pigment to hair. The reason that littermates in Wolff's colony could have dark or yellow hair was because the IAP element was variably influencing the expression of the *nonagouti* gene. In exploring the mechanism of the IAP effects, DNA methylation at the IAP element was found to be present when the element was silenced and had no effect on *nonagouti* and absent when the IAP element was active and causing abnormal regulation of this pigment gene, leading to the addition of yellow pigment to the hair.

Although this on its own was a striking model of DNA methylation being causally associated with an obvious phenotypic outcome, overriding the DNA sequence information present, George Wolff took it a step further in the 1990s and asked whether altering the diet of the mother of such a litter during pregnancy could influence whether the IAP element was silenced by DNA methylation. He fed pregnant mothers bearing the viable yellow mutation a diet designed to increase DNA methylation. The mothers fed these diets had an increased proportion of pups born with the brown hair phenotype, indicating silencing of the IAP element.

There is one more twist to add—the phenotype of these yellow mice was not limited to their fur color. They also gained weight to a much greater extent than their genetically identical, brown littermates and had a metabolic phenotype resembling type 2 diabetes mellitus in humans. When Wolff fed the pregnant mothers the diet designed to promote DNA methylation, he was influencing not just the cosmetic outcome of fur color but also a disease resembling an increasingly prevalent human disorder.

These findings resonated across multiple domains of research. The idea that a simple dietary change during pregnancy could modify a phenotype that included the risk of adult obesity provided striking support for the DOHaD model proposed by Barker almost a decade earlier. It also suggested that an unbiased test of DNA methylation across the genomes of the animals discordant for the phenotype would have picked out the causative locus as differentially methylated, not only finding the mechanism for a memory of a past (dietary) exposure but also the gene (*nonagouti*) mediating the phenotype. The finding also supported the idea that interventions could not only be harmful but could also be helpful, by mitigating genetic risk. All of these lessons continue to resonate today, reflecting the substantial influence of this mouse model.

Cigarette Smoking

As we developed the epigenomic technologies to survey DNA methylation and other presumed regulators of gene expression, an obvious place to start in human cohort studies was the large number of samples stored away in freezers from genetic studies. Almost universally these samples were collected from peripheral blood cells and were stored as DNA. DNA methylation is a very stable biochemical modification, so it was easy to defrost these samples and start asking questions about the patterns of DNA methylation in these cells and how they correlated with the characteristics of the donors.

A very reproducible finding that has emerged from these studies is the distinctive pattern of DNA methylation in people who smoke cigarettes, looking at individual genes (Breitling et al. 2011; Monick et al. 2012) or throughout the genome (Shenker et al. 2013). It was also shown that your DNA methylation profile in blood cells reflects whether your mother smoked during pregnancy (Richmond et al. 2015).

This excellent example of a robust epigenetic biomarker represents another reason why people became interested in epigenetics. If past exposures, even to your mother during pregnancy, can be reflected by DNA methylation patterns in your peripheral blood, you have the potential to develop numerous biomarkers that could be helpful in both research and in medical care.

Epigenetic Clock CpGs

The second major epigenetic biomarker to be developed associated DNA methylation with age in humans. Steve Horvath from the University of California in Los Angeles took the conceptually relatively straightforward approach of mining data in public databases representing the DNA methylation in individuals of different ages. By studying almost 8000 samples from 51 different cell types or tissues, he could test which loci changed their DNA methylation most consistently with chronological age. As methylation of cytosines typically occurs when the cytosine is followed by a guanine in the DNA sequence, a so-called CpG dinucleotide, he used that term to describe the 353 loci that change DNA methylation with age as "epigenetic clock CpGs." This biomarker has been tested repeatedly since its first description and has proven to be very reproducible.

What this association between DNA methylation and aging highlights is an assumption that is pervasive in epigenetics research. When a molecular marker like DNA methylation is found to be associated with something (e.g., a phenotype or an exposure), it is very tempting to assume that the relationship is causal. We know that DNA methylation can be associated with a gene being silenced, so when we see a change in DNA methylation associated with age, we assume the DNA methylation changes actually mediate the cell's aging. In fact, as will be discussed in Chapter 7, this is a very difficult conclusion to make, given what we now know about why DNA methylation can differ between individuals or over time.

The second assumption that follows is that by intervening and thus changing DNA methylation, we can alter the rate of biological aging. An association does not have to involve causation; our updated perspective on DNA methylation is that it may, in fact, reflect a footprint of where transcriptional regulatory proteins bind in the genome, as discussed later in this book. Altering DNA methylation may therefore be less valuable as an intervention than altering the control of the transcriptional regulatory proteins binding at the distinctively methylated loci.

Epigenetic Mechanisms in Gliomas

In Chapter 8 we will dive deeply into the fascinating field of cancer epigenetics, which has provided us with many reasons to be excited about epigenetics, especially because the potential for therapeutic intervention is much more tangible. One example of a story linking epigenetic events with the mechanism of malignancy was developed by Brad Bernstein at the Massachusetts General Hospital. His group was studying gliomas, tumors derived from the glial cells of the brain. His group linked together a number of steps in a cascade of events leading to the activation of an oncogene. First, they found that a mutation in a gene involved in metabolism started to generate an unusual metabolite, which interfered with the function of a gene regulating DNA methylation. The net effect was to increase the amount of DNA methylation in these glioma cells, which in turn led to the addition of methylation to sequences where a protein called CTCF usually bound. The binding of CTCF to DNA can be influenced by methylation of the target site, causing CTCF to have difficulty in its role to partition the regulatory regions of the genome away from each other at the *PDGFRA* gene. The outcome was to increase the expression of *PDGFRA*, a known glioma oncogene, driving the malignant transformation of these cells. This glioma example demonstrates nicely how the study of epigenetics now encompasses everything from metabolism to the three-dimensional organization of chromatin in the cell nucleus and the recognition of interventions that may be helpful in human diseases.

FROM RESEARCH TO CLINICAL INTERVENTION

From the earliest days of epigenetics research, there has been the hope that if we can define an abnormal pattern of epigenetic regulation in a disease, the inherent malleability of epigenetic regulatory processes should allow these to be targeted for reversal. This remains a high priority, with cancer the focus for much of the drug development to date.

Drugs in oncology have been developed to target DNA methylation and modifications of the proteins in chromatin as major priorities. Clinical trials have shown that inhibitors of enzymes adding DNA methylation to the genome improve outcomes in a blood cancer called myelodysplastic syndrome (Cabezón et al. 2021). Other tumors have high expression of bromodomain and extraterminal motif (BET) proteins, which bind to chromatin where acetyl groups have been added to the histones in chromatin. The use of BET protein inhibitors is being evaluated as a potential aide to conventional therapies (Sun et al. 2020; Trojer 2022).

In the current era of immunotherapy of cancers, another avenue has opened up for epigenetic drugs. The relatively nonselective actions of drugs that target genome-wide regulators like DNA methylation or histone modifications have the effect of reactivating the transposable elements, mentioned earlier in the story of the viable yellow mice. These reactivated elements make the cancer cell more likely to produce molecules that can be recognized by the immune system. An area of scrutiny for epigenetic drugs is therefore focused on the possibility that pretreating the patient with these drugs before activating the immune system to seek out the cancer cells may be a useful addition to immunotherapy (Jones et al. 2019).

Cancer is a serious condition, which has allowed exploration of epigenetic therapies that may have toxic effects because the risk–benefit ratio remains favorable even when the use of the drug influencing the epigenome has significant side effects (Sun et al. 2020). Applying similar therapies in other, less dangerous conditions believed to involve epigenetic changes is not justifiable, which is the reason why epigenetic interventions remain focused on cancer. As drugs influencing the epigenome become less toxic, the range of conditions that could be explored for positive effects in clinical trials will broaden accordingly.

PROBLEMS THAT HAVE EMERGED IN EPIGENETICS

Hopefully the case is being made as to why there is currently so much excitement about epigenetics and the possibility that this area of research will allow us insights into the mechanisms of disease and into possible novel therapies.

A challenge for this book is to maintain that spirit of excitement while digging deeply into some reasons why we need to temper this excitement with caution. Possibly the biggest reason for dampening our enthusiasm is the overinterpretation of differences between samples in patterns of epigenetic regulators like DNA methylation. Although this will be addressed in detail in Chapter 7, in essence the problem boils down to the fact that DNA methylation is influenced by a number of factors. If you are pursuing the idea that DNA methylation changed in the cells studied, and you have not excluded the effects of confounding influences upon the DNA methylation patterns, then you cannot interpret positive findings as supportive of your starting hypothesis. There is no study to date that has accounted for all known confounding influences, so therefore it follows that no study to date can be said to be fully interpretable. This issue is not confined to DNA methylation but involves any functional genomic output, including gene expression and chromatin studies.

There are other issues that are causing scepticism in epigenetics research. The idea that epigenetic mechanisms can propagate memories from parent to daughter cells has led logically to the concept that memories of adverse events can be propagated through generations. This is where the strength of the science is most mismatched with the power of how the story captures the imagination. We have seen a major interest in the popular press and from the lay public in how the diet, habits, or stresses of prior generations pass on their effects to children or grandchildren. Sometimes this is prompted by epidemiological observations and sometimes by model organism work, but the same interpretability issues described above remain when attempting to link molecular changes with the phenotype. In plants, as mentioned earlier, there are indeed molecular mechanisms for epigenetic changes to be induced in one generation and propagated in subsequent generations, but their molecular regulators of transcription are quite distinct from those in animals, and it remains to be demonstrated in animals that our molecular mediators can perform the same way.

Annoyingly but perhaps inevitably for a loosely defined and exciting field like epigenetics is the rise of misuse of the term in commercial products and services. As mentioned earlier, the idea that interventions can either increase or decrease one's risk of developing a phenotype to which one is genetically predisposed has fostered ideas of genetic nondeterminism and has allowed a favored intervention to be touted as the way of protecting against disease. The upshot is a circus of epigenetic diets, epigenetic face creams and other cosmetics, epigenetic yoga and meditation, and even epigenetic antidandruff shampoo, all to be found with a search of the internet. The foundation in evidence for any of these interventions is, at best, extremely weak. The resulting collective eye roll from the broader scientific community demands a response from epigenetics researchers, who individually may be performing rigorous research but who collectively suffer from what would be described in the world of marketing as "brand damage."

Finally, there has been the tendency to treat epigenetic properties of an individual in odd ways. One occurs when a researcher succumbs to the temptation to treat epigenetic information the same way as genetic information, as if it is a fixed property of an individual. DNA methylation of peripheral blood leukocytes is a phenotype, and it varies as any phenotype can, undermining some of the assumptions that are foundational in the field of epigenetic epidemiology, for example. A second weird tendency was our enthusiastic embrace of every biochemical process occurring in the nuclear genome as epigenetic, but initially excluding transcription factors for some reason. Furthermore, our relentless focus on molecular processes ends up creating a blind spot—a failure to think in terms of how these molecular events reflect cellular properties. The bases for these tendencies become understandable when we delve into the history of our field, as will be the focus of the next couple of chapters and a theme throughout the book.

THIS BOOK

Why write a book in this day and age? An internet search allows most questions to find sources of information that can potentially provide answers, whereas review papers in journals offer deep dives into scientific topics. For epigenetics, the problem is that even a comprehensive review is only informative about one segment of what has become a sprawling field. This book is designed to be like a set of connected reviews—delving deeply into as many components as possible of today's field of epigenetics while cross-referencing between these areas to create a coherent overall picture.

It is worth acknowledging that a book written by a single author is inherently biased. The research for this book involved reading monographs by Conrad Waddington and others, so apparently this kind of behavior used to be at least tolerated, if not encouraged. The challenge is to overcome your biases as you write so that the reader is not left with something resembling an opinionated rant. There are plenty of podcasts out there if that's your thing. The hope is that this book will challenge the reader to explore

their preconceptions about epigenetics, reflecting the bias of a single author but expressed politely, and hopefully based on solid evidence. When
Waddington and others wrote their books, the value of their work was in
part because they injected some interpretation and opinion into the discussion. In this book, the goal has been to do likewise, a strange writing style to
learn for one trained in the current era.

The breadth of the field of epigenetics today is daunting and requires
some degree of focus even in a book trying to maintain a broad perspective. The choice of emphasizing mostly mammalian organisms and disease
phenotypes in this book was based on several factors, including how these
reflect the author's major interests, but also the degree to which mammalian
and disease studies appear to dominate public narratives and publications in
this field. However, what represented the most compelling rationale for this
focus was the need to link what we call epigenetics today with the original
definition of the term, which was based on cellular differentiation and lineage commitment and how this is influenced by the action of genes.

There will therefore be an overt goal in the chapters to come to refocus
our attention on cellular properties as a primary way of thinking about epigenetics. The molecular mediators that have been studied enthusiastically
and given the description epigenetic do not go to waste—they remain the
means by which the epigenetic properties of cells are mediated but do not all
need to be lumped into a single vague category of being "epigenetic" in their
actions. Having used the words epigenetic and epigenomic promiscuously in
this first chapter, the remainder of the book will attempt to avoid using the
terms when a more accurate alternative way of saying the same thing—such
as "transcriptional regulation" or "cellular memory"—can be substituted.

This clarity of terminology and the refocusing on cellular properties
should make it easier for the newcomer to the field to create a framework to
learn about epigenetics, which is otherwise an intimidating field. The goal is
to allow better insights by specialist and nonspecialist alike, more productive and interpretable research, and a foundation for therapeutic interventions for human patients.

CHAPTER 2

A History of Epigenetics

To understand how we have come to develop today's sprawling, splintered field of epigenetics research, it is essential to start with some insights into the history of embryology. In fact, as historical questions about embryology were linked closely to those about conception and sex determination, it is difficult to separate out these fields, so they will all receive attention in the timeline below. In addition to yielding insights into our main question about the history of epigenetics, we can see that throughout this history scientists have eagerly pursued hypotheses that were subsequently disproven and that seem inexplicably popular in retrospect, embryology being no different to the rest of the world of science. We must assume that the future will show that we today are making comparable mistakes in some of our current assumptions about epigenetics.

What will be strikingly apparent below is how the field of epigenetics underwent a dramatic change at the turn of the twenty-first century, not just in terms of shifting radically away from its developmental biology roots, but fracturing to become several distinct fields, each of which refers to itself as the study of epigenetics. These definitions will be explored in the next chapter, providing insights into the conflicting uses of the term and the confusion that this causes today.

ANTIQUITY

Possibly the earliest text describing a theory of embryogenesis dates from sometime between 1000 and 600 BCE. This text is attributed to Sushruta, living in the Benares state in Northern India. Sushruta is described as the father of plastic surgery, describing in detail how to perform procedures such as rhinoplasty.[1] In his book *Sushruta Samhita* (Sushruta's Compendium),

---

[1] Apparently not for vanity, but to replace noses when they had been cut off as punishment.

he departs from talking about surgery to describe the embryo being formed by the mixing of semen and blood and to provide a description of sequential development of different organs and body parts. The model of heritability proposed was one of the hard body parts being from the father and the soft body parts from the mother. In Needham's excellent *A History of Embryology* (Needham 1934), he attributes to Sushruta the idea of semen and blood mixing to solidify like the skin on milk, a model that we will see used repeatedly during antiquity.

Whether the writings of Sushruta or other Indian scholars made it to Greece is uncertain, but this was to be the next major center for writings about embryology for several centuries. Pythagoras (c. 570–495 BCE) had a few words to say about conception and the vexing question of how the offspring acquires a soul. His pronouncement (*Volume 2, Pythagorean Fragments: Theages, On the Virtues*) on this issue reflect a mindset that would be difficult to defend today:

> In the conjunction of animals, the female subsists for the sake of the male; for the latter sows, generating a soul, while the former alone imparts matter to that which is generated. In the soul, the irrational subsists for the sake of the rational part.

Not too long after Pythagoras lived the physician Hippocrates (c. 460–370 BCE), who provided the first record of writings about embryological experimental research. Needham translates Hippocrates (or, more likely, someone from his school of followers) referring to embryological development in terms of getting "nourishment from the food and breath introduced into the mother," and the formation of the body as a drying of humidity by fire to generate solid structures.

However, what is most intriguing about his school's writings is how he appears to believe in all parts of the embryo being present from the outset, the theory later called preformationism:

> Everything in the embryo is formed simultaneously. All the limbs separate themselves at the same time and so grow, none comes before or after other, but those which are naturally bigger appear before the smaller, without being formed earlier. Not all embryos form themselves in an equal time but some earlier and some later according to whether they meet with fire and food, some have everything visible in 40 days, others in 2 months, 3, or 4. They also become visible at variable times and show themselves to the light having the blend (of fire and water) which they always will have.

The Hippocratic writer also describes the first systematic experiment in embryology, examining development serially over time in chicken eggs:

> Take 20 eggs or more and give them to 2 or 3 hens to incubate, then each day from the second onwards till the time of hatching, take out an egg, break it, and examine it.

As will be seen below, chicken development was to become a mainstay of embryological research over the next two millennia and influenced many scientists and their hypotheses about development. Hippocrates can probably be credited with the first experiments performed to study embryological development.

Moving to one generation later, and from Hippocrates' school on the island of Kos to the Lyceum in Athens on the mainland of Greece, Aristotle (384–322 BCE) now takes up a prominent part in the story. A former student of Plato, Aristotle was extraordinarily prolific and encompassed embryology among his broad range of interests. In his book *On the Generation of Animals*, as he describes ideas about conception, he mentions something that remains of interest today, the possibility that the acquired characteristics of a parent can be inherited by the offspring:

> ...it would seem to be reasonable to say that as there is some first thing from which the whole arises, so it is also with each of the parts, and therefore if semen or seed is cause of the whole so each of the parts would have a seed peculiar to itself. And these opinions are plausibly supported by such evidence as that children are born with a likeness to their parents not only in congenital but also in acquired characteristics for before now, when the parents have had scars, the children have been born with a mark in the form of the scar in the same place, and there was a case at Chalcedon where the father had a brand on his arm and the letter was marked on the child, only confused and not clearly articulated.

The more startling passages come from his book *History of Animals*. He provides an extremely detailed description of the development of the chicken embryo, taking the experimental model proposed by Hippocrates much further in terms of insights. For example, he describes what had to be observations with the naked eye about early development:

> ... the heart appears, like a speck of blood, in the white of the egg. This point beats and moves as though endowed with life, and from it two vein-ducts with blood in them trend in a convoluted course...

A little afterwards the body is differentiated, at first very small and white. The head is clearly distinguished, and in it the eyes, swollen out to a great extent. This condition of the eyes lasts on for a good while, as it is only by degrees that they diminish in size and collapse. At the outset the under portion of the body appears insignificant in comparison with the upper portion.

When the egg is now ten days old the chick and all its parts are distinctly visible. The head is still larger than the rest of its body, and the eyes larger than the head, but still devoid of vision. The eyes, if removed about this time, are found to be larger than beans, and black; if the cuticle be peeled off them there is a white and cold liquid inside, quite glittering in the sunlight, but there is no hard substance whatsoever. Such is the condition of the head and eyes. At this time also the larger internal organs are visible, as also the stomach and the arrangement of the viscera; and the veins that seem to proceed from the heart are now close to the navel.

These careful observations led Aristotle to a theory that rejected preformationism emphatically:

There is a considerable difficulty in understanding how the plant is formed out of the seed or any animal out of the semen... Either all the parts, as heart, lung, liver, eye, and all the rest, come into being together or in succession... That the former is not the fact is plain even to the senses, for some of the parts are clearly visible as already existing in the embryo while others are not that it is not because of their being too small that they are not visible is clear, for the lung is of greater size than the heart, and yet appears later than the heart in the original development.

This idea that embryogenesis is an ordered sequence of events, with one organ driving the formation of others, was one reason why Aristotle thought the preformationism theory was "strange and fictitious" and that "[p]lainly, then, while there is something which makes the parts [of the developing embryo], this does not exist as a definite object, nor does it exist in the semen at the first as a complete part." Aristotle's legacy included the alternative model to preformationism, to be resurrected and given a name almost 2000 years later by William Harvey, as described later in this chapter.

THE UNPRODUCTIVE MILLENIUM

Sometimes scientists have days or weeks when we don't get a lot done at work. However, we can feel relatively virtuous that at least we didn't let 1000 years go by without making much of a dent in the world of science. After

Aristotle, there was a long period of history during which there were very few advances in embryological research. Galen of Pergamon (130–210), considered to be history's first pathologist (van den Tweel and Taylor 2010), was one exception, talking about a constant reshaping of the embryo during development, very much the Aristotlean model. He used the word "genesis" when talking about embryo formation:

> Genesis is not a simple activity of Nature, but is compounded of alteration and of shaping. That is to say, in order that bone, nerve, veins, and all other tissues may come into existence, the underlying substance from which the animal springs must be altered; and in order that the substance so altered may acquire its appropriate shape and position, its cavities, outgrowths, and attachments, and so forth, it has to undergo a shaping or formative process. One would be justified in calling this substance which undergoes alteration the material of an animal, just as wood is the material of a ship and wax of an image.

During the prolonged fallow period of few advances in embryology research from Aristotle to William Harvey in the 1600s, it should be borne in mind that major educational sources for centuries in the Western world and beyond were the Abrahamic religious texts, which would have been influential and widely read by many people, including scientists. It is intriguing how religious text parallels the writings of Aristotle so closely. From the Book of Job 10: 10–11:

> Didst Thou not pour me out like milk and curdle me like cheese; Clothe me with skin and flesh and knit me together with bones and sinews?

Aristotle, in his *On the Generation of Animals*, describes the effect of semen to create an embryological structure using very similar terms of curdling of cheese:

> This material of the semen dissolves and evaporates because it has a liquid and watery nature. Therefore we ought not to expect it always to come out again from the female or to form any part of the embryo that has taken shape from it; the case resembles that of the fig-juice which curdles milk, for this too changes without becoming any part of the curdling masses.

It is estimated that the Book of Job was written sometime between the sixth and fourth centuries BCE (Vargon 2001), which potentially predates Aristotle, who was writing at the end of the fourth century BCE. The Babylonian Talmud also cites the text of Job, translated by Kottek as follows (Kottek 1981):

When the womb of the woman is full of retained blood which then comes forth to the area of her menstruation, by the will of the Lord comes a drop of white-matter which falls into it: at once the embryo is created. [This can be] compared to milk being put in a vessel: if you add to it some lab-ferment [drug or herb], it coagulates and stands still; if not, the milk remains liquid.

The common theme between Aristotle and these texts used by Christians and Jews—describing semen curdling like milk to form a structure—allows us to conclude that there was at least an exchange of ideas between religious texts and the ancient Greek philosophers. Furthermore, the same idea was proposed by Sushruta in India long prior to the writing of the Book of Job. It is difficult today to distinguish between ideas developing independently, rediscovering the same relatively simple model of the curdling of milk and an exchange of ideas between scholars from different times and countries in the far less mobile world of the time.

Digging more deeply into religious texts, we find that the Talmud makes conjectures on the heritability of traits and was a bit less sperm-centric and more open to the idea that the female was contributing something of importance:

Can we say that the organ [of the father] produces the [corresponding] organ [of the embryo]? ... Or do we say that the seeds [of the two parents] are mixed together? The answer is: of course the seeds mix together. If not, a blind would produce a blind child, and a cripple would generate a cripple. It is thus evident that the seeds are mixed.

The Qu'ran was more emphatic in giving equal prominence to the contributions of the male and the female in conception. From verse 76:2:

Verily, We have created man from Nutfah [a small drop of fluid of mixed origins, in this case both male and female], in order to try him; so We made him hearer and seer.

It is likely that a combination of the writings of Aristotle and the occasional mentions of conception and development in religious texts were the major influences for the next millennium. The notable contributor during the Middle Ages would have been St. Thomas Aquinas (1225–1274), whose writings reflect the continued male domination of society (from *Summa Theologica*):

The generative power of the female is imperfect compared to that of the male; for just as in the crafts, the inferior workman prepares the material and the more skilled operator shapes it, so likewise the female generative virtue provides the substance but the active male virtue makes it into the finished product.

The goal here is not to focus on male chauvinism, but to reiterate the point that scientific insights can be influenced and damaged by current societal and personal prejudices, a lesson from the Middle Ages with ongoing relevance.

Aquinas was preoccupied with issues of spirituality and religion, specifically how and when the soul entered into the organism. His theory was quite convoluted, involving an initial "vegetative soul" in the foetus that would expire to be replaced by a "sensitive soul," which in turn died and was replaced by a "rational soul," this time supplied by divinity. As a theory, it lacked the elegance of simplicity, and he abandoned it later in life.

During the Renaissance a couple of centuries later, the contributor of note to embryology was Leonardo da Vinci (1452–1519). He described briefly the first quantitative studies of the developing embryo:

> ...the length of the umbilical cord always equals the length of the foetal body in man though not in animals.

His major legacy was, however, the well-known, beautiful illustrations he created of the developing foetus, which remain iconic to this day (Fig. 2.1).

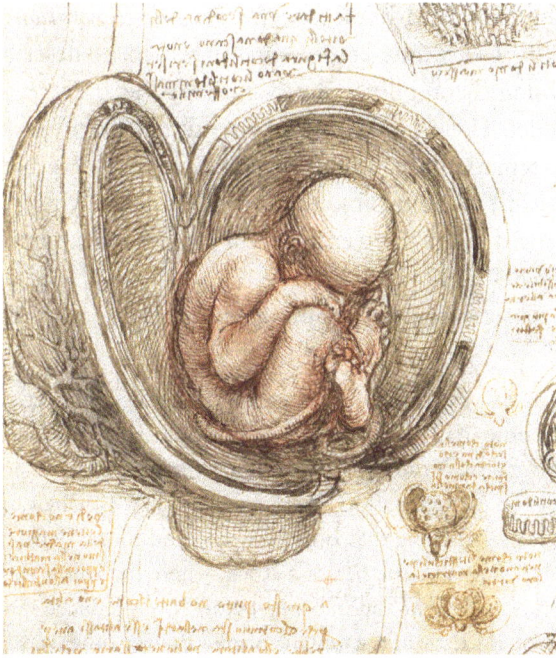

Figure 2.1. *The foetus in the womb,* pen, ink and chalk drawing (1511) by Leonardo da Vinci, Royal Collection, London, UK. © Royal Collection Enterprises Limited 2025 | Royal Collection Trust.

EMBRYOLOGY REBORN

The small contributions since Aristotle had little significant effect on the field of embryology. It took until the seventeenth century for embryological research to make any substantial new steps forward, prompted by the work of William Harvey (1578–1657). Harvey had studied in Cambridge University but traveled to the University of Padua in Italy to learn medicine, mentored by Hieronymus Fabricius (1537–1619). The Bursa fabricii, the site of B lymphocyte development beside the gut of birds, bears his name, recognising his major contributions to the field of anatomy.

To place Harvey in a historical time frame, also on faculty while Harvey was a student at the University of Padua was Galileo Galilei, teaching mathematics, physics, and astronomy, and Harvey's time in Italy meant that he was missing the latest performances back in England of a prolific playwright by the name of William Shakespeare. Toward the end of his life, English society was riven by the civil war between the Royalists and the Parliamentarians (1642–1651). Harvey was a Royalist, acting as a personal physician to King Charles I, accompanying him on hunting expeditions that allowed him to perform dissections of the deer killed, one of the sources of material for his 1651 book *Exercitationes de Generatione Animalium* (*Exercises on the Generation of Animals*) in which he described his ideas about embryology, compiling observations made over several decades. Apart from deer, Harvey also returned to the chicken egg experimental system for systematic observations of embryogenesis. He had the advantage over Aristotle of being able to use simple magnifying glasses, which had been invented in the thirteenth century in what is modern-day Italy by Salvino D'Armate (1258–1312).

Harvey's observations of embryogenesis led him to the following conclusions:

> There is no part of the future foetus actually in the egg, but yet all the parts of it are in it potentially... I have declared that one thing is made out of another two several wayes and that as well in artificial as natural productions, but especially in the generation of animals. The first is, when one thing is made out of another thing that is pre-existent, and thus a Bedstead is made out of Timber, and a Statue out of a Rock, where the whole matter of the future fabrick was existent and in being, before it was reduced into its subsequent shape, or any tittle of the designe begun. But the other way is when the matter is both made and receiveth its form at the same time... So likewise in the Generation of Animals, some are formed and transfigured out of matter

already concocted and grown and all the parts are made and distinguished together *per metamorphosin*, by a metamorphosis, so that a complete animal is the result of that generation; but some again, having one part made before another, are afterwards nourished, augmented, and formed out of the same matter, that is, they have parts, whereof some are before, and some after, other, and at the same time, are both formed, and grow... These we say are made *per epigenesin*, by a post-generation, or after-production, that is to say, by degrees, part after part, and this is more properly called a Generation, than the former... The perfect animals, which have blood, are made by Epigenesis, or superaddition of parts, and do grow, and attain their just future ... after they are born... An animal produced by Epigenesis, attracts, prepares, concocts, and applies, the Matter at the same time, and is at the same time formed, and augmented... Wherefore Fabricius did erroneously seek after the Matter of the chicken (as it were some distinct part of the egg which went to the imbodying of the chicken) as though the generation of the chicken were effected by a Metamorphosis, or transfiguration of some collected lump or mass, and that all the parts of the body, at least the Principall parts, were wrought off at a heat or (as himselfe speaks) did arise and were corporated out of the same Matter.

What Harvey is doing is distinguishing between his two proposed models of development, *metamorphosis*, in which the formless embryo received a transformative signal to allow it to take shape, and *epigenesis*, in which morphogenesis occurs in a sequentially ordered process. Oddly enough, with all the influence of the Greek philosophers mentioned up to now, you might think that they had originally created the word epigenesis, but instead it was this English scientist who coined the word. Harvey's book became a respected and powerful influence on scientists of the era and included another striking insight into his mindset. The frontispiece of the book was a depiction of Zeus holding an egg in his hands with the inscription *Ex ovo omnia*, or "From the egg, everything." Probably influenced by his chicken experiments, Harvey put the egg from the female in a position of central importance in animal generation, a departure from many prior models.

Unfortunately, many of Harvey's scientific notes on insect biology were destroyed during the English Civil War by Parliamentarian soldiers, limiting his legacy in this parallel area of his research. His interest in insects may explain why his alternative to an epigenesis mechanism was not preformationism but instead metamorphosis, a process in insect development but otherwise an unusual choice. Although Harvey's proposals on embryological development mechanisms in *Exercises on the Generation of Animals* were necessarily limited by the technologies available at the time to magnify developing structures, he reinvigorated the field for the first time since Aristotle and left it poised to build on his ideas.

Harvey's emphasis on the role of the egg in generation of a new offspring prompted a renewed focus on the female reproductive tract. In 1672, 21 years after Harvey published *Exercises on the Generation of Animals*, Regnier de Graaf (1641–1673) described follicles on the ovaries that he believed to be human ova, not yet recognizing that the follicles were larger, multicellular structures that contained and released the single cell that is the ovum (de Graaf 1672). This publication caused a rift with his friend and collaborator Jan Swammerdam (1637–1680), who accused him of plagiarism. Mourning the recent death of his 3-week-old son, and in conflict with his former friend, de Graaf died in 1673 at the age of 32, a year after publishing his work on the female reproductive tract. His name lives on today in the description of ovarian follicles as "Graafian" follicles.

Swammerdam was one of the first scientists to use microscopy extensively, making a remarkable series of contributions to science by combining his expertise in anatomical dissection with microscopy (Cobb 2000). Although he trained as a physician in Leiden's then-new university, from childhood he had been passionately interested in insects and never ended up practicing medicine, instead immersing himself in basic science research with a major focus on the study of insect development. For example, he was the first person to recognize that the "king bee" in a colony had ovaries and was, in fact, a queen. His influential contribution to the world of embryology came from his revelation that structures like the wings, legs, and antennae of the butterfly were present in the mature caterpillar, simultaneously damaging the idea of metamorphosis (defined as a complete change in morphology) while prompting speculation supporting the idea of preformationism. Ironically and tragically, he died at the early age of 43 of malaria, an insect-borne disease still prevalent in Northern Europe at that time.

Swammerdam was one of the early adopters of the new technology of microscopy. The idea of taking more than one lens and placing them in series within a tube was explored by the Dutch spectacle makers Hans Janssen and his son Zacharias (c. 1580–1638), working in the 1590s. They thus invented the first compound microscope, albeit with limited magnification power, about the same time that the first patent was being awarded by the government of The Hague to Jacob Metius (c. 1571–1628) in 1608 for the similar technology of the telescope. Robert Hooke (1635–1703) further developed the compound microscope and published his observations in his influential book *Micrographia* in 1665. Contemporaneously with Swammerdam, another early adopter of microscopy was the Italian physician–scientist Marcello Malpighi (1646–1694), whose name remains eponymously linked to skin and renal structures today, reflecting his groundbreaking work in human histology. However, it was an erroneous observation by Malpighi that was taken to support the idea of the egg containing a preformed embryo. As reviewed by Waller (Waller 2004), Malpighi studied fertilized but unincubated chicken eggs and found them to contain rudimentary embryos. What he did not appreciate was that the warm conditions in which the eggs were maintained in the several days prior to his analysis had led to the advancement of development of the embryos. The conclusion that the egg, rather than the sperm, contained the elements of a preformed embryo was referred to as the "ovist" model for preformationism.

Ovism only dominated for a few years before the inevitable competing theory emerged—that the homunculus was not contained within the egg but was, instead, in the sperm. Anton van Leeuwenhoek (1632–1723) is credited with the most significant early advances in the field of microscopy, developing instruments capable of ~270× magnification. His advances did not involve compound microscopy but instead exceptionally well-crafted single lenses. He started to describe the world of microorganisms for the first time, and in 1677 reported (with his trainee Johan Ham in the *Philosophical Transactions of the Royal Society*) that semen contained tiny worm-like creatures that they named spermatozoa. This was the catalyst for speculation that the homunculus was present in the sperm, among whose proponents Dutch biologist Nicolaas Hartsoeker (1656–1725) was prominent. In Hartsoeker's *Essay de Dioptrique* published in 1694, he showed an iconic image of a homunculus in the head of the sperm (Fig. 2.2). This image has been taken to represent what Hartsoeker was claiming to have observed, but instead he was illustrating what he believed should be possible to observe with optimal imaging techniques (Hill 1985). This prompted ridicule from the French astronomer François de Plantade (1670–1741), the secretary of the Montpellier Academy of Sciences, who in 1699 published under the pseudonym Dalenpatius a mocking depiction of the supposed homunculus in the sperm head (Prasad 2012). Leeuwenhoek made, in retrospect, a strategic error in reproducing Plantade's drawings in a refutation sent to the Royal Society of any claim that he had observed any such structures,

230   ESSAY DE DIOPTRIQUE.

que la tête feroit peut-être plus grande à proportion du reste du corps, qu'on ne l'a deffinée icy.

ART. XC.
Ce que c'est
que l'œuf de
la femme, &
comment un
enfant vient
ordinairement
au monde.

Au reste, l'œuf n'est à pro-
prement parler que ce qu'on
appelle *placenta*, dont l'enfant,
aprés y avoir demeuré un certain temps tout courbé & com-
me en peloton, brife en s'éten-
dant & en s'allongeant le plus
qu'il peut, les membranes qui le
couvroient, & pofant fes pieds
contre le *placenta*, qui refte atta-
ché au fond de la matrice, fe
pouffe ainfi avec la tête hors de
fa prifon ; en quoi il eft aidé par
la mere, qui agitée par la dou-
leur qu'elle en fent, pouffe le
fond de la matrice en bas, &
donne par confequent d'autant
plus d'occafion à cet enfant de
fe pouffer dehors & de venir
ainfi au monde.

L'experience nous apprend
que beaucoup d'animaux for-
tent à peu prés de cette maniere

ART. XCI.
Que l'on peut
pouffer bien
plus loin cette
nouvelle pen-
fée de la gene-
ration, &
comment.

des œufs qui les renferment.
L'on peut pouffer bien plus
loin cette nouvelle penfée de la
generation, & dire que chacun de ces animaux
mâles, renferme lui-même une infinité d'autres

Figure 2.2. A pencil sketch of a homunculus enclosed in a human spermatozoon by Nicolaas Hartsoeker. From Hartsoeker N. 1694. *Essay de Dioptrique*. Jean Anisson, Paris. CC BY-NC-SA 3.0

thereby linking himself inextricably to the images. The net effect of the well-intentioned speculation by these groundbreaking microscopists was to intensify the interest in the competing preformationist theories of ovism and spermism/animaliculism for decades to come.

◆ ◆ ◆

At this point the impatient reader is probably asking why advances in microscopy were not resolving the conflict between the preformationists and the epigenesists, in favor of the latter. Surely all that was needed was to look down one of these new microscopes and you would see the absence of a homunculus and the formation of organs from precursor structures that don't resemble the mature organ? This represents an example of how authoritative voices coupled with an attractive concept (preformationism) can influence how their field thinks about a scientific idea, representing a triumph of dogma over observation, a source of bias always worth considering by contemporary readers. In fact, it took more than half a century before preformationism was dealt a significant blow by a young medical student.

THE RISE OF EPIGENESIS

In 1759, at the age of 26, Caspar Friedrich Wolff (1733–1794) was finishing the thesis required by the University of Halle for his doctoral degree in medicine. He entitled it *Theoria generationis*, a work written in Latin, translated into German in 1764. In 1766, having had to spend time serving in the Prussian army during the Seven Years' War and having incurred the displeasure of the senior German academics of the day with his impertinent support of epigenesis, he found himself unable to get a teaching position in Germany and instead accepted an invitation from Catherine the Great (1729–1796) to become a member of the Russian Academy of Sciences in St. Petersburg (with the support of the famous mathematician Leonhard Euler [1707–1783], he of diagram fame). Wolff's *Theoria generationis* is described by Needham as written in a style that was "very formal, logical, and unreadable" (Needham 1931) but represented his strong defense of epigenesis. Wolff argued that the preformed embryo and all its organs should be visible from the outset if the preformationist theory was correct, but that epigenesis would be supported by the changing of shapes within the developing embryo. He also used the well-tested model of chicken embryogenesis, performing detailed microscopic studies of the blood vessels of the blastoderm, finding a coalescence and separation of clumps of cells leaving spaces filled with clear liquid in between, spaces that would then fill with the red liquid of blood before becoming covered by membranes to form blood vessels. He argued that the blood vessels therefore did not exist from the outset but were formed in a manner consistent with epigenesis.

Wolff sent a copy of his thesis to the major authority in the field at the time, Albrecht von Haller (1708–1777), a Swiss physician who studied anatomy and botany and worked at the University of Göttingen. Haller is described as the father of physiology, having made new findings about the properties of nerves and muscles and the role of bile in digestion. A year before Wolff forwarded his thesis, Haller had published his own defense of preformationism, *Sur la formation du coeur dans le poulet (On the Formation of the Chicken Heart*), making him probably even less receptive than he might otherwise have been to a medical student's contrary viewpoint. Politely invited by Wolff to give feedback in the form of corrections to his thesis, Haller did so by publishing them publicly in *Göttingische Anzeigen von gelehrten Sachen* in 1760. What follows is a cautionary tale for more established scientists when challenged by their early-stage peers. Haller dismissed Wolff's findings on the basis that the structures must have been present all along but were initially transparent. Haller supported this contention with his own experimental observation that when he immersed the blastoderm in alcohol, structures started to appear. However, in reality this represented an unrecognized artefact of coagulation by the alcohol, understandably misdirecting Haller toward confirming his belief in preformationism. The courteous Wolff responded with thanks and a promise to look further into the issues raised.

In 1769, now in St. Petersburg, Wolff published his second work, *De formatione intestinorum (On the Formation of the Intestine)*. In it he focused on the development of the chicken gut, finding that it formed from a folding process of a previously flat membranous structure. This removed any issues about transparency, as structures were visible before and after the folding process. He also described the transient mesonephros (still called the Wolffian bodies) and blood being formed in extraembryonic mesoderm remote from the heart. These findings were more destructive of a theory of preformationism than they were instructive about a model of epigenesis but should have served to rebut the original criticisms of Haller effectively. Wolff and Haller continued to correspond, with Haller publishing responses to *De formatione intestinorum* in 1770 and 1771, continuing to insist stubbornly that Wolff's observations could not account for transparent structures.

It should be noted that although Wolff's work is now regarded as a set of landmark studies in the history of embryology, it was largely ignored at the time. Perhaps his distance in St. Petersburg from the other major European academic centers left him relatively marginalized, and certainly his views were not well-received by his more established scientific peers. Although Needham describes the impact of Wolffe publishing *De formatione intestinorum* as having "ruined preformationism" (Needham 1931), it did not appear to change the thinking of most scientists for a number of years. There was one notable exception: By 1789 it was clear that Wolffe's work had influenced J.F. Blumenbach's (1752–1840) *Über den Bildungstrieb (On the Formative Impulse*, 1781). Blumenbach, a convert from preformationism, was notable for being the originator of the term Caucasian to mean white European

(Bhopal 2007) and was another member of the faculty of the University of Göttingen. Blumenbach's advocacy for the epigenesis model was influential enough that it attracted the attention of Kant, the major German philosopher of the time.

Immanuel Kant (1724–1804) weighed in on epigenesis in his book *Critique of Judgment (Kritik der Urteilskraft)* in 1790. He distinguished preformationism and epigenesis in a useful way, describing the former as the "educt" and the latter as the "product" of the being generating the offspring. The word educt indicates that the being is already there and needs to be brought forth, whereas the word product indicates the generation of something new.

He was aware of the recent work by Blumenbach and appeared to be attempting to be supportive when he wrote the following (Kant 2006):

> No one has done more by way of proving this theory of epigenesis than Privy Councilor *Blumenbach* and by way of establishing correct *[echt]* principles for applying it, which he did in part by avoiding too rash a use of it. Whenever he explains any of these structures physically he starts from organized matter. For he rightly declares it contrary to reason that crude matter on its own should have structured itself originally in terms of mechanical laws, that life could have sprung from the nature of what is lifeless, and that matter could have molded itself on own into the form of a self-preserving purposiveness. Yet by appealing to this principle of an original *organization,* a principle that is inscrutable to us, he leaves an indeterminable and yet unmistakable share to natural mechanism. The ability of the matter in an organized body to [take on] this organization he calls a *formative impulse.* (It is distinguished from the merely mechanical *formative force* that all matter has, [but] stands under the higher guidance and direction, as it were, of that formative force.)

Kant was clearly trying to have it both ways, retaining some preformationist ideas while attempting to voice enthusiasm for epigenesis. By describing an original model and dismissing the idea that lifeless crude matter could in some way mold itself, he is making the case for some sort of preexisting template, a preformationist position.

So why was Kant putting himself through such contortions? It should be noted for perspective that preformationism was felt to be more reflective of the actions of a divine being, placing epigenesis in the light of being less respectful of religious beliefs. As religious faith was pervasive in society and undoubtedly of great personal importance to many of the scientists of the day, it is understandable that anything that could be construed as irreverent would be uncomfortable for scientists to support. This concern caused the preformationism versus epigenesis question to be maintained without resolution for far longer than might be assumed to be reasonable, especially as the ability to visualize developing embryos continued to improve.

THE EMERGENCE OF DEVELOPMENTAL GENETICS

We are seeing that the use of the word epigenesis predates the concept of genes, for which we had to wait another century. The nineteenth century was the time of Gregor Mendel (1822–1844) performing his experiments on plant hybrids, published in 1866 (Mendel 1866). As Holliday points out, an underappreciated aspect of Mendel's work was that he realized that heritability could be studied without also studying development (Holliday 2006). This represents the first divergence between embryology and what would come to be called genetics—a split that would grow over the next century.

Predating Mendel was Charles Darwin (1809–1882), whose *On the Origin of Species* was published in 1859 (Darwin 1859). In that work, Darwin described an idea that remains topical today in the field of epigenetics research:

> It is commonly assumed, perhaps from monstrosities often affecting the embryo at a very early period, that slight variations necessarily appear at an equally early period. But we have little evidence on this head—indeed the evidence rather points the other way; for it is notorious that breeders of cattle, horses, and various fancy animals, cannot positively tell, until some time after the animal has been born, what its merits or form will ultimately turn out. We see this plainly in our own children; we cannot always tell whether the child will be tall or short, or what its precise features will be. The question is not, at what period of life any variation has been caused, but at what period it is fully displayed. The cause may have acted, and I believe generally has acted, even before the embryo is formed; and the variation may be due to the male and female sexual elements having been affected by the conditions to which either parent, or their ancestors, have been exposed. Nevertheless an effect thus caused at a very early period, even before the formation of the embryo, may appear late in life; as when an hereditary disease, which appears in old age alone, has been communicated to the offspring from the reproductive element of one parent.

What Darwin was describing was the heritability of traits and disease, but by invoking the "conditions to which either parent, or their ancestors, have been exposed" and events occurring during development, he was laying the foundation for a major area of contemporary epigenetics research, the developmental origins of health and disease (DOHaD) within a life course, and inter- and transgenerational inheritance of acquired traits (Chapter 9). This issue of heritability of acquired characteristics had been proposed decades earlier by Jean-Baptiste Lamarck (1744–1829), who believed the idea to be so self-evident that he didn't feel it necessary to write much about it (Burkhardt 2013) but described it to be governed by two laws:

First Law: In every animal that has not reached the end of its development, the more frequent and sustained use of any organ will strengthen this organ little by little, develop it, enlarge it, and give to it a power proportionate to the duration of its use; while the constant disuse of such an organ will insensibly weaken it, deteriorate it, progressively diminish its faculties, and finally cause it to disappear.

Second Law: All that nature has caused individuals to gain or lose by the influence of the circumstances to which their race has been exposed for a long time, and, consequently, by the influence of a predominant use or constant disuse of an organ or part, is conserved through generation in the new individuals descending from them, provided that these acquired changes are common to the two sexes or to those which have produced these new individuals.

This theory was challenged by August Weismann (1834–1914), who performed a simple experiment to test it, cutting the tails off mice for each of six generations and allowing the animals to breed. He proposed that the Lamarckian model would predict that the mice would eventually stop forming tails, the phenotype of lacking a tail a phenotype already observed in Manx cats (later shown to be due to T-locus mutations, which will come up again in Chapter 3), and that by failing to demonstrate the inherited acquisition of the phenotype, he had invalidated the Lamarckian model. His motives in testing the hypothesis of a fellow scientist were not wholly altruistic, as he was making a case favoring his own alternative model instead: that events in the germline and soma were independent, with continuity of information through the germ cells, unaffected by somatic events like tail mutilation.

♦ ♦ ♦

Getting back to genetics, Gayon has nicely summarized the convergence of ideas as the nineteenth century drew to a close (Gayon 2016). Mendel was rediscovered independently by the botanists de Vries, Correns, and von Tschermak in 1900 and was the focus of William Bateson's (1861–1926) book in 1902, *Mendel's Principles of Heredity: A Defence,* in which he introduced the words "allelomorph" (which later became "allele"), "homozygote," and "heterozygote" and began to use the word "genetics" at that time to describe his own research. The Danish botanist Wilhelm Johannsen (1857–1927) is credited with introducing the words "gene," "genotype," and "phenotype" in his 1906 book *Elemente der exakten Erblichkeitslehre* (Johannsen 1909). A major technical breakthrough came from the independent work of two scientists, the American Walter Sutton (1877–1916) and the German Theodor Boveri (1862–1915). Boveri was working at the University of Munich, doing groundbreaking work in the field of cytology, observing chromosome numbers during meiosis. Studying a nematode called *Ascaris megalocephala*, Boveri showed that the number of chromosomes halved during meiosis; he subsequently worked on sea urchin and found evidence that fertilization involved the haploid chromosome number following meiosis being restored to diploidy. Sutton took this observation a step further and suggested that these events could help to explain aspects of Mendelian inheritance.

An initial sceptic of this Boveri–Sutton chromosomal theory was the American biologist Thomas Hunt Morgan (1866–1945). Like Sutton, Morgan worked at Columbia University in New York, which did not preclude his asking some difficult questions of his colleague's theory (Morgan 1910):

> If Mendelian characters are due to the presence or absence of a specific chromosome, as Sutton's hypothesis assumes, *how can we account for the fact that the tissues and organs of an animal differ from each other when they all contain the same chromosome complex...* For on such a view the chromosomes should be sorted out in the soma until each region gets its proper kind. The facts are the reverse. However important therefore the chromosomes are in transmitting the full quota of hereditary traits, we must be prepared to admit that the evidence is entirely in favor of the view that the differentiation of the body is due to other factors that modify the cells in one way or in another. This consideration is, to my mind, a convincing proof that we have to deal with two sets of factors—*the common inheritance of all the cells to produce all the kinds of tissues and organs in the body, and the limitation of that property in the course of development.* If the former is due to the chromosomes and the unspecialized parts of the cytoplasm, the latter may be due to the local changes that the relation of the parts to each other calls forth. It might even be argued that since in the development we find no evidence of a sorting out of the chromosomes that produce special parts, the individual chromosomes can not stand each as the representative of those parts, but rather that each part needs the entire set of chromosomes for its normal life. *[emphasis added]*

Morgan is now defining a question that became a central driver in the need to define epigenetics as a new term later in the twentieth century, the question why genes, as the regulator of development, do different things in different cell types. He goes on to take a position on the topic:

The modern literature of development and heredity is permeated through and through by two contending or contrasting views as to how the germ produces the characters of the individual. One school looks upon the egg and sperm as containing *samples* or *particles* of all the characters of the species, race, line, or even of the individual. This view I shall speak of as the *particulate theory of development*. The other school interprets the egg or sperm as a kind of material capable of progressing in definite ways as it passes through a series of stages that we call its development. I shall call this view the *theory of physico-chemical reaction*, or briefly the reaction theory. The resemblance of this comparison to the traditional theories of preformation and epigenesis is obvious, and I should willingly make the substitution of terms were it not that the terms preformation and epigenesis have certain historical implications, and, as I wish to emphasize certain things not necessarily implied in the historical usage, I prefer descriptive terms other than these overladen with so many traditions.

Even then, it appears, the word epigenesis was so loaded with ambiguous meaning that Morgan felt the need to skirt it by calling it "reaction theory" instead. He was representing the prevailing view of developmental biologists of the time: If a chromosome contained information for how multiple different cell types should develop, these should be encoded in different parts of the chromosome, and these parts should break away from the chromosome as development progresses, distributing selectively to the cell lineages they control. The observation that mature cells had just as much genetic material as progenitor cells was not reconcilable with this theory, leading to doubt that genetic material influenced development.

It took experimental data from his own laboratory to change his mind. He was the first to use *Drosophila* as a model organism, establishing that these fruit flies have four pairs of chromosomes, of which one pair determines sex—the female having two large X chromosomes, the male only one X and a smaller Y chromosome. A chance observation of a male fly lacking the normal red eye pigment was the starting point for Morgan's epiphany. By breeding the fly's progeny in thousands of matings, his group showed the white-eyed phenotype was likely to be due to a gene located on the X chromosome. This finding converted Morgan to the belief that genes were, in fact, embedded within chromosomes and could influence normal development.

How Morgan changed his position on a fundamental and public tenet of his scientific beliefs is described in detail by Benson, a fascinating account of how a great scientist was willing to accept challenging new ideas and, thus, admitting his past opinions to be wrong (Benson 2001).

THE SCHISM BETWEEN EMBRYOLOGISTS AND GENETICISTS

Morgan was not alone in resisting the idea of genes mediating morphogenesis. In retrospect, the schism within the world of embryology appears amusing, but at the time appears to have been quite heated, with the traditional embryologists believing that the information for developmental programs was located in the cytoplasm, and the upstart geneticists espousing instead the nucleus. Morgan had been firmly in the cytoplasmic camp before his own observations prompted him to switch to the nuclear insurrectionists. He was undoubtedly influenced by the work of Edmund Beecher Wilson (1856–1939), who was Walter Sutton's graduate thesis advisor, and Nettie Stevens (1861–1912), a groundbreaking woman at the dawn of the field of genetics. Stevens and Wilson both published similar findings at the same time—that there appeared to be a pair of chromosomes that were morphologically distinctive and were found in combinations that corresponded to the sex of the organism. This finding indicated a genetic mechanism for sex determination, the foundation upon which Morgan would then interpret his own observations in fruit flies.

Morgan's apostasy must have dismayed the traditional embryologists, who regarded the geneticists as impertinent and ill-informed about developmental biology, neither of which may have been an unfair characterization. The embryologists were also influenced by the discovery, reported in 1924 (Spemann and Mangold 1924), of the "organizer." This work was carried out by graduate student Hilde Mangold (1898–1924) working in the laboratory of Hans Spemann (1869–1941) at the University of Freiburg and described a cluster of mesodermal cells in newts that altered the differentiation of (or organized) the adjacent ectodermal cells. This cell-autonomous property was difficult to reconcile with the idea that nuclear processes influenced development, part of the reason for the resistance of the embryologists to these new ideas. Spemann had been a student of Boveri but was clearly sufficiently independent a thinker to make a break from the teachings of one of the founders of the theory of chromosomal inheritance, embracing instead the viewpoint of the embryologists.

The emotion associated with the disagreements between the embryologists and the geneticists is reflected by the hyperbolic language used by Ross Harrison (1870–1959), the inventor of animal tissue culture, in 1937 (Harrison 1937). He described the threat of the encroachment of geneticists into the field of embryology in terms of "invasion" and using German words like *"Wanderlust,"* sending a message that likened geneticists to the then-current aggressive militarism of Nazi Germany:

> The location of genes in the chromosomes, the proof of their linear order, the association of somatic characters with definite points in the chromosomes, in short, the whole development of the gene theory is one of the most spectacular and amazing achievements of biology in our times. The embryologist, however, is concerned more with the larger changes in the whole organism and its primitive systems of organs than with the lesser qualities known to be associated with genic action. As Just remarked in the symposium this morning, he is interested more in the back than in the bristles on the back and more in eyes than in eye color. Now that the necessity of relating the data of genetics to embryology is generally recognized and the *"Wanderlust"* of geneticists is beginning to urge them in our direction, it may not be inappropriate to point out a danger in this threatened invasion.
>
> The prestige of success enjoyed by the gene theory might easily become a hindrance to the understanding of development by directing our attention solely to the genom [sic], whereas cell movements, differentiation and in fact all developmental processes are actually effected by the cytoplasm. Already we have theories that refer the processes of development to genic action and regard the whole performance as no more than the realization of the potencies of the genes. Such theories are altogether too one-sided.
>
> The importance of substances of genic origin lies in their continuous source of supply and in their transmissibility through generations. Organizers have come into prominence through the dramatic manner in which they have demonstrated epigenetic development at a time when the tendencies of thought were in the direction of preformation. Their most striking action, still veiled in mystery, lies not in the induction of a particular organ here or there, but in making plastic material form a harmoniously constructed embryo.

The problem remained that it was difficult for embryologists to believe that chromosomes conferred developmental properties when the chromosomes looked the same in cells and tissues with strikingly different appearances and properties. Furthermore, what the geneticists were describing as heritable phenotypes were relatively subtle, like eye color, whereas the embryologists were trying to figure out how the eye itself formed. Also seemingly failing to be encompassed by genetic mechanisms were some puzzling situations such as environmental regulation of sex determination (Gilbert 2000). How were these fields of research to be reconciled?

CONRAD HAL WADDINGTON

Conrad Hal Waddington (1905–1975) is the pivotal player in this historical narrative, as the originator of the "epigenetic landscape," the foundation for the modern use of the term "epigenetic." As his father was employed in India as a tea planter, young Conrad and his sister were brought up by Quaker relatives on a farm in rural Worcestershire in England, where he spent a lot of time collecting snails and fossils, reading books about natural science, and being inspired by the owner of the local chemist's shop, a distant relative with a passion for science in all its forms (Robertson 1977). Waddington went on to study at the University of Cambridge, initially focusing on geology, moving on to a relatively diffuse postgraduate career nominally studying paleontology, while having the time to become an expert in Morris dancing (Robertson 1977). His studies of philosophy led to an interest in evolution, from which he moved to a new focus on genetics. He was reading about the Spemann–Mangold organizer and was prompted to seek out his Cambridge colleagues Joseph Needham (1900–1995) and Dorothy Moyle Needham (1896–1987), a husband and wife team who were both biochemists. Joseph Needham was focused on the biochemistry of embryology, writing a classic book that reviewed the history of the subject (Needham 1931). Although Waddington did not complete his original doctoral thesis in paleontology, his research output was such that by 1938 he was awarded the Sc.D. higher degree based on his published work in embryology.

In 1930, Waddington received a grant to allow him to work in Freiburg for 6 months in the laboratory of Hans Spemann, learning techniques to study the organizer, techniques he took back to Cambridge to demonstrate the inductive effect of these cells on other parts of the embryo. In 1938–1939, he was funded to visit the United States, where he worked with two former trainees of T.H. Morgan, Alfred Sturtevant (1891–1970) and Theodosius Dobzhansky (1900–1975), learning the new field of *Drosophila* biology. In 1939 he published the astonishingly comprehensive and lucid textbook *An Introduction to Modern Genetics* (Waddington 1939). In this he discussed the difficulties of reconciling epigenesis, genotype, and phenotype. He used eye pigment as an example:

> The main difficulty in defining the concept of phenotype is caused by the fact that animals change in time. In its original sense, the word referred to the characters, both anatomical and physiological, of the adult. But clearly if we have an animal whose eye colour darkens with age there is no essential difference between the light eye of the young animal and the dark eye of the older one; both must be included in the phenotype. But if this is allowed, there is no reason to exclude from the phenotype the processes (about which we usually know very little) by which the eye pigments are synthesized during development. The phenotype in fact must be used as a name for the whole set of characters of an organism, considered as a developing entity. Phenotypic differences between two organisms may be caused by genotypic differences or may be produced by different environments acting on the same genotype.

> The concepts of genotype and phenotype are defined in the first place in relation to differences between whole organisms. They are not adequate or appropriate for the consideration of the development of differences within a single organism. In this connection we do, indeed, require the concept of the hereditary (chromosomal) constitution of the zygote, and we can without danger extend the meaning of the word genotype to cover this. But the difference between an eye and a nose, for instance, is clearly neither genotypic nor phenotypic. It is due, as we have seen, to the different sets of developmental processes which have occurred in the two masses of tissue; and these again can be traced back to local interactions between the various genes of the genotype and the already differentiated regions of the cytoplasm in the egg. One might say that the set of organizers and organizing relations to which a certain piece of tissue will be subject during development make up its "epigenetic constitution" or "epigenotype" then the appearance of a particular organ is the product of the genotype and the epigenotype, reacting with the external environment. In transplantation experiments, such as those described above, it is the epigenotype which is altered.

This description illustrates how he could combine his interest in the organizer with the *Drosophila* genetics he learned with Sturtevant and Dobzhansky. He appears to have been unusual in that era for having a foot in both the embryology and genetics camps, providing him with a unique perspective. This fusion of epigenesis as a developmental process with genetic influences on development is Waddington's major contribution, emerging in a 1939 textbook while he was still in academic limbo as a trainee without a permanent position. He went on in this book to describe a concept to which he would return later, the idea of representing development as a landscape of branching valleys:

> In a species, certain types of morphological variation may occur particularly frequently, and from diverse genetical and environmental causes. In terms of the "geological" model of gene reactions, we could say that the landscape defined by the genotype includes certain definite valleys branching downwards, so that any slight variation in the upper part of the main valley may divert the gene reaction into one or other of these already existing side channels. A very good example of this type of behaviour is the Minute group of genes in *D. melanogaster*. The Minutes are all dominant (and lethal when homozygous), and they all shorten the bristles to various extents and have a characteristic effect on the eyes, wings, etc. Their loci are scattered throughout the chromosomes, but although they are definitely different genes, they all produce the same syndrome of effects, though with different strengths. Schultz showed that in compounds (double heterozygotes) two Minutes do not reinforce one another, and thus do not perform the same primary reaction. Each Minute has a different effect on the early part of the valley along which the bristle-forming process is moving, but any of these effects is sufficient to divert the course of the process out of the main valley into the Minute branch valley. The course of the branch valley is not absolutely fixed by the rest of the genotype, but is also effected to some extent by the particular Minute used, since the final effect can be more or less extreme. Schultz discovered some of the modifiers which determine the course of the Minute branch valley; these are factors (Delta, Jammed, etc.) which will increase or decrease the effects of Minutes, and which modify all Minutes in the same way, and therefore must affect the Minute valley itself and not the different primary effects which originally diverted the process into that valley.

He credits a founder of the field of population genetics, Sewall Wright (1889–1988), with having used a similar landscape analogy in describing population fitness, possibly revealing an inspiration for his model for developmental biology. It was in Waddington's 1947 book *Organisers and Genes* that Waddington published his first picture of what he called The Epigenetic Landscape, a drawing he commissioned from his friend, the British artist John Piper (1903–1992).

This picture (Fig. 2.3) is puzzling, as it shows a river cutting more deeply into a landscape as it flows toward the sea, implying that the land remains as a plateau or rises as it reaches the coast. His written description of the epigenetic landscape idea in the book is more intuitive, describing "valleys diverging down an inclined plane."

Figure 2.3. From a drawing by John Piper, frontispiece to Waddington CH. 1940. *Organisers and Genes*. Cambridge University Press, Cambridge. Description from book: Looking down the main valley toward the sea. As the river flows away into the mountains it passes a hanging valley, and then two branch valleys, on its left bank. In the distance the sides of the valleys are steeper and more canyon-like. CC BY-NC-ND 4.0

The depiction of the epigenetic landscape that became famous was published in Waddington's 1957 book *The Strategy of the Genes: A Discussion of Some Aspects of Theoretical Biology.* This illustration (Fig. 2.4) is much more intuitively representing the description of valleys on an inclined plane and usefully includes a ball at the highest point to remind us of a trajectory with choices occur at each bifurcation of the valleys, or creodes (a word he invented from biblical Greek, combining the Greek words χρή [chre], "it should, it is necessary, it may be useful," and ὁδός [hodos], "road, path") as he called them.

The less-heralded part of the picture is the depiction of guy-ropes underneath the landscape, pulling a floating surface into its contours to form the creodes. This part of the model is very awkward and requires some suspension of belief in the laws of physics, as the tethering by the guy-ropes implies that gravity is acting from above, but the rolling of the ball down the hill implies the opposite. The equivalent of tent poles would probably have been a better approach.

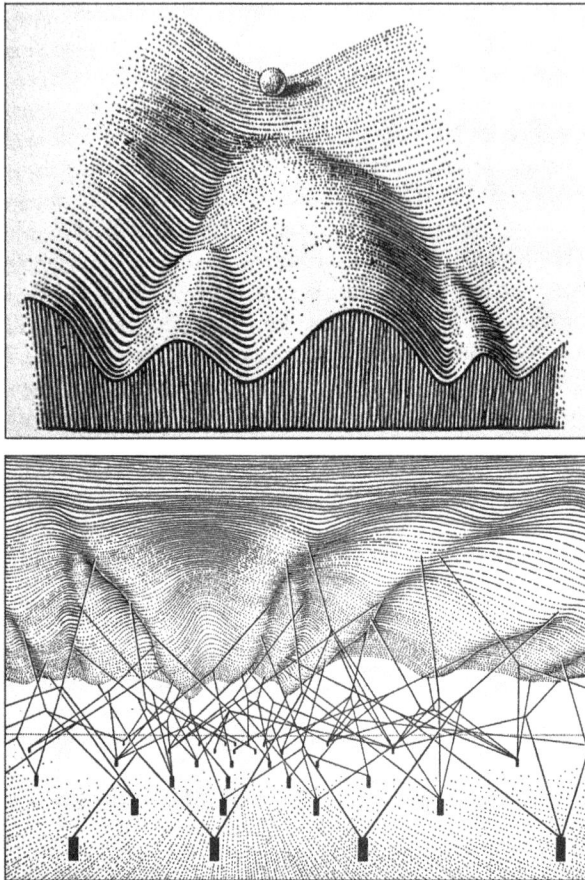

Figure 2.4. (*Top*) Part of an epigenetic landscape. (*Bottom*) The complex system of interactions underlying the epigenetic landscape. From Waddington CH. 1957. *The Strategy of the Genes: A Discussion of Some Aspects of Theoretical Biology.* Allen and Unwin, United Kingdom. Wellcome Collection. CC BY-NC 4.0

The implication of the model becomes clear—if you snap one of the guy-ropes tethering the landscape, the landscape will lose its shape, and the choice of creodes available to the rolling cell will change. In calling his first book *Organisers and Genes*, Waddington was referring to both cellular differentiation and genetics, and this landscape model attempts to integrate the two. The series of bifurcations in the landscape represents cell fate decisions and lineage commitment, or epigenesis. The role of the guy-ropes is the equivalent of the genes involved in development. If you snap one (mutate a gene), the repertoire of differentiation decisions available to the cell changes, resulting in altered morphogenesis. He referred to this combination of epigenesis and genetics as the "epigenetic landscape."

Waddington, in proposing this simple idea, created a conceptual template for the field of developmental genetics. It is worth bearing in mind that the existence of genes driving cell fate decisions (now recognized to be transcription factors) was not known at that point in history. His major goal appears to have been to reconcile the hostility between the embryology and genetics communities, pointing toward a way that both viewpoints could be accommodated.

His prior studies as a geologist may also have influenced his use of terrain and valleys as analogies. The description "epigenetic" was being used in geology as early as 1914 to describe drainage of a terrain (Whitney and Smith 1914) (Fig. 2.5). Also described as "inherited" drainage, it referred to how the drainage of a region sometimes has no clear relationship to the local rock

drainage, *n.*—**Adjusted drainage**, a drainage-system in which the streams and valleys have come, by spontaneous changes, to follow chiefly the belts of weak rock, while the ridges and divides follow the belts of resistant rock. Rivers and divides also are similarly adjusted.— **Antecedent drainage.** See ★*antecedent.*—**Arterial drainage**, that part of drainage which is effected through large open channels which are either artificial watercourses or natural ones improved: opposed to *minor* ★*drainage.*—**Autogenetic drainage.** See ★*autogenetic.* —**Deep drainage**, Same as *thorough* ★*drainage* (a).— **Drainage cycle**, the initiation, development, and maturity of drainage of any given region to the time of interruption introduced by new conditions.

"The old drainage of this basin presents some interesting peculiarities, and the interpretation of these will enable us to determine some of the deformations of the basin during the development of this old *drainage cycle.*" *W. G. Tight*, U. S. Geol. Surv., Professional Paper 13, [p. 76.

Dumb-well drainage. Same as *sink-hole* ★*drainage.* —**Epigenetic drainage.** Same as *inherited* ★*drainage.* —**Essex system of drainage**, an earlier British method of drainage in which drains of the bush and straw type (which see, under ★*drain*) were placed under each, or each second or third, water-furrow: hence, also called *furrow drainage.*—**Inherited drainage**, streams the courses of which have been determined by the slope of a once overlying series of strata now removed by erosion so as to disclose rock structures of another arrangement with respect to which the streams manifest no sympathy. Also called *superposed* or *epigenetic drainage.*—**Minor drainage**, deep or thorough drainage as opposed to arterial ★*drainage* (which see).—**Mole drainage**, drainage with mole ★*drains* (which see).—**Parallel drainage.** Same as

Figure 2.5. Epigenetic or inherited drainage, a definition from 1914. Whitney WD, Smith BE. 1914. *The Century Dictionary and Cyclopedia*. The Century Company, New York.

formations, because the drainage patterns had been established in a covering layer that had eroded over time to expose the strata below. Waddington may have been completely unaware of this, but at the other extreme of possibility he could have been mischievous in using this old geology term with his new colleagues in embryology and genetics. In contemporary terms, his behavior could be described as "trolling" the field of biology.

In proposing the epigenetic landscape model in 1937, Waddington's timing was excellent. Mutational effects on phenotypes had started to encompass model organisms beyond *Drosophila*. In 1938, a year before Waddington's *An Introduction to Modern Genetics* was published, a 29 year old was publishing the results of her unpaid work in the laboratory of L.C. Dunn at Columbia University in New York. Salome Gluecksohn-Waelsch (1907–2007) had fled Nazi Germany in 1933 with her husband. She had received her Ph.D. in 1931 in the laboratory of Hans Spemann in Freiburg, where she encountered the visiting Waddington during his 6-month visit (Solter 2008). Dunn was a groundbreaker for demonstrating genetic linkage in mammals and for identifying mutations causing malformations in rodent embryos (Anon. 1974), findings attributed to Gluecksohn-Waelsch by Davor Solter (Solter 2008). Gluecksohn-Waelsch's 1938 publication described how the Brachyury *T* mutation led to abnormal formation of the notochord, leading to secondary malformations of the neural tube. This model established "the conceptual basis for the field of developmental genetics" (Solter 2008) and led to her developing a research program combining genetic mutation studies with the organizer concept of Spemann (Solter 2008). Gluecksohn-Waelsch's work represented that of an emerging group of biologists linking mutations with malformations at the time, putting pressure on the embryologists to incorporate genetic influences into their thinking about development.

Waddington was not content with his use of the epigenetic landscape idea for reconciliation of the dispute between embryologists and geneticists with a model for modern developmental genetics. In the same *The Strategy of the Genes* book, Waddington included a much less heralded illustration that depicted variability of the creode shapes. He was using this to describe how he believed the Baldwin effect (Baldwin 1896) could occur (Fig. 2.6). The Baldwin effect has been defined by Maynard Smith (Smith 1987) as follows: "If individuals vary genetically in their capacity to learn, or to adapt developmentally, then those most able to adapt will leave most descendants, and the genes responsible will increase in frequency. In a fixed environment, when the best thing to learn remains constant, this can lead to the genetic determination of a character that,

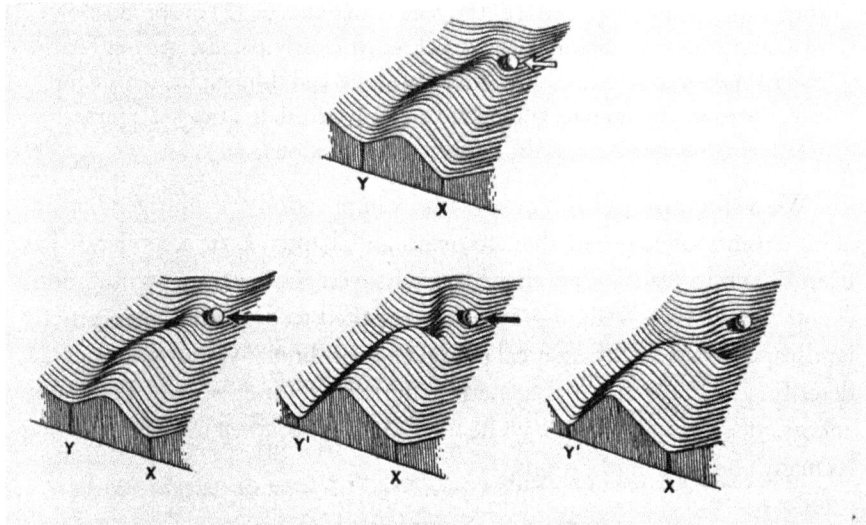

Figure 2.6. "Organic selection" (the Baldwin effect) and genetic assimilation. Waddington saw cell fate as something that could be influenced. The hollow arrow in the *top* panel is meant to represent an environmental influence. In the *bottom* section, the filled arrows represent a "mutant allele," with varying contributions of "the genotype as a whole" (represented by the landscape) to permit new cell fates. Waddington was attempting to fit an evolutionary model for acquisition of new heritable characteristics of the organism. From Waddington CH. 1957. *The Strategy of The Genes: A Discussion of Some Aspects of Theoretical Biology*. Allen and Unwin, United Kingdom. Wellcome Collection. CC BY-NC 4.0

in earlier generations, had to be acquired afresh in each generation." Maynard Smith explicitly referred to this as a Lamarckian hypothesis and called out Waddington for being "curiously reluctant" to refute Lamarckism.

Waddington also used the epigenetic landscape model as the foundation for further ideas. For example, he saw the potential for genetic influences on epigenesis to cause specific phenotypes to be modifiable by the environment, describing some supportive data from *Drosophila* experiments:

In some organisms, belonging to very different groups, experimental treatment has produced changes which are maternally inherited for some generations and then gradually lost. They seem to be caused by persistent, but not unalterable, changes in the cytoplasm. They were discovered in Protozoa, where they were produced by heat and chemicals; and it was found that repeated treatments increased the effects produced. Similar persistent modifications have also been obtained in *Drosophila* by heat. A culture heated for some hours to just below the lethal temperature (i.e., to about 37°C) shows three types of effects: (1) a rise of the gene-mutation rate; (2) the production of non-hereditable abnormalities which often closely parallel gene effects; and (3) persistent modifications which are inherited through the female for a few generations and then gradually return to normal. In *Drosophila* further treatment does not increase the degree of modification.

We will return to this model, with its implication for multigenerational heritability of acquired characteristics, in Chapter 9. As a spoiler, it has been shown in recent years that his results were, in fact, due to mutations (Fanti et al. 2017). Waddington also attempted to describe the epigenetic landscape idea as a mathematical model in a later book (Waddington 1962), describing concepts like "epigenetic temperature" and "epigenetic kinetic energy," demonstrating the ways he built upon his original model, exploring its many potential applications.

◆ ◆ ◆

If the reader is getting the impression that the epigenetic landscape model is just not that exciting, merely representing the idea that genes can influence cell fate, that appears to be how others were also regarding it at the time; it did not seem to have much traction in the scientific community. The model was probably destined for extinction when it underwent the first of several definitional mutations that it would undergo through history, this time by David Nanney, corrupting and prolonging the life of the word epigenetic.

DAVID NANNEY AND THE RE-INTERPRETION OF EPIGENETIC PROPERTIES AS THOSE OF CELLS

Once again, it is worth broadening the perspective to understand where influences were coming from in other fields of science. Jan Sapp has described the conflict in the first half of the twentieth century as that between people who believed that the cellular information needed for development resided in the cytoplasm and those who believed that it resided in the nucleus. This was another way of portraying the ongoing conflict between the embryologists and the geneticists, the embryologists favoring the cytoplasm and the geneticists the nucleus (Sapp 1987).

A major figure in this discussion was David Nanney (1925–2016). Nanney was the scientist who first developed *Tetrahymena* as a model eukaryotic unicellular organism. One of the discoveries he made in *Tetrahymena thermophila* was of their mating type loci. These organisms

can mate sexually, but only with other *Tetrahymena* with a different mating type. Once a *Tetrahymena* had chosen its mating type, that decision was
maintained in offspring. Similar observations had been made by Nanney's
Ph.D. mentor, Tracy Sonneborn (1905–1981), in another ciliate organism,
Paramecium, while comparable discoveries were being made at the same time
in *Saccharomyces cerevisiae*, indicating that this was a broadly used process.

Nanney saw this as an example of "The developmental paradox—the
chromosomal equivalence of cells with hereditary differences" (Nanney
2004). It appeared to him that the nuclear genome could not possibly account for all cellular properties, as it appeared to be the same DNA present
in cells with different properties like mating types. Instead, in the 1940s and
1950s, researchers like Nanney and other prominent scientists of the time like
Sonneborn, Max Delbrück (1906–1981), and Boris Ephrussi (1901–1979)
began to support the idea of plasmagenes, definable as some sort of entity
within or property of the cytoplasm that could mediate heritability of characteristics in the next generation of daughter cells. In a sense, the plasmagene
theory had to emerge to account for the seeming inability of nuclear genes to
explain inherited cellular traits. In proposing the plasmagene idea, they were
accomplishing two goals—challenging what had become the mainstream,
nuclear gene-centric paradigms of the time and highlighting the capability
of the cytoplasm to mediate heritability. Waddington's reconciliation of embryology and genetics may have been served well by his epigenetic landscape
model, but a version of the same conflict had broken out on another front.

Nanney wanted heritability of cellular properties to be thought of as
mediated by a cellular control system that included cytoplasmic regulatory
influences. He had decided to describe this as "paragenetic," giving the sense
of something that exists beside or accompanying genetic processes. In 1957,
he flew 14 hours in a propeller plane from the United States to France to
participate in a meeting, organized in Gif-sur-Yvette outside Paris by Boris
Ephrussi, of scientists sympathetic to the idea of cytoplasmic inheritance. It
was his first time meeting many of his European colleagues, including Guido
Pontecorvo (1907–1999), an *Aspergillus* geneticist at Glasgow University
whose education in Pisa in Italy had been focused on the classics, "so by the
age of 16 he could, for example, read and enjoy Aristophanes in Greek"
(Cohen 2000). Pontecorvo persuaded Nanney that "paragenetic" was et-
ymologically less appropriate than "epigenetic." They both felt that they
should really get input and permission from Waddington to use the term,
but unfortunately Waddington had been unable to attend the meeting, and

so the travel-fatigued Nanney relented and used the word "epigenetic" in his presentation to describe nongenetic properties of heritability of the cell.

Nanney recounts (Nanney 2004) how, after the Gif-sur-Yvette meeting, Tracy Sonneborn was infuriated that Boris Ephrussi appeared to appropriate Nanney's ideas as his own and persuaded the editor of the *Proceedings of the National Academy of Sciences* to allow Nanney to publish these ideas under his own name before they were attributed to Ephrussi. This became the 1958 paper "Epigenetic Control Systems" that gave visibility to Nanney's ideas (Nanney 1958). In this review, he describes epigenetic properties in several ways. One is what he called "persistent homeostasis," properties of the cell "perpetuated in the absence of the inducing conditions." He was open to the idea that some epigenetic control systems could be located in the nucleus and presciently warned that the assumption in such cases must be one of genetic rather than epigenetic control.

We now recognize that the mating type choice in *Tetrahymena* is due to small RNAs produced by the micronucleus using Dicer, passed through the cytoplasm to influence DNA elimination in daughter cells (Noto et al. 2015). This will be explained further in Chapter 4. However, the principal impact of Nanney's use of the word "epigenetic" was to equate it with cellular memory, which is a relatively tenuous link with Waddington's epigenetic landscape concept. We have an insight into what Waddington thought of Nanney's redefinition of the word "epigenetic" from the published proceedings of a meeting (Abercrombie 1967). In this 1967 paper, Michael Abercrombie from University College London argues that cellular and tissue differentiation is the phenotypic expression of Nanney's epigenetic systems. He recognized that the term "epigenetic system" as used by Nanney differed from its longer-standing use by Waddington and others to mean "developmental." Abercrombie referred to "epigenotype" as "the set of self-reproducing regulatory mechanisms that characterizes each of the different tissue types of an organism," acknowledging that Waddington had used the same word as early as 1939 but infrequently and with a different meaning. This must have been a nerve-racking talk for Abercrombie to give, explaining to the venerable Waddington, sitting in his audience, how to use the words epigenetic and epigenotype.

In responding to Abercrombie, Waddington was extremely courteous. "When I first used the term 'epigenotype'" he said, "it was in a different sense from yours, as a general word for the epigenetic characteristics of the organism as a whole, namely its particular system of developmental interactions, whether it has organizers or gradients or some other system. The term is not much needed today in that sense and I am perfectly willing to give it up to somebody else!" He was less convinced with Nanney's use of the term, which he attributed to Max Delbrück's "idea that differentiation depends on cells being switched between alternative steady states," which he believed he had already discussed in his *An Introduction To Modern Genetics* (Waddington 1939) in describing the switching between alternative pathways of differentiation. Waddington was, in essence, graciously passing on the words epigenetic and epigenotype to a new generation of scientists who had uses for the words different from those he originally intended.

A DEFINITION THAT SURVIVED BY MUTATING

We see how that the word epigenesis, itself a product of history, became a part of the portmanteau word epigenetic, fusing epigenesis with genetic, to describe Waddington's idea for how genetic mutations could cause morphological changes. We also find the word epigenetic to be reused by Nanney to help him make his case for nonnuclear heritability through cell division, representing the first case of a rederivation of the meaning of the word that would become a repeated event. However, both Waddington and Nanney's uses of the word epigenetic were forgotten by the time we emerged into the modern molecular era, when it would undergo an even more dramatic evolution.

RECOMMENDED FURTHER READING

Goldberg B. 2017. Epigenesis and the rationality of nature in William Harvey and Margaret Cavendish. *Hist Philos Life Sci* **39**: 8.

Great historical context for the era in which Harvey was developing his ideas of epigenesis.

Needham J. 1931. *Chemical embryology*, Volume 3. MacMillan, New York.

An excellent and comprehensive history of embryology from ancient times through to the eighteenth century.

Sapp J. 1987. *Beyond the gene: cytoplasmic inheritance and the struggle for authority in genetics*. Oxford University Press, New York.

An excellent overview of the twentieth century schism between those arguing for cytoplasmic roles in heritability and the geneticists focused on nuclear DNA. Provides insights into the political and social influences of the time.

Waddington CH. 1939. *An introduction to modern genetics*. MacMillan, New York.

In which a young Waddington, still a trainee, not only teaches about the entire field of genetics of the time but presents for the first time the ideas that became the epigenetic landscape hypothesis.

CHAPTER 3

The Multiple Definitions of Epigenetics

EPIGENETICS, SILENCED

After the late 1950s, when Nanney was re-appropriating the word "epigenetic" for his own purposes, there were two decades during which there was very little discussion of epigenetics, despite intense interest and advances in molecular development, the biochemistry of DNA, and heritability in general. In Figure 3.1, the resurrection of the term occurred in the 1970s, despite minimal use of the word in publications since 1950.

In the words of Marilyn Monk (2018, pers. comm), who went on to define the ontogeny of DNA methylation in mammals (Monk 1990):

> We all knew in the 60s that all cells had the same genes and it seemed obvious to me that different differentiated cell types meant that genes were regulated differently. Methylation was an obvious thing to look at for those of us [with] some molecular background in prokaryotes as we knew that differential methylation and its recognition was the means by which bacteria resisted viral attack (restriction/modification work—the basis of so much molecular biology of higher organisms).

The restriction/modification work referred to by Monk was a phenomenon described originally by Salvador Luria and others in the early 1950s (Anderson and Felix 1952; Luria and Human 1952; Bertani and Weigle 1953). They noted that bacterial viruses (phage) would infect some bacterial species more effectively than others, leading to the discovery a decade later that this was due to an interaction between host enzymes digesting DNA at specific sequences (restriction) that were sensitive to the addition of methyl groups in the nucleic acid of the phage (modification) (Gold et al. 1963). Restriction–modification systems and the enzymology of DNA methylation will be addressed in more detail in Chapter 4.

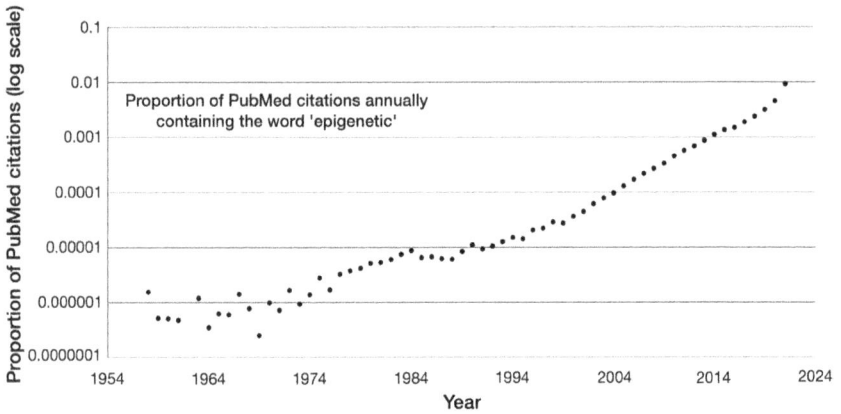

Figure 3.1. The rise of the word "epigenetic" in publications in PubMed. Note the log scale, without which we would not be able to appreciate the beginning of the sustained contemporary use of the term in the 1970s. In 2021, there were 15,202 citations with the word "epigenetic" out of a total of 1,648,629 publications indexed in PubMed (https://catalog.data.gov/dataset/pubmed-total-records-by-publication-year-fcf4a), representing almost 1% of all indexed publications.

A leader in the field of these restriction–modification systems was Ruth Sager (1918–1997). Working with *Chlamydomonas*, she was interested in the same kinds of questions of cytoplasmic inheritance as Nanney— identifying the agent transmitting information in cytoplasmic chloroplasts to be a unique DNA, and that the intriguing maternal inheritance pattern in *Chlamydomonas* was due to maternal but not paternal DNA being methylated and, therefore, differentially susceptible to restriction enzyme digestion (Burton et al. 1979).

Sager was not content with having made these extraordinary discoveries in a eukaryotic algal species and expanded her interests to cancer biology, spending a sabbatical at the Imperial Cancer Research Fund in 1972–1973 (National Academy of Sciences 2001). Her primary goal was to define tumor-suppressor genes, whose expression was downregulated in cancer, and explored mammalian paradigms like X-chromosome inactivation to gain insights into the normal regulation of gene expression. Her *Chlamydomonas* background prompted her to consider the role of DNA methylation in such processes. This work became the fuse that led to the explosion of epigenetics in the 1970s, reawakening the field from its dormancy. A critical link was an inquisitive graduate student, working in a laboratory focused on an entirely separate question, the mechanisms of DNA recombination.

JOHN PUGH GETS ON HIS BIKE

On February 5, 1973, a graduate student called John Pugh climbed onto his bicycle at the National Institute for Medical Research in Mill Hill, in the north of London, and prepared to cycle the 11 miles into the center of the city to hear a lecture by a visiting scientist, Dr. Ruth Sager, at the Imperial Cancer Research Fund laboratory in Lincoln's Inn Fields. Sager was giving two talks that week on chloroplast genetics and DNA inheritance in the alga *Chlamydomonas*. Pugh wanted to hear the talks, as he wished to follow up on his previous year's graduate student seminar on X-chromosome inactivation, which he had concluded was a frustratingly complex biological problem.

Pugh had been an undergraduate student at Oxford, where he had been assigned to John Gurdon as his "moral tutor" from 1968 to 1971.[1] Gurdon had received his D.Phil. degree with Michael Fischberg at Oxford, working on nuclear transfer in frogs, and had spent a year at CalTech before returning to Oxford. The nuclear transfer technique used by Gurdon was later called cloning and prompted the idea that differentiated cells could revert to a pluripotent state.

Another influence on Pugh during undergraduate life at Oxford was a book cowritten by Kenneth Roderick Lewis, a lecturer in the Botany Department, with Bernard John. In this book, *Chromosome Marker*, they described the work of Helen Crouse (1914–2006) on *Sciara coprophila*, an insect with an interesting pattern of elimination of its paternal but not maternal chromosomes in the developing embryo. Crouse proposed that this was due to a distinctive marking of the chromosomes in each germline, or "chromosomal imprinting" (Crouse 1960).

This all suggests that Pugh had interests other than his assigned thesis project, to characterize the hypothesized inducible system of genetic recombination in the smut fungus and maize pathogen, *Ustilago maydis*. Pugh's mentor was Robin Holliday (1932–2014), who had been appointed head of a new Division of Genetics at Mill Hill, and who had made major breakthroughs in understanding recombination of DNA using the *U. maydis* system. However, having made these discoveries, Holliday was interested in

---

[1] The role of moral tutor at Oxford is defined as the person in a college to whom a student can turn with concerns about their teaching or general welfare, allowing close interactions with faculty who might be otherwise inaccessible.

broadening his research to encompass other scientific questions like aging and was receptive to novel ideas from trainees and colleagues. In the words of his former collaborator Tom Kirkwood (2018, pers. comm.):

> There was a strong sense of community and collegiality within the group which extended to include collaborators like me and a regular stream of scientific visitors. Ideas were openly discussed and there was a vibrant journal club and seminar program... Robin's gift was for treating scientists—senior or junior—as equally qualified to join a discussion, provided they had something to contribute. He could be challenging and enjoyed debate.

So, when John Pugh began his journey into London in early 1973, he could be confident that any new ideas he developed would be received positively by his mentor. The talk on the Monday was focused on Sager's *Chlamydomonas* work, but his second trip two days later was rewarded by Sager expanding her ideas to propose that restriction/modification systems similar to those in *Chlamydomonas* might be acting in higher eukaryotes to mediate phenomena like genetic imprinting and heterochromatin formation.

To Pugh's initial chagrin, he only realized after his 44 miles of cycling that Sager was scheduled to visit Mill Hill a month later. During this lecture, while she was presenting the same ideas about restriction–modification as Pugh had heard several weeks earlier, he started to muse about the mechanism that would be needed for enzymatic modification of DNA to mediate the kinds of processes that were interesting Sager. It was during the talk that Pugh "hit on the idea of maintenance methylation," raising it as a possibility in the questions and answers session with Sager after her talk and suggesting that it may be involved in "X-chromosome inactivation, genetic imprinting, and the stability of the determined state" (J Pugh, 2018, pers. comm.).

The next day, Pugh brought his theory to Holliday, who, according to Pugh, "had been thinking about the two cycles of enzyme- and sequence-led transition mutations and the DNA clocks." This reflects how Holliday was more focused on DNA mutational processes, prompting his interest in DNA methylation as a potential mediator of mutation, as previously proposed by Scarano (Scarano 1971), a mechanism described further in Chapter 4. Both student and mentor were interested in the idea of a molecular memory mediated by maintenance DNA methylation (Fig. 3.2) from distinct starting points: Holliday focused on how the memory decayed over time to produce aging and cancer, whereas Pugh was more intrigued by how maintenance DNA methylation allowed developmental order to be created in

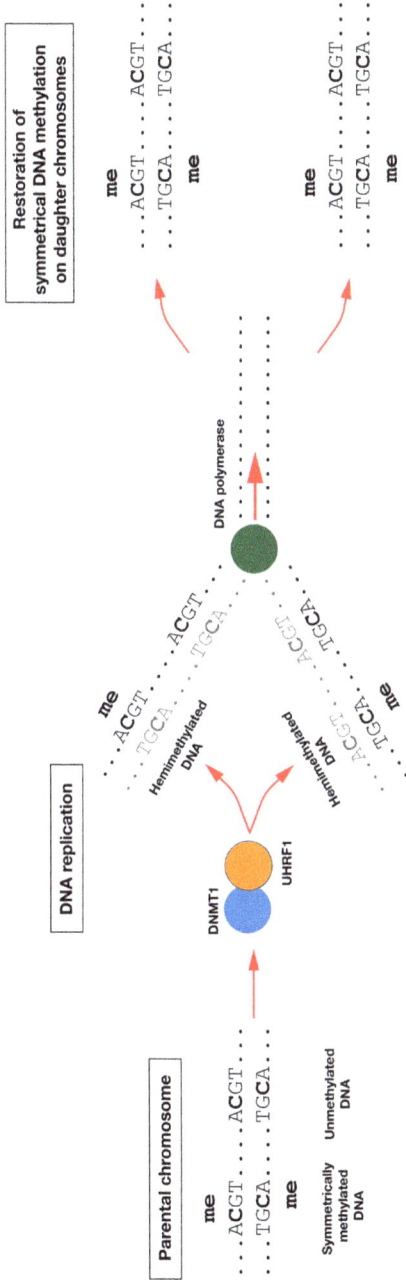

Figure 3.2. Why was there this interest in restriction–modification systems? The biochemistry of DNA methylation during DNA replication is now well-understood, as shown, with the maintenance of the symmetrical DNA methylation propagated from parent to daughter chromosomes. This permits DNA methylation to mediate a memory of past states, without altering the DNA sequence, providing a model for phenomena like X-chromosome inactivation, which initiates early in development and is maintained thereafter through cell division and differentiation.

DO CHEMICAL CARCINOGENS ACT BY ALTERING EPIGENETIC
CONTROLS THROUGH DNA REPAIR RATHER THAN BY
MUTATIONS?

J. E. PUGH and R. HOLLIDAY

National Institute for Medical Research, Mill Hill, London NW7 IAA

It has been debated whether carcinogenesis has a mutational or epigenetic basis (*e.g.*
J. H. Coggin and N. G. Anderson, *Adv. Cancer Res.*, *19*, 106, 1974). Recent work shows a
correlation between cancer and mutation production by a wide range of chemicals. (J.
McCann and B. N. Ames, *Proc. Nat. Acad. Sci. USA*, *73*, 950, 1976). A proponent of a strict
epigenetic mechanism of carcinogenesis has to explain how chemical damage to DNA can
alter an epigenetically regulated state of gene expression without requiring mutation.
    Holliday and Pugh (*Science*, *187*, 226, 1975) proposed that the methylated state of
particular DNA sequences could stably control gene expression. During development,
specific bases are methylated on both strands of these sequences by transiently-active switch
methylases. Half-methylated sequences arise after DNA replication and, to maintain the
methylated state, the newly replicated strand is methylated by constitutive enzymes specific
for half-methylated sequences.
    If a DNA lesion occurs in or near one of the methylated sequences and is repaired, by
excision or recombination mechanisms, just before or after DNA replication, a sequence
unmethylated on both strands may be generated; this would no longer be a substrate for a
maintenance methylase and would result in a stable change in genetic control. The change
is not mutational since the sequence of base-pairing specificities is unaltered. The weak
mutagen ethionine, which is a potent liver carcinogen, could act by inhibiting DNA
methylation. Further aspects of these ideas will be discussed.

Figure 3.3. The abstract in which Pugh and Holliday use the word "epigenetic" for
the first time. Reprinted from Pugh JE, Holiday R. 1978. *Heredity* **70**: 329, with permission from SpringerNature.

the first place. Pugh and Holliday started to apply this hypothesis to several
biological questions, including normal physiological processes like developmental programming, prompted by Gurdon's nuclear transfer experiments,
X-chromosome inactivation, and Crouse's genetic imprinting ideas, as well
as pathological processes like aging, cancer, and transdifferentiation. The result of this discussion was a landmark review paper in *Science* by Holliday
and Pugh in January 1975 (Holliday and Pugh 1975).

Although this was an extremely influential publication in the history of
epigenetics, it is noteworthy that it does not include the word "epigenetic" at
all. In fact, the first use of the word by Pugh and Holliday was in an abstract
(Fig. 3.3) submitted to a meeting on the biology of chemical carcinogenesis,
held at the Royal Society in May 1976 (Dulbecco 1977). As John Pugh remembers it (2018, pers. comm.):

I was not trying to define epigenetics. My preference had not been to use
that term as it was mixed up in my mind with epigenesis and Aristotle.[2] I

---

[2] It has been a common misperception that Aristotle, a Greek philosopher who studied
development, was the source of the word epigenesis, or its adjectival form epigenetic. In
fact, as described in Chapter 2, it was an Englishman (William Harvey) in a book written
in Latin (*Exercitationes de Generatione Animalium*) who coined a Greek word *epigenesis*.
Which is indeed confusing.

Figure 3.4. John Pugh (second from *left*) during his graduate student training, at an EMBO Symposium on Developmental Genetics, April 30–May 3, 1975, at Hirschhorn, near Heidelberg in Germany. In the foreground on the *left* is Werner Arber, who went on to receive the 1978 Nobel Prize in Physiology and Medicine with Hamilton Smith and Daniel Nathans for the discovery of restriction endonucleases. The bearded face of Robin Holliday is visible immediately over John Pugh's left shoulder. (Photograph courtesy of John Pugh.)

preferred the term developmental genetics as just one view or part of the wider field of developmental biology. What I was trying to do was define mutation in such a way that the new mechanism I was proposing would not be encompassed in it.

So, a definition of epigenetic as not involving a change of sequence of base pairing specificities is really a "photographic negative" of my definition of mutation. I would not have thought it was a sufficient definition for all of the field of epigenetics—if indeed you chose to use that term. I invented the idea of a "strict epigeneticist" as a device to get the background debate between epigenetic and mutational origins of cancer into the short abstract.

The critical distinction that Pugh and Holliday (both pictured in Fig. 3.4) were trying to make here was to define two distinct ways that information could be altered in the cancer cell: one by mutation, in which one nucleotide in replaced by a different nucleotide, and another distinct process in which the nucleotide remains the same but its methylation status changes, which they called "epigenetic" to distinguish it from genetic, or sequence-based, mutation. Furthermore, as Pugh clarifies, "I was not using epigenetic as a shorthand for maintenance methylation."

In his obituary for Robin Holliday (Holloman 2014), Bill Holloman independently describes the Pugh–Holliday relationship in a way that is consistent with the story recounted by Pugh:

> A number of very talented students interested in recombination mechanisms joined Robin's laboratory at Mill Hill to work on *U. maydis* genetics. One of these was John Pugh, a student who was exceptionally bright and enthusiastic. John's project was to search for mutants with constitutively elevated levels of recombination in mitotic cells. But, in his perusal of the literature, John became more and more attracted to investigating other biological phenomena. He was intrigued by X chromosome inactivation in eutherian mammals, and in frequent discussions with Robin about possible mechanisms for maintaining an active versus an inactive state, they speculated about epigenetic control in embryogenesis and development. The notion occurred to them that methylation of DNA could be responsible for gene regulation and that gene expression patterns so regulated could be stably inherited. Robin and John formulated these ideas into a theoretical framework for developmental regulation. The paper was published in *Science* (1975) and became the foundation of another focus for Robin's research efforts in parallel with his work on ageing. He continued investigation of epigenetic control for the rest of his scientific career.

By 1979, Holliday was using the word "epigenetic" in his influential proposal, "A new theory of carcinogenesis," published in the *British Journal of Cancer* (Holliday 1979). In this paper, Holliday provided the following footnote at the outset:

> This paper is in large part based on the ideas of my colleague J.E. Pugh. He first discussed the theory at a meeting of the Genetical Society of Great Britain in November 1977 (see Pugh and Holliday 1978) but did not consider that its publication in full was justified.

The 1978 Pugh and Holliday reference that he cites is the abstract shown in the figure. Holliday's use of the word "epigenetic" in this review begins as follows:

> The view that cancer is essentially a developmental aberration or a disease of differentiation is, of course, not a new one (see, for example, Markert 1968). It is well known that tumour cells often gain new surface antigens or produce proteins or tRNAs which are normally present only in embryonic cells (see Coggin and Anderson, 1974; Medawar 1977). In principle, gene mutation could produce multiple or abnormal changes in gene expression; however,

for the reasons just summarized, a strong case can be made for an epigenetic or non-mutational origin of cancer. The problem is then to explain how the initial damage to DNA by carcinogens can lead to heritable changes in gene expression, in the absence of mutation. I shall suggest that the accurate repair of DNA lesions may have the side effect of causing these epigenetic changes in somatic cells.

This paper represents an interesting transition in the definition of epigenetics. It starts by being consistent with the definition used in the 1978 Royal Society abstract, using the word epigenetic as a nonmutational mechanism involving DNA methylation, but it progresses to the further idea that DNA methylation regulates gene expression as another epigenetic property. The abstract ends up focusing on the latter idea, describing "epigenetic changes in gene expression" and how "the probability of epigenetic changes in gene activity will depend on the activity of methylating enzymes and the rate of excision repair."

MEANWHILE, IN CALIFORNIA

The idea that maintenance of DNA methylation in daughter strands following DNA replication could mediate a memory of the state of the parent cell occurred simultaneously and independently to Arthur (Art) Riggs (1939–2022). The work that Riggs performed links, for the first time in this narrative, the development of ideas about epigenetics with the *lac* repressor work of Jacob and Monod (Jacob and Monod 1961). Riggs' postdoctoral work involved purifying large amounts of the *lac* repressor protein (Riggs and Bourgeois 1968) and studying its DNA interactions (Riggs et al. 1968). The *lac* operon appeared to him to be a good example of feedback loops and mass action chemistry (Fig. 3.5), but he was intrigued by the more stable, long-term model of X-chromosome inactivation, which attracted him to join the City of Hope National Medical Center in 1969 for his first faculty appointment so he could interact with Susumu Ohno, a leader in the study of X-chromosome inactivation (Riggs 2002).

Riggs describes in an illuminating 2002 publication how he developed his early insights (Riggs 2002). Four years into his faculty appointment, he spent a short sabbatical working with Herbert Boyer, taking advantage of the Boyer laboratory expertise in cloning DNA into plasmids using the new restriction enzyme digestion approaches being developed by Boyer. He had already identified an effect of methylation of thymidine on *lac* repressor

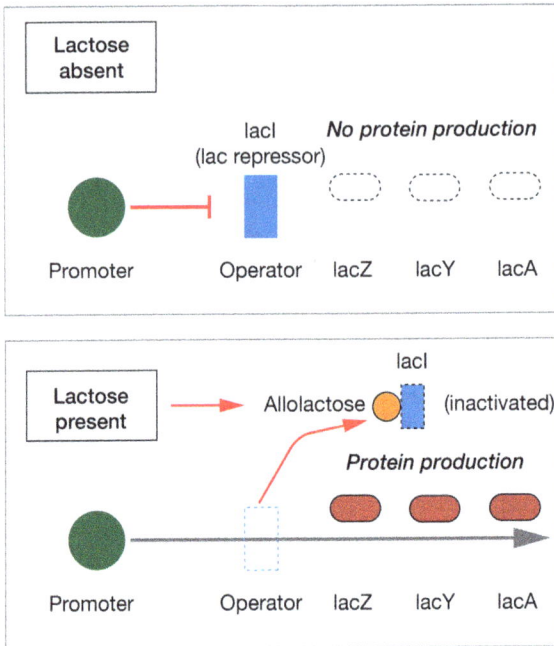

Figure 3.5. The *lac* operon: In the quest to find out how genes could be selectively expressed, a central puzzle for the mechanism of differentiation (epigenesis) in multicellular organisms, the corollary approach was taken by bacterial geneticists, asking how a single-celled organism switches gene expression in different conditions of nutrient availability. This was the work performed by Jacques Monod (1910–1976) and colleagues. What Monod first noted was that bacteria responded to the presence of a nutrient by switching on the genes required for its metabolism and switched these genes off just as quickly when the nutrient was removed. Focusing on the metabolism of lactose by *Escherichia coli*, they observed ß-galactosidase, a lactose permease and a lactose transacetylase to be switched on within minutes of lactose exposure. They found that these three genes mapped in proximity in the *E. coli* genome, and that a mutation that allowed the genes to be expressed constitutively, even without lactose being present, mapped close by. This was the first mutation ever found for a regulator of gene expression.

The group then went on to deduce that the normal regulation of this three-gene locus was mediated through the separate *I* locus in the genome, encoding a protein that bound to a sequence called the "operator" immediately adjacent to the three genes. The operator and the three genes were collectively called the *lac* operon, kept silenced at all times by the *I* protein, unless that protein was bound by the nutrient inducer (lactose), which caused its inactivation and the derepression of the *lac* operon genes.

This work allowed a molecular mechanism to be proposed for regulation of gene expression—not only in prokaryotes but also multicellular eukaryotic organisms—through selective activation and repressive mechanisms responding to environmental and other cues.

binding and was now working with restriction endonucleases like EcoRI, whose digestion was sensitive to methylation of adenines in its recognition sequence. Furthermore, he was exposed, while at the Boyer laboratory, to the characteristics of the DNA methyltransferase subunit of a type I restriction enzyme, and how it worked only on double-stranded DNA, poorly on unmethylated DNA, and efficiently only on hemimethylated DNA (when there is a methyl group already on one but not the other strand). All of these observations combined to give Riggs a sudden realization that these properties found in prokaryotes could explain "XCI [X-chromosome inactivation], as well as other somatically heritable events that do not involve changes in DNA sequence" (Riggs 2002). He discussed the ideas with Ohno, who encouraged him to publish them, but his submission to the journal *Cell* was rejected, leading him to publish it instead in the journal *Cytogenetics and Cell Genetics*, which had been founded by Ohno's collaborator Harold Klinger (Riggs 1975). In 2002, Riggs retrospectively summarized the components of his 1975 theory as follows (Riggs 2002):

1. Mammalian cells will have a DNA methyltransferase(s) with a strong preference for hemimethylated DNA;

2. Methylation will be found in symmetrical sites;

3. DNA methylation will be involved in the maintenance of XCI and imprinting; and

4. DNA methylation at the 5-position of cytosine will affect protein–DNA interactions and, thereby, chromatin structure and gene expression.

Separated by eight time zones, Riggs was developing ideas about a potential role for DNA methylation independently of Pugh and Holliday in the north of London. Both Riggs and Pugh were coming from a starting point of interest in XCI, but Riggs was more intrigued by the potential effects of DNA methylation on binding of sequence-specific proteins, whereas Holliday and Pugh pursued the idea of DNA methylation as a type of information that could establish order in cells, and whose loss could lead to aging and cancer.

Riggs (2018, pers. comm.) did not use the word "epigenetic" in his 1975 publication:

> At that time, I was unaware that Waddington had introduced the word much earlier. For that matter, I was not even aware of Waddington... I don't recall ever hearing of Waddington or the word epigenetic until well after 1975...

When I did learn of Waddington (from Adrian Bird, who got his Ph.D. from Edinburgh) and read some of his papers...I thought his use of epigenetics was significantly different from that which I favored. I liked a meaning more like that suggested by Nanney, not Waddington. I don't have any special insight into Holliday's thinking. I don't recall ever discussing it with him.

Therefore, although both Riggs and Holliday/Pugh were thinking the same way about maintenance of DNA methylation as a mechanism for maintaining regulatory information through cell division, neither initially described the phenomenon as epigenetic or perceived any influence from Waddington's epigenetic landscape in their ideas, although Riggs was possibly influenced to a minor extent by the idea proposed by Nanney, the property of persistent homeostasis and cellular memory. A common theme was that the cellular memory had to survive cell division, which became the core meaning associated with John Pugh's use of the word "epigenetic" as shorthand for nonmutational memory in his 1976 Royal Society abstract. Holliday's use of the word epigenetics in his influential, high-profile reviews helped to popularize the word in an era when a number of discoveries were being made that appeared to involve nongenetic information in the inheritance of information by cells, such as the example of X-chromosome inactivation. These paradigms also began to be described as epigenetic, expanding the use of the term and leading to the surge in publications using the word epigenetics following the 1970s.

EPIGENETIC PARADIGMS AND THE HIGHER LEVEL OF REGULATION

The increased use of the word "epigenetic" was being driven by several areas of research, but three are worth a focus, as they were probably the most influential in mammalian epigenetics, which is where Pugh, Holliday, and Riggs, despite having backgrounds in studying the biology of prokaryotic or single-cell organisms, were applying their new ideas. Two have been mentioned earlier, X-chromosome inactivation and "chromosomal" (which became "genomic") imprinting, whereas the third, cancer, became the major human disease focus for epigenetics research.

X-Chromosome Inactivation

It was known since 1949 that something distinctive was happening with the chromatin in the cell nucleus in female compared with male cells. Murray Barr (1908–1995) and his graduate student Ewart (Mike) Bertram from the University of Western Ontario published a letter in *Nature* (Barr and Bertram 1949) that described what they called a "nucleolar satellite," an area of condensed chromatin sitting just beside or obscured by the nucleolus, that was present in the neurons of female but not male cats. The story of how Barr and Bertram made this observation (Barr 1988) provides a great example of why we should be careful to understand the sources of variation in experiments. Barr wanted to test whether neuronal activity induced cytological changes. He chose cat as a model because of the accessibility of the hypoglossal nerve and location of its central nucleus near the midline of the medulla, allowing the ready comparison between the stimulated and nonstimulated sides. Mike Bertram dutifully noted the changes in the stimulated neurons, including "chromatolysis,... enlargement of nucleoli, and movement of a normally juxtanucleolar chromatin mass, later termed the 'nucleolar satellite', away from the nucleolus..." (Barr 1988).

They were puzzled why they could not find this nucleolar satellite in a couple of the animals. It occurred to Barr that they should check the sex on the animals to see whether this was associated with the difference and found that the two cats lacking the nucleolar satellite were both male and all the others were female.

Barr was skeptical that such an obvious cytological difference should have gone unrecognized to date and pursued the finding further, using nervous tissue samples from cats of known sex, finding the same result. He then studied slides of human tissue from the neuropathology collection at Toronto General Hospital and found that in "sections in which there was good nuclear detail" he could discriminate male and female samples by "the presence or absence of a larger-than-usual mass of chromatin" (Barr 1988). The observation became translated to clinical practice soon thereafter, initially using fibroblasts and subsequently buccal epithelial smears, allowing testing for female- or male-type nuclei in individuals being evaluated for intersex conditions and conditions of unusual sex chromosome numbers.

Barr and Bertram's "nucleolar satellite" became commonly known as "sex chromatin," but it was unclear why such a structure formed in female and not male cells. In 1959, Barr was in bed in a dormitory at Pennsylvania State University, the night before he was due to give a keynote address at a meeting of the Genetics Society of America, when he heard a knock at his door. The visitor was a young scientist called Susumu Ohno, who had arrived late following his trip from California, and could not wait to tell Barr,

"I know the origin of the sex chromatin." Barr stayed up and shortened his keynote address, generously giving space for Ohno to use some of Barr's allotted time to present his new findings. Ohno used this opportunity to show beautiful photomicrographs of female rat liver cells, demonstrating how the mass of sex chromatin disappeared during mitosis to be later reformed by one of the sex chromosomes (Barr 1988).

Susumu Ohno (1928–2000) had been born in Seoul, South Korea, to Japanese parents. He was the only one in his family to emigrate from Asia, initially moving to the University of California, Los Angeles, and subsequently to the City of Hope National Medical Center in Duarte. He developed skills visualizing cells in the living bone marrow and then began to use these techniques on cytogenetic preparations. One observation that remains associated with Ohno arose from his documentation of the varying amount of chromosomal material between organisms, in a way that was not predicted by their relative positions on the phylogenetic tree. He attributed this to the accumulation of what he called "junk DNA," which did not encode genes (Ohno 1972), a description that is still taken as a challenge by those of us studying the function of noncoding DNA.

His conclusion that one of the X chromosomes formed the sex chromatin body was inferred from studies of the appearance of the chromosomes during mitosis. In female rat liver cells, he described pycnosis (a cytological term meaning the appearance of shrinking and becoming more dense) of one but not the other X chromosome, but no pycnosis of the male's single X chromosome (Ohno et al. 1959). This was the observation that prompted Ohno to knock on Murray Barr's door late that night in the dormitory at Penn State, proposing that the sex chromatin was not, as then speculated, due to the juxtaposition of both X chromosomes, but the selective condensation of one entire X chromosome. As Riggs later wrote (Riggs 2002):

> Colleagues who worked with him at that time told me that Ohno thought that the heterochromatin-like X chromosome of the Barr body would be genetically inactive, but he was persuaded to leave this idea out of the first paper, as it was just speculation.

If Susumu Ohno was known as the father of X inactivation (Beutler 1998), the mother of the field was about to take Ohno's ideas substantially further. Mary Lyon (1925–2014) published a letter in *Nature* in 1961 that succinctly assembled several lines of evidence to propose the theory of what she described as X-chromosome inactivation. Lyon was a University of

Cambridge graduate, where she read Zoology, Physiology, and Biochemistry and was the Turle Scholar and recipient of the Crewdson Prize. That was the time when women represented less than 10% of the students in the University and could only receive a titular degree, as women could not be full members of the university. Following her Ph.D. in 1948, she worked for a period in the laboratory of Waddington and then moved on to work on the genetic effects of radiation, a health concern in the new era of nuclear war. Her eventual long-term research home was the Medical Research Council Radiobiological Research Unit at Harwell, where she went on to head the genetics section.

Lyon was influenced not only by Barr and Bertram and Susumu Ohno but also by the finding that mice with a single X chromosome and no Y chromosome were viable and fertile, Ernest Beutler's finding that females carrying X-linked G6PD deficiency had a subset of cells that were G6PD-deficient, and Liane Russell's observation of a similar "variegation" of gene expression in mice with translocations between the X chromosome and autosomes. Lyon also noted the variegation effect of X-linked genes manifested by cats, and that the sex chromatin body was present from as early as the blastocyst stage, prompting the model that inactivation chose an X chromosome early and persisted thereafter (Lyon 1992). She modified the theory when additional sex chromatin bodies were noted in females with more than two X chromosomes, reversing the emphasis to focus on one X remaining active in the presence of other X chromosomes. As phenotypic abnormalities were noted in individuals with unusual numbers of X chromosomes, she postulated that there were some genes on the X chromosome that did not undergo the same inactivation and whose expression would be dependent on the number of X chromosomes present. These predictions all ended up being correct.

X-chromosome inactivation gave us a paradigm for an early inactivation event that persisted even as cells progressed through many cell lineage decisions. To those like Art Riggs, who had come from the *lac* operon background, and John Pugh, struggling to deliver a punchline for his graduate student symposium, X inactivation was an enormously challenging model of gene regulation. The maintenance of DNA methylation through cell division allowed a molecular mechanism to be invoked and suggested a mysterious kind of information that overrode the normal transcriptional program, so that even though both X chromosome alleles are exposed to transcriptional regulators, only one can respond.

Genomic Imprinting

The unusual behavior of chromosomes inherited from the paternal germline was first described by Charles Metz, who worked on the fungus gnat, *Sciara coprophila*. Metz ended up studying *Sciara* following a combination of disappointment and resourcefulness, as recounted by *Sciara* geneticist Susan Gerbi (2018, pers. comm.), who met him in 1970 in Woods Hole where Metz had retired, while she was completing her Ph.D. at Yale:

> He was a grad student at Columbia. As a first-year grad student he suggested to his supervisor (Thomas Hunt Morgan) that he would like to see if *Drosophila* had giant polytene chromosomes similar to those described by Balbiani in 1881 for the gall midge *Chironomus*. However, he was still taking courses and had not yet begun his thesis research. Morgan thought it seemed like an interesting idea and squashed some *Drosophila* salivary glands and discovered their polytene chromosomes! (A little under two decades later, in 1933 Heitz and Bauer and also Painter published their findings on *Drosophila* polytene chromosomes.) Disheartened that his idea had already been done, Metz was searching for another topic for his Ph.D. thesis and a friend suggested that he should study the fly *Sciara* that was in the pigeon house at Cold Spring Harbor—and that started Metz's career with this fly.

Sciara had two shots at becoming a major model organism in modern genetics. Susan Gerbi described the history of this model organism in a 2017 interview (Gelling and Genetics Society of America 2017):

> Around 1914 Charles Metz decided to study *Sciara* for his PhD thesis at Columbia. He captured it in the pigeon house at Cold Spring Harbor Laboratory on the suggestion of a friend. It took him quite a number of years to figure out the chromosome mechanics, but he ultimately succeeded and ended up dedicating his career to studying *Sciara*.

> In the 1930s geneticists had a meeting at Cold Spring Harbor and realized that they would make more progress if they all worked on the same organism. They discussed which to choose, and the two finalists were *Sciara* and *Drosophila*. We all know who won! The reason *Drosophila* was chosen was because geneticists of the 1930s relied on making mutations by X-irradiation, and *Sciara* turns out to be extremely resistant to X-irradiation. This is another of its unique biological features, but it was not good at the time. *Sciara* surfaced again in 1970–71 when Sydney Brenner spent two years in the library trying to figure out a good model system for developmental biology. *Sciara* made his final shortlist of six organisms, but the winner of that competition was the nematode worm *C. elegans*.

Charles Metz had a laboratory member by the name of Helen Crouse (1914–2006), who eventually became the driver of the major discoveries about chromosomal imprinting in *Sciara*. Crouse moved to the University of Missouri to perform her graduate studies, where she met Barbara McClintock, who became her Ph.D. thesis advisor remotely from her new position at Cold Spring Harbor Laboratory. Of the only three Ph.D. students supervised by McClintock, Crouse was the sole student who remained in science as a career.

The phenomenon of chromosomal imprinting in *Sciara* is illustrated in Figure 3.6. During male meiosis, there are three separate times when chromosomes from the paternal origin do something distinctive. Male *Sciara* have two X chromosomes, females one, and sperm brings two X chromosomes to add to the one from oogenesis, creating a trisomic embryo. The first event is the recognition and elimination of one of the X chromosomes derived from sperm. There is a stage that follows when the other paternal/sperm-derived chromosomes decondense cytologically, and then a third event during the first meiotic division of spermatogenesis in male offspring in which the remaining three autosomes and a single X chromosome from the paternal source fail to segregate on the spindle and are lost to daughter cells. The remaining three autosomes and a single X are now all maternally derived and have to undergo a process of nondisjunction to create a second X chromosome for mature sperm.

This ability to recognize paternal chromosomes was described by Crouse as reflective of some sort of marking left by recently being in sperm cells, a "chromosomal imprint." As Crouse put it (Crouse 1960):

> First, the dramatic chromosome unorthodoxies in *Sciara* are clearly unrelated to the genic make-up of the chromosomes: a chromosome which passes through the male germ line acquires an "imprint" which will result in behavior exactly opposite to the "imprint" conferred on the same chromosome by the female germ line. In other words, the "imprint" a chromosome bears is unrelated to the genic constitution of the chromosome and is determined only by the sex of the germ line through which the chromosome has been inherited.

It is a pity that *Sciara* was considered unworthy of mainstream model organism status by the geneticists convened at Cold Spring Harbor Laboratory and later by Sydney Brenner, as this dramatic parent-of-origin effect would have become a more mainstream area of research. Instead, we

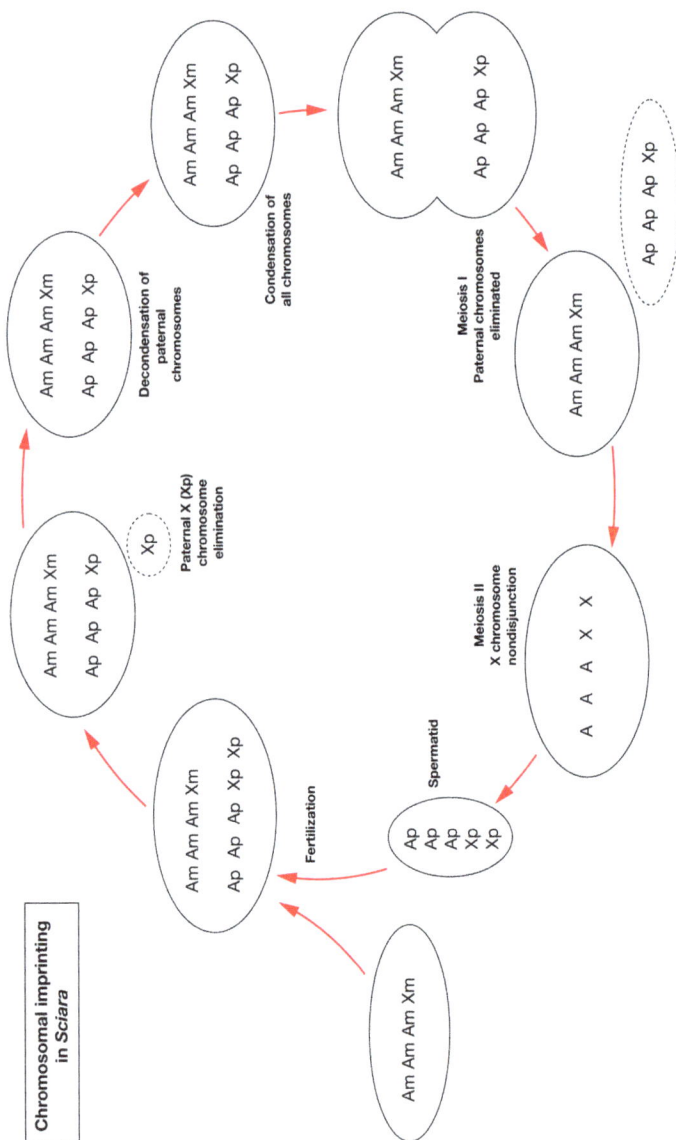

Figure 3.6. The distinctive way that paternal chromosomes are treated during male meiosis in *Sciara* represents a model for chromosomal imprinting, the memory of the last germline in which the chromosome was propagated. Somehow the cell knows which chromosomes came from the paternal germline, selectively excluding on of the two paternal X chromosomes, decondensing the remaining chromosomes, and then eliminating the four paternal chromosomes completely when making a haploid gamete during meiosis I. This parent of origin effect was later found to exist for domains within mammalian genomes, but nothing as dramatic as the elimination of chromosomes in *Sciara*.

had to wait until the 1980s for the demonstration of a similar memory of parental origin, this time in mammals, to heighten interest in this biological phenomenon.

Davor Solter didn't expect to find parent of origin effects. As he put it in 2015 (Solter 2015):

> We wanted to explain why parthenogenesis doesn't work in mammals. We were expecting that an embryo to which you'd transferred two male or two female pronuclei would develop perfectly well because they were diploid embryos that had all the necessary chromosomes. The usual problems involved in parthenogenic failure were eliminated because the zygotes were fertilized normally, so any contributions from the sperm were already there. To our surprise, they never developed. It took us, Jim McGrath and me, probably two or three years to convince ourselves that these embryos were not going to develop and that there must be something different between the male and the female genome. That was the beginning of imprinting. The results were completely unexpected because none of the classical Mendelian genetics suggested that something like that should exist.

Solter's graduate student of the time Jim McGrath (1952–2024) added more perspective (2018, pers. comm.):

> While the question of parthenogenesis was prominent in our thinking, it was not the only reason we questioned the possibility of imprinting. There was also the curious case of the T-hairpin mice—T^{hp}. This is a deletion on proximal mouse Chromosome 17 and part of the T complex. D.R. Johnson published a paper in 1974 (Johnson 1974) where he showed that the deletion is viable when paternally inherited and lethal when maternally inherited. I was pretty familiar with this paper when it first came out as I did my graduate work on the T locus with Nina Hillman at Temple. Johnson's observation was hard to understand. This was a heterozygous deletion; the first meiotic division in the oocyte occurs shortly before fertilization and the second division occurs just after fertilization and the chromosomes are condensed so it's not likely there could be gene transcription, yet zygotes that retained the nondeleted wild-type chromosome were viable and those that retained the maternal T^{hp} deletion were not. This made a cytoplasmic deficiency in all the eggs from T^{hp} mice unlikely as an explanation. Also around that time I became aware of work that George Snell had published in the mid-1940s (Snell 1946). While working with translocations, he described "noncomplementation" for offspring that inherited two paternal copies of proximal Chromosome 17. In other words, he never saw progeny that had paternal uniparental disomy for

proximal 17. You couldn't be sure that perhaps the products of male meiosis might not create such gametes, but his results were also consistent with a maternally expressed only gene on proximal Chromosome 17.

So, the first two experiments we wanted to do were (1) pronuclear transfer between T^{hp} and wild-type zygotes, and (2) to create androgenones and gynogenones, and we were working on both experiments in 1982. We finished the T^{hp} project first and showed that maternal T^{hp}-deleted fetuses still died despite wild-type egg cytoplasm and sent that paper off in 1983 (McGrath and Solter 1984a), and the gynogenones and androgenones paper we sent the next year. We now know that the *Igf2r* gene is in the T^{hp} deletion, and that explains the maternal T^{hp} lethality.

For the gynogenone and androgenone experiments, McGrath was harvesting embryos from matings of BALB/c females with C57BL/6J males. He would remove the pronuclei (the separate nuclei from the oocyte and that introduced by the sperm) and then re-introduce either two maternally derived pronuclei (to generate "gynogenones"), two paternally derived pronuclei (creating "androgenones"), or one from each parental origin, as controls. Because BALB/c and C57BL/6J differed by coat color and glucose phosphate isomerase-1 (GPI-1) alleles, the genotype of the offspring could be confirmed.

McGrath and Solter found that the only embryos that survived were those which had received pronuclei from each parent, and that none of the gynogenones or androgenones completed pregnancy. They recognized that one-quarter of the androgenetic embryos should have a lethal YY sex chromosome complement, but this was not enough to explain the universal loss of these embryos. As they wrote, "We conclude that the maternal and paternal contributions to the embryonic genome in mammals are not equivalent and that a diploid genome derived from only one of the two parental sexes is incapable of supporting complete embryogenesis" (McGrath and Solter 1984b). Simultaneously at the University of Cambridge, Azim Surani was showing the same phenomenon and used the word "imprinting" to describe the memory of the chromosomes of their last gametogenesis, whether spermatogenesis or oogenesis (Surani et al. 1984).

It took until 1991 for Denise Barlow (1950–2017) to identify the first imprinted gene. She was studying the T-associated maternal effect (Tme) locus on Chromosome 17 in mouse. The odd thing about this locus was that when a deletion of the locus was inherited from the mother, the embryos died in utero, but paternally inherited deletions had no such viability consequences, as had been noted earlier by Snell. When Barlow mapped genes to the deletion region, she found the insulin-like growth factor type-2 receptor (*Igf2r*) to be one of those present, and that embryos only express the copy on the maternal chromosome (Barlow et al. 1991). This created a molecular basis for the nonequivalence of the maternal and paternal genomes in mice, if some genes were expressed in a manner that depended on which germline they last experienced.

Although the physical loss of chromosomes in *Sciara* is obviously a much more dramatic process than a molecular signal switching off a gene, genomic imprinting gave us a paradigm other than X-chromosome inactivation in which there was a repressive signal, probably established as early as gametogenesis, also overriding the normal transcriptional program, resulting in only one allele being capable of responding to transcriptional regulatory factors.

Cancer Epigenetics

There will be a dedicated discussion of cancer epigenetics in Chapter 8, but it is worth adding some historical context here, as we dissect the evolving meaning of the word epigenetic. Robin Holliday was an early and major contributor to the possibility that cancer could result from epigenetic mutations (Holliday 1979), whereas Melanie Ehrlich (Gama-Sosa et al. 1983) and Andrew (Andy) Feinberg and Bert Vogelstein (Feinberg and Vogelstein 1983a) noted the loss of DNA methylation in cancer, linking Holliday's speculation to a specific molecular mechanism. However, attention began to focus on a genomic feature described by Adrian Bird as the CpG island (Bird 1986), a region unusually rich in cytosine-guanine (CG, CpG) dinucleotides. Bernhard Horsthemke showed that the cancer genome is not uniformly demethylated but has loci such as the CpG island at the promoter of the retinoblastoma suppressor (*RB*) gene that gain DNA methylation in cancer (Greger et al. 1989). With the development of methylation-specific polymerase chain reaction (PCR), which allowed a number of loci to be tested simultaneously in the same sample, it began to be appreciated that multiple CpG islands were simultaneously gaining DNA methylation in specific tumors (Herman et al. 1996), eventually leading to the recognition that a subset of colorectal tumors were characterized by dramatic patterns of acquisition of DNA methylation by CpG islands, referred to as the CpG island methylator phenotype (CIMP) (Toyota et al. 1999a).

Although today's sequencing technologies have led to the insight that CIMP is a consequence of mutations activating KRAS (Serra et al. 2014) or BRAF (Fang et al. 2014) or by mutating *IDH1* (Turcan et al. 2012) for a long period of time, it was assumed that DNA methylation was, in effect, making its own decision to target these loci, leading to the silencing of the genes nearby. With X-chromosome inactivation and genomic imprinting also involving the selective DNA methylation of CpG islands, we were faced with the possibility of the existence of an as-yet undiscovered higher level of organization of the genome that was not only critical for normal development but could go awry in human disease.

THE RISE OF THE GENOMIC BIOCHEMISTS

If there was a higher level of organization of the genome, it should be mediated by molecular processes. The discovery of genomic imprinting was followed by the identification of the imprinting of *Igf2r*, providing a mechanistic basis for the failure of androgenesis and gynogenesis (also referred to as parthenogenesis, indicating "virgin" conception). X-chromosome inactivation was also being linked mechanistically to DNA methylation, revealed by exposure of cells to the DNA demethylating drug 5-azacytidine, causing reactivation of genes on the silent X chromosome (Mohandas et al. 1981).

Adding to the excitement was the enormously fertile period of study by what could be described as "genomic biochemists," starting in the 1980s, when David Allis was gaining insights into histone biology in *Tetrahymena*, insights subsequently brought to mammalian systems in the 1990s and since. Tim Bestor was cloning and characterizing mammalian DNA methyltransferases and showed, in a collaboration with En Li and Rudolf Jaenisch in the early 1990s, that the mutation of the DNA methyltransferase 1 (*Dnmt1*) gene in mice was not compatible with survival past very early embryogenesis (Li et al. 1992). In 2000, Thomas Jenuwein characterized the mammalian homologs of fly and yeast histone methyltransferases, showing how they catalyzed post-translational modifications of mammalian histones (Rea et al. 2000) and setting the stage for the proposal, in 2000, by David Allis and his postdoctoral fellow Brian Strahl, that these modifications exist on individual histones in specific combinations that represent a "histone code" (Strahl and Allis 2000).

These only represent the tip of the iceberg of new insights from scientists using biochemical approaches to understand mammalian transcriptional regulation. A more comprehensive discussion of these findings will follow in Chapter 4. As these new insights about histone modifications, histone variants, chromatin accessibility, and DNA methylation were being made, scientists were exploring their favorite epigenetic paradigms and finding associations. The histone variant macroH2A was being observed to be enriched on the inactive X chromosome (Costanzi and Pehrson 1998), the chromatin insulator protein CTCF was found to be instructed by DNA methylation to guide where it should bind at loci undergoing genomic imprinting (Bell and Felsenfeld 2000), and tumor-suppressor genes were being found to acquire DNA methylation at promoter CpG islands, resulting in their silencing (Herman et al. 1994).

The upshot for all these exciting advances was crystallized in a 2000 paper that reflected what had become the pervasive new definition of epigenetics. As the author put it (Stern 2000):

> The Greek prefix means 'above', and therefore epigenetics might be translated as 'above genetics', or 'above the genes'. Were it not in current use in other contexts, it might now be interpreted as 'mechanisms that control gene expression', and could therefore be appropriate as a description of some developmental mechanisms.

This is a clear back-translation of the word epigenetics, splitting it into epi- and -genetics to signify the mysterious higher level of regulation that kept one of the two X chromosomes and imprinted alleles silent because of events very remotely during development and causing cancer cells to switch off tumor-suppressor genes that did not have any mutations. As such, it captures the excitement in the field that we may be able to understand a regulatory process that can override our genetic makeup, potentially mediating environmental influences, and alter our susceptibility to diseases. This epi+genetics definition became enormously popular as a result.

There are, however, major problems with this epi+genetics back-translation. One is that it is historically untethered—the word epigenetics was originally a fusion between epigenesis and genetics as envisaged by Waddington and was never intended to be split into epi- and genetics. Second, it only refers to a higher level of regulation of genomic function and does not encompass the idea of memory passed from parent to daughter cells, as emphasized when the word epigenetics was resurrected by Pugh and Holliday. Third, it is overencompassing—any putative transcriptional regulatory influence is now being described as epigenetic, such as the addition of a methyl group to

RNA, leading in turn to the even more corrupted term "epitranscriptomics" to describe the study of such modifications (Saletore et al. 2012).

As it is obvious that definitions matter in science, to ensure that we reduce the room for ambiguity when two researchers use the same term, there were several attempts to impose order on this definitional flux. Art Riggs in 1996 defined epigenetics as follows (Russo et al. 1996):

> The study of mitotically and/or meiotically heritable changes in gene function that cannot be explained by changes in DNA sequence.

A few years later, Adrian Bird shifted the definition toward a more biochemical footing, but maintaining the memory component by using the word "perpetuate" (Bird, 2007):

> ...the structural adaptation of chromosomal regions so as to register, signal or perpetuate altered activity states.

With all this ambiguity comes challenges. When Riggs included "mitotically and/or meiotically heritable changes in gene function...," he was being a careful scientist, noting that there was no reason why what applies to mitosis should not also apply to meiosis. This in turn has led to a further definition of epigenetics that sits awkwardly with the others—the possibility that, in multicellular organisms, information not encoded in genomic DNA is passed from parent to offspring. The field that studies this possibility also refers to its area of research as epigenetic, even though a chromatin biologist or someone studying cell fate decisions would find it difficult to reconcile such a definition with their own areas of research.

FOUR MAJOR, DISCONNECTED DEFINITIONS OF EPIGENETICS

Reviewing the definitions of the last two chapters, we find that each really had very little foundation in the previous definition and were largely de novo creations. To put each definition simply:

1. Waddington's epigenetic landscape
   He was trying to illustrate how cell fate decisions (epigenesis) could be influenced by genetic mutations, extending the model to include environmental perturbations, leading to altered phenotypes.

2. Nanney's persistent homeostasis
   Nanney had two uses for the word epigenetic: one referring to extranuclear (cytoplasmic) inheritance of information by the daughter cell, and

the second describing information that could then be maintained in the absence of the inducing condition, a persistent homeostasis. Neither component of his definition is rooted in the cell fate decisions highlighted by Waddington, probably because it didn't apply to the single-celled organism on which Nanney worked.

3. The epi+genetics definition
   Any influence on transcription regulation. Purists continue to require that the biochemical process be heritable from parent to daughter cells, but this influence (reflecting Nanney's idea of persistent homeostasis) is largely abandoned today.

4. Multigenerational memory
   The inclusion of parent to daughter cell transmission of information, expanded to include meiosis, allows the multigenerational definition of epigenetics to be proposed.

These four definitions are depicted in Figure 3.7. Chapter 5 will attempt to show the value of a definition of epigenetics that is focused on cellular properties, to amalgamate the most coherent and useful aspects of the fragmented definitions into something that Waddington might recognize today. With the emergence of single-cell genomic technologies, bringing a new cellular perspective to what we call epigenetic mechanisms and cellular or organismal phenotypes is probably timely.

WHY ARE MULTIPLE DEFINITIONS OF EPIGENETICS A PROBLEM?

The anguish above about the fragmented definitions of epigenetics is not new. As Art Riggs put it (2018, pers. comm.):

> In the 80s the "chromatin field" discovered epigenetics and started using it for just about any change in chromatin. I thought this was too broad. I thought that epigenetics should be used only when there was some form of heritability, at least somatic heritability. This led to my definition that has been often quoted from the book on *Epigenetic Mechanisms of Gene Regulation* (1996, Volume 32), which I edited with V.E.A. Russo, and R.A. Martienssen, writing the introduction and a chapter. The broader usage seems to have prevailed; I have given up trying to keep it confined to just persistent heritable changes.

So, what is the problem? Surely people are allowed to have different ideas about the same area of science? Why impose a single definition, and who is in the exalted position to direct everyone in this way? It has been said that

Figure 3.7. The four major historical definitions of what it meant for something to be epigenetic. Oddly enough, Waddington's original definition is the only one that required DNA sequence changes.

Waddington's epigenetic landscape

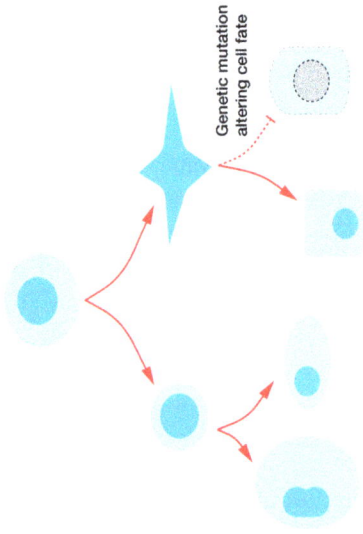

Genetic mutation altering cell fate

Nanney's persistent homeostasis

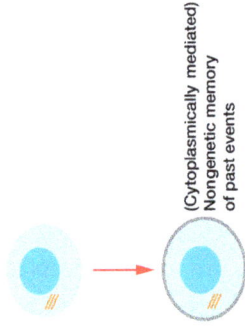

(Cytoplasmically mediated) Nongenetic memory of past events

The epi+genetics definition

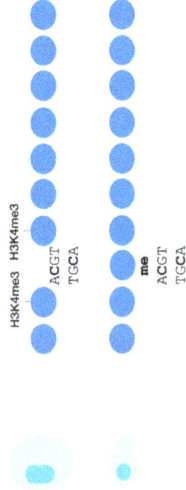

H3K4me3 H3K4me3

me
ACGT
TGCA
me

me
ACGT
TGCA
me

The same DNA sequence, but different organization of biochemical regulators

Multigenerational memory

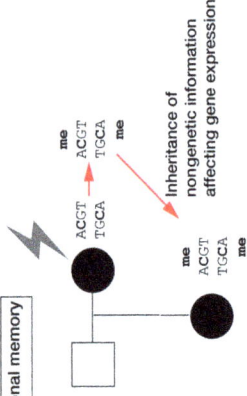

me
ACGT
TGCA
me

me
ACGT
TGCA
me

Inheritance of nongenetic information affecting gene expression

scientists would rather use each other's toothbrushes than each other's hypotheses; any attempt to unify or limit the definition of epigenetics would inevitably be met with stubborn resistance.

A major issue at present is that the definitions of epigenetics end up bleeding into each other. When a chromatin regulatory process is described as epigenetic, using the epi+genetic definition, someone studying multigenerational memory may infer that it must therefore be capable of mediating phenotypes across generations. Without such a definitional escape hatch, we are forced to ask specific questions instead. For example, how the addition of a methyl group to DNA in lymphocytes (epi+genetic) could find its way to gametes and have that methyl modification be propagated to the lymphocytes of the next generation (multigenerational memory), which is much more challenging when you could instead use a vague, ambiguous term to link the phenomena.

We should be aware that, in describing our research in publications, talks, or funding proposals, when we use a word like epigenetics with its multiple meanings, what may be clear in our own minds might be resonating with a substantially different interpretation of the word by the colleagues receiving the message.

As mentioned earlier, the extremely loose epi+genetics definition, which includes all candidate regulators of transcription, leads to the implication that any such regulator can propagate itself from parent to daughter cell, when such self-propagation is, in fact, biochemically implausible for most regulators.

Finally, we have the very weird situation that the extremely permissive epi+genetics definition traditionally selectively excluded one major type of regulatory element, the transcription factor. In Chapter 4 the historical reasons for this omission are described, as well as the need to bring back this group of regulators centrally into the discussion of epigenetics.

Our lack of rigor in the use of the word epigenetics has consequences for the broader public outside the specialized scientific community. The idea that there is a higher level of regulation acting above our genetic sequence is extraordinarily attractive. If a nonmutated tumor-suppressor gene can be silenced in cancer, or an allele can be silenced by X-chromosome inactivation or genomic imprinting whatever the sequence of that allele, and the mediators of these regulatory processes could be potentially influenced by the environment (Chapter 7), then we have opened up the possibility that interventions can change our health, positively or negatively, and overcome our innate genetic susceptibilities. That genetic nondeterminism exists is reflected by the discordance of monozygotic twins for certain conditions; there is nothing inherently wrong with this train of logic.

However, the use of epigenetics as a putative mechanism for genetic nondeterminism also attracts pseudoscience like an overripe banana attracts fruit flies. We have seen an explosion of health products come to market being described as epigenetic, whether yoga, diets, face cream, meditation, or whatever other grim realities have come to pass since the time of writing. Whenever we perform a study associating epigenetic changes with phenotypes, we become proponents of genetic nondeterminism, at least to some extent, so that I and my other colleagues in this field must accept some responsibility in fostering the misuse of the term epigenetics in this context.

Which leads to the final, entertaining possibility. If you are a proponent of the environment affecting your "epigenome," the genetic nondeterminism hypothesis, you have no common ground with those who propose that your health risks are locked in because of the misbehaviors of your parents or grandparents, the adherents of epigenetics as the mediator of multigenerational memory (Chapter 9). If I drink a glass of red wine, the resveratrol it contains has histone deacetylase inhibition properties—does this modify my own health risks, or am I doomed because of a grandparent who was unusually stressed early in life? For some reason, nobody has yet put the proponents of these irreconcilable hypotheses into what would be an entertaining debate with each other, illustrating how the word epigenetics is now not only ambiguous, it is self-contradictory.

A SOLUTION: WHEN YOU SAY EPIGENETICS, GO ON TO SAY WHAT YOU MEAN BY THE TERM

There is no way that the constantly evolving definition of epigenetics will coalesce to a single, universally approved definition. What we should do, therefore, is to continue to use the term, but to add each time our specific, contextual definition of what we mean. For example, we may be talking about "...the epigenetic processes at work in my system...," to which we add the phrase "...by which I mean..." and then an explanation of what we are trying to communicate (Greally 2018).

That is how this book is written from this point forward. When discussing, for example, DNA methylation, it is relevant in terms of Waddington's cell fate decisions, Nanney's propagation of cellular memory, and information over and above DNA sequence and is also sought as a mediator of multigenerational inheritance. These definitions, plus the integrated, cellular definition of epigenetics, will all be maintained as threads through all the subsequent discussions. It is difficult to write a textbook about a topic with multiple meanings; this solution is cumbersome but the only practical approach that suggests itself.

RECOMMENDED FURTHER READING

Papers that come in pairs, showing how the same discoveries and insights were happening simultaneously in different parts of the world.

Holliday R, Pugh JE. 1975. DNA modification mechanisms and gene activity during development. *Science* **187:** 226–232.

Riggs AD. 1975. X inactivation, differentiation, and DNA methylation. *Cytogenet Cell Genet* **14:** 9–25.

These 1975 papers describe the independent development of the idea of DNA methylation being heritable and regulatory.

McGrath J, Solter D. 1984. Completion of mouse embryogenesis requires both the maternal and paternal genomes. *Cell* **37:** 179–183.

Surani MA, Barton SC, Norris ML. 1984. Development of reconstituted mouse eggs suggests imprinting of the genome during gametogenesis. *Nature* **308:** 548–550.

These two 1984 papers independently describe the discovery of genomic imprinting in mammals.

CHAPTER 4

Molecular Mechanisms of Epigenetic Properties

The prior chapter should not be read as dismissive of the achievements made in the field of genomic biochemistry. The advances made in understanding chromatin biology, DNA modifications, the functions of noncoding RNAs, and other regulatory processes have been astonishing and serve as the foundation for understanding the mechanisms of whatever definition of epigenetics one might be using. The concern raised in the last chapter was that, by calling all such processes epigenetic—defined as a higher level of information regulating how genomic DNA is used—we have created a new source of ambiguity about what epigenetics means.

In this chapter, we will delve more deeply into the many molecular processes described as epigenetic, understanding what each does in the cell and allowing us to interpret how each may mediate cell fate decisions and cellular memory, the more deeply rooted definitions. The further definition of epigenetics—that describing multigenerational inheritance—will be described here only to a limited extent, as this complex question will be separately dealt with as the focus of Chapter 9. For the remainder of this current chapter, we will dive deeply into the complex and fascinating world of the biochemistry of genomic regulation. However, because this topic is big enough to fill an entire book by itself (Allis et al. 2015), the chapter will be focused on identifying those aspects of genomic regulation that can mediate "epigenetic" properties—that is, those defined in the preceding chapter.

THE NECESSARY PROPERTIES OF EPIGENETIC REGULATORS

Fresh from considering the competing definitions of epigenetics in Chapter 3, we can now regard each of the molecular processes below in terms of what is required of it to mediate each definition. To recap, it is easy to call any molecular process epigenetic if it acts "above" or "upon" the genome, based on the back-translation of epi- and -genetics. Waddington would be pleased if the process involved cell fate decisions, whereas Nanney would be gratified if it mediated long-term cellular memory.

Condensing the prior chapter, the following appear to be reasonable, necessary properties of molecular regulators of epigenetic states.

1. They allow the same DNA to be used in different ways.
   This makes the top of the list because it represents the pervasive epi+genetics definition, anything acting above/upon the DNA sequence. Whether this is the same sequence in different cells or on homologous alleles in the same cell, anything that can distinguish these sequences in molecular terms would probably be considered epigenetic by many.

2. They maintain a molecular state stably over time and through cell division.
   This is no longer the fundamental requirement that it was in the past but reflects Nanney's idea of persistent homeostasis and the interest of Pugh/Holliday and Riggs in a molecular message being stably transferred to daughter cells.

3. They can be targeted to specific genomic regions.
   This is almost never considered to be a necessary property of epigenetic regulators but is critically important—something initially sent the regulatory process to that region in the genome, even if we typically focus on its "epigenetic" capacity to maintain itself locally thereafter through cell division. If the same sequence is being used differently in different cells or alleles, as described in the first property above, this reflects differences in targeting. We will include this property of targeting as part of the description of the candidate epigenetic regulators below.

There is an important further component to the second property above, that of stability, a feature originally described by Adrian Bird (a successor to Conrad Waddington as Buchanan Professor of Genetics at the University of Edinburgh) in his 2007 definition of epigenetics (Bird 2007):

> ... the structural adaptation of chromosomal regions so as to register, signal or perpetuate altered activity states.

What has gone generally unnoticed was the intent Bird (2018, pers. comm.) had in using the word "perpetuate" (Bird 2007):

> I was indeed thinking of non-dividing cells when using the word "perpetuate" in my definition. As I say elsewhere in the same article: "The restrictiveness of

the heritable view of epigenetics is perhaps best illustrated by considering the brain. A growing idea is that functional states of neurons, which can be stable for many years, involve epigenetic phenomena, but these states will not be transmitted to daughter cells because almost all neurons never divide."

Many cells in multicellular organisms are postmitotic and would be candidates for mediating the phenotypes studied when testing epigenetic changes in human diseases. What Bird describes is a way of thinking about long-term transcriptional regulatory changes in postmitotic cells capable of mediating cellular memory of past events, and it is therefore consistent with Nanney's idea of a cell's epigenetic state reflecting a "persistent homeostasis." It is therefore reasonable to separate out what we mean by long-term changes in cells that are still undergoing division and those that are postmitotic. The broader epi+genetics definition gains relevance in postmitotic cells, as a persistent homeostasis can be linked to any molecular changes that persist in the cell.

What follows is a sampling of many of the molecular genomic processes referred to as epigenetic. Although no such review can be fully comprehensive, what's included below should represent a substantial proportion of the candidate transcriptional regulatory systems in the mammalian genome. Many but not all mechanisms also occur in plants, but it should be noted that the organization of genes and transposon-rich heterochromatin in plants is quite distinctive from animals, so the information presented here with a mammalian focus should not be considered universally applicable.

Addressing the multiple definitions of epigenetics, it should be apparent that any candidate transcriptional mechanism will fall under the current epi+genetic definition, so this will not be discussed as a property of these genomic regulators. However, for each one we can also ask the questions whether it is known to be *stable* through cell division and whether or how it can be *targeted* to specific loci in the genome. By addressing these two additional properties, we will be able to scrutinize each regulatory process to assess whether it could be considered epigenetic beyond the pervasive epi+genetics definition.

VARIABILITY OF DNA

The 5-Methylcytosine Cycle

As should be clear from the preceding chapters, the recognition of the maintenance of DNA methylation through cell division ended up being the catalyst for the resurrection of the term epigenetics in the 1970s. So, what do we know about DNA methylation?

The start of the story of DNA methylation was in unicellular organisms, with the recognition of the restriction–modification systems by researchers like Ruth Sager (systems described in more detail below). It was proposed by Pugh and Holliday and by Riggs that an equivalent process in mammalian cells could mediate X-chromosome inactivation and other current mysteries of cell regulation. The later demonstration of DNA methylation in the genomes of multicellular organisms helped to propagate interest in this reversible modification of DNA, with some striking observations in the field of plant epigenetics research, which continues to be an extraordinarily exciting area of investigation (Wang and Yamaguchi 2024). As plant transcriptional regulatory systems are not all directly comparable to those in animals, to avoid losing coherence the focus will be maintained here on mammalian insights.

By far the most common animal form of DNA methylation is of cytosine at the carbon at position 5 in the six-atom ring (5-methylcytosine, 5mC) (Fig. 4.1). This modified nucleotide was originally found in vivo in "tuberculinic acid," the name used for the nucleic acid of *Mycobacteria*. To pass on the 5mC "mark" following DNA replication, it needs to propagate to each daughter cell. It is therefore essential that both the DNA strands have the 5mC mark, as one strand is passed on to each daughter cell. This requires that DNA methylation be symmetrical, occurring on both strands of DNA. The smallest possible sequence that can have symmetrical methylation of cytosines is where there is a cytosine on one strand and a cytosine immediately adjacent on the opposite, complementary strand. As cytosine pairs with guanine, this means that the methylated cytosine on one strand has to sit beside a guanine on that same strand, and the cytosine on the other strand complementary to that guanine also undergoes DNA methylation. In theory, this cytosine and guanine combination could be either a cytosine followed by a guanine or a guanine followed by a cytosine, when reading in the usual 5′ to 3′ direction along the DNA; evolution chose the former, the CG (or CpG) dinucleotide. Some cytosine methylation occurs when the guanine is two bases distant, with a non-guanine nucleotide in between. With non-G nucleotides referred to as "H" by the International Union of Pure and Applied Chemistry (IUPAC) nucleotide code, this is referred to as CHG methylation. In plants, and to a modest extent in mammals, especially in cells of the central nervous system, cytosines are also methylated in a CHH context, an asymmetrical pattern of methylation on one strand only (Schultz et al. 2015).

Something needs to put the methyl group onto the cytosine on each strand during cell division. In 1998, having previously isolated proteins from

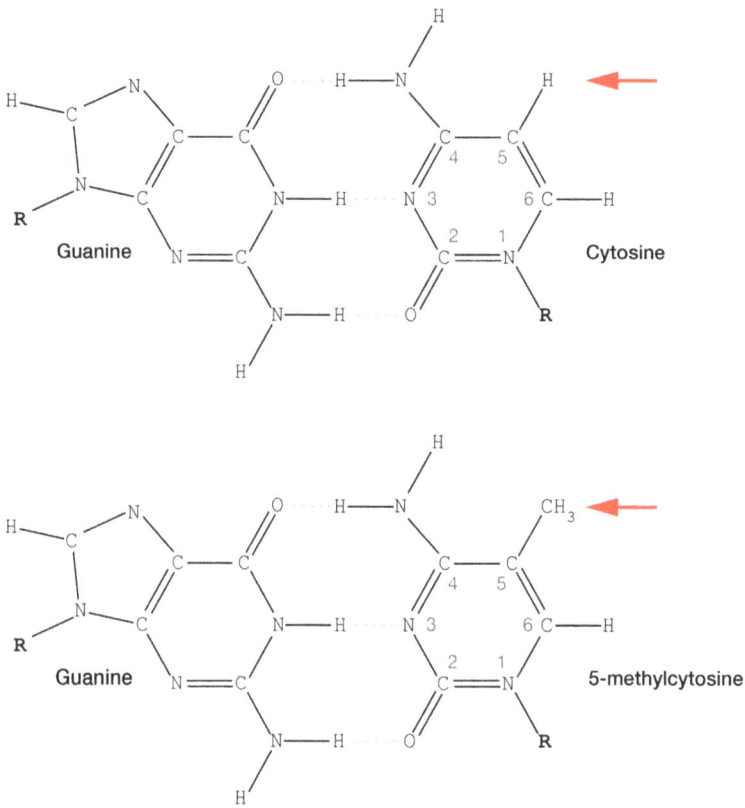

Figure 4.1. The difference between unmodified cytosine and 5-methylcytosine is shown. The 5 refers to the position within the six-atom ring. The three hydrogen bonds pairing cytosine or 5-methylcytosine with guanine are also shown as dashed lines.

mouse erythroleukemia (MEL) cells with DNA methyltransferase activity, Tim Bestor cloned a complementary DNA (cDNA), the DNA copy of the spliced messenger RNA (mRNA), from the MEL cells and showed it to have sequence homology with bacterial type-II DNA methyltransferases (Bestor 1988). We now appreciate the family of mammalian DNA methyl-transferases (DNMTs) to include this DNMT1 protein, which acts specif-ically during cell division to propagate DNA methylation to daughter cells. When DNA replication occurs, unmethylated cytosine is incorporated, so that what used to be symmetrically methylated DNA in the parent cell now creates DNA in the daughter cells that is methylated on one strand but unmethylated on the other, referred to as hemimethylated DNA (Fig. 3.2). This hemimethylated state is recognized by a protein called UHRF1, which in turn recruits DNMT1 during replication, allowing DNMT1 to restore symmetrical methylation to the locus in daughter cells.

This seems risky, right? The DNA replication machinery is copying at least tens of nucleotides per second running through the genome, and UHRF1 and DNMT1 have to find the hemimethylated DNA and selectively add a methyl group to the cytosine on one strand. Is there not a risk of adding methylation to the majority of cytosines that don't undergo methylation? As it turns out, the mammalian DNMT1 enzyme only works efficiently on hemimethylated DNA, as had been predicted by Art Riggs (Riggs 2002). The risk of ectopic DNA methylation is therefore reduced. There is also a second wave of remethylation well after the DNA synthesis phase of the cell cycle, which could help to increase the fidelity of the transmission of 5mC patterns, but this appears to involve only a subset of loci (Gowher and Jeltsch 2018). If, on the other hand, the concern is that UHRF1 and DNMT1 will miss the occasional hemimethylated site, that can indeed happen and is likely to be a mechanism for passive loss of DNA methylation in dividing cells (Fig. 3.2).

For DNA methylation to be propagated to daughter cells, it needs to have been put there in the first place. Two other so-called de novo DNA methyltransferases were identified that have the ability to add DNA methylation to a locus where it did not exist before: DNMT3A and DNMT3B. It appears that DNMT3B is more active in early development and DNMT3A later, and that both enzymes are less stringent than DNMT1 when it comes to avoiding adding DNA methylation to cytosines not located at CG dinucleotides. DNMT3B tends to add methylation when the cytosine is in a CAG context, whereas DNMT3A acts on the first cytosine in the CAC context, consistent with why we tend to see more CAG methylation early in development and CAC in mature neurons (Gowher and Jeltsch 2018).

There is also an interesting member of the DNMT3 family called DNMT3L, which lacks any catalytic activity. However, two units of DNMT3L can form a complex with two units of DNMT3A. This tetramer places the catalytic regions of DNMT3A a specific distance apart from each other, favoring the co-methylation of CGs located 8–10 bp apart on the DNA double helix (Suetake et al. 2004).

The story gets even more strange in the yeast *Cryptococcus neoformans*. This species lacks any de novo DNA methyltransferases but has DNA methylation in the heterochromatic regions of its genome and a maintenance

DNA methyltransferase called DNMT5. The *DMX1* gene encoding the DnmtX de novo methyltransferase of the fungal family *Tremellaceae* that includes *C. neoformans* was lost sometime between 150 and 50 million years ago (Catania et al. 2020). This indicates that the DNA methylation present in the genomes of this species has been faithfully maintained over tens of millions of years since the loss of the *DMX1* gene and evolution of derived *Tremellaceae* species. We went straight from DNMT1 to the DNMT3s; what happened to DNMT2? As it turns out, this family member does not seem to methylate DNA well at all but instead prefers to act upon RNA, specifically the cytosine at position 38 in the transfer RNA for aspartic acid (Goll et al. 2006). As such, the enzyme should probably be called an RNA methyltransferase. It had been suggested that the limited amount of DNA methylation in *Drosophila melanogaster* is mediated by its DNMT2 ortholog called MT2, but more recent studies show this DNA methylation to be independent of MT2 and probably therefore mediated by another, undiscovered insect methyltransferase (Takayama et al. 2014).

Once DNA methylation is present at a locus, how do you get rid of it? This was a mystery for some time but was again a conundrum solved by genomic biochemists with the discovery in 2009 that 5-methylcytosine could be converted to 5-hydroxymethylcytosine (5hmC) by the TET enzymes (Tahiliani et al. 2009). TET stands for ten-eleven translocation, referring to how the *TET1* gene on Chromosome 10 was originally identified as a fusion gene with the *KMT2A* (originally known as *MLL*) gene on Chromosome 11 in an adult patient with acute myeloid leukaemia (AML) (Ono et al. 2002). Two other genes in the family were then identified—*TET2* on Chromosome 4 and *TET3* on Chromosome 2—all producing enzymes with the same properties of adding a hydroxyl group to the methyl group creating 5-methylcytosine and requiring iron, α-ketoglutarate, and vitamin C (ascorbic acid) as co-factors. It was then shown that 5hmC could be oxidized further by the TETs to 5-formylcytosine (5fC) and then 5-carboxylcytosine (5caC) (Ito et al. 2011).

This observation does not by itself explain how 5mC demethylates—all that is happening is that the position 5 modification is changing with each oxidation step. What completes the cycle are thymine DNA glycosylases (TDGs), which preferentially excise 5fC and 5caC from the DNA double helix; this leaves a gap (what is called an abasic site) in the DNA that is then recognized by the base excision repair machinery, which adds an unmodified cytosine to complement the guanine on the other strand (Fig. 4.2). This completes the process

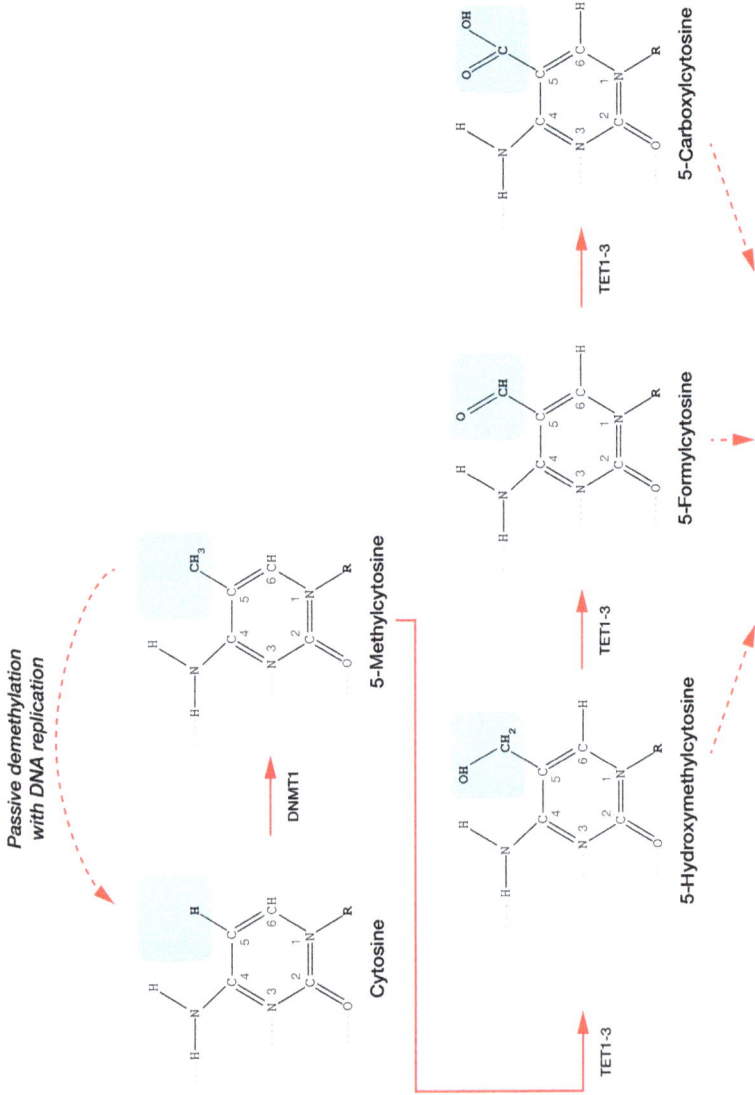

Figure 4.2. How 5-methylcytosine is removed at a locus. One way is passively, by failing to propagate the methylated cytosine through DNA replication, as shown in Figure 3.2. The other is an active process that involves the TET enzymes (TET1, TET2, and TET3, shown as TET1-3 above). The progressive oxidizing of 5-methylcytosine creates nucleotide variants (5-hydroxymethylcytosine, 5-formylcytosine, and 5-carboxylcytosine) present on one strand of the DNA, which are recognized by thymine DNA glycosylase (TDG), which removes them from the DNA. The result is a gap in the nucleotide sequence on one strand, with a G on the complementary strand. This is repaired by adding an unmodified cytosine where the base was excised, in effect reversing the cytosine methylation.

of replacing 5mC with unmodified C, a process that does not require DNA replication but does, intriguingly, involve enzymes that depend on co-factors (iron, vitamin C) that are potentially influenced by dietary deficiencies.

The Putative 6-Methyladenine Modification of DNA

Cytosine is by far the more modified locus in mammalian cells, but there is one other DNA modification thought to be present in small amounts that also occurs—methylation of adenine at position 6 (6-methyladenine, 6mA) (Luo et al. 2015). This modification is potentially added by N-6 adenine-specific DNA methyltransferase (N6AMT1) and removed by the AlkB homolog 1 (ALKBH1) demethylase (Xiao et al. 2018). The concern has been raised that 6mA occurs at such low levels in metazoan genomes that measurements are difficult to distinguish from background, and that "mammalian cells incorporate exogenous methylated nucleosides into their genome, suggesting that a portion of 6mA modifications could derive from incorporation of nucleosides from bacteria in food or microbiota" (O'Brown et al. 2019). At present, the functional significance and physiological regulatory associations for 6mA, and even whether it is genuinely present in mammalian cells, remain to be ascertained.

RESTRICTION–MODIFICATION SYSTEMS

The resurrection of the word epigenetics in the 1970s was prompted by the discovery of restriction–modification systems in single-cell organisms (e.g., Ruth Sager's work on *Chlamydomonas*). To close this circle, it is worth recounting here what she was describing that caused John Pugh to become inspired (Chapter 2).

Restriction–modification (R-M) systems are estimated to be encoded in the genomes of ~90% of bacteria and have been described to be akin to a bacterial immune system (Vasu and Nagaraja 2013). Bacteria can be infected by bacterial viruses, known as phage, which introduce their own DNA into a host bacterium to propagate themselves. If the host expresses a restriction enzyme that cuts a DNA sequence motif present in the phage DNA, then the introduced phage DNA can be destroyed before it has the chance to cause harm, thus "restricting" the range of phage that can infect that bacterium. However, expressing a restriction enzyme that cuts DNA should also cut the DNA of the bacterium itself in places where this motif occurs, which would be self-defeating. That is where the modification aspect of the system comes in—if the restriction

Continued

enzyme cannot cleave when the DNA at the cut site has become "modified" by DNA methylation, and the bacterium produces a second enzyme that methylates this specific sequence, then the host DNA is protected from the restriction enzyme. R-M systems do not represent the only bacterial mode of defense against foreign DNA—the clustered regularly interspaced short palindromic repeats (CRISPR) system provides another such protective process. Just as restriction enzymes became essential tools for molecular biology in the era when Art Riggs was learning about their use while on sabbatical with Herbert Boyer (Chapter 3), CRISPR is revolutionizing molecular and cellular biology today, another gift from bacterial immune systems.

Variation of DNA Structure

The idea that DNA structure can be added to the list of potentially "epigenetic" mechanisms is going to raise a few eyebrows. One thing we may tend to forget is that DNA is not a blandly uniform double helix throughout the genome and in all cell types. It can exist in multiple conformations, to some extent driven by local sequence composition, and in some cases responding to local binding of proteins and RNA molecules. Although our insights into this potential regulator of genomic activity remain relatively limited, it is worth providing an overview here as another type of information that could potentially influence gene regulation.

DNA can exist in multiple forms (Fig. 4.3). When the complementary strands are bound together as double helices, the most common conformation is the right-handed B form, but when the local DNA sequence alternates between purines and pyrimidines (e.g., CGCGCG), it can form a left-handed Z (zigzag) form. If a strand of DNA is complexed with a strand of complementary RNA, the conformation of the nucleic acid at these RNA:DNA hybrids resembles that of double-stranded RNA, a thicker right-handed duplex called A-form DNA. If one strand of DNA is complexed with RNA, the remaining strand can therefore exist as single-stranded DNA in vivo. The combination of an RNA:DNA hybrid on one strand and a single-stranded DNA on the other is referred to as an R-loop. When multiple guanines occur with a certain spacing on one strand of DNA, four of them can complex at a time within that strand to form a G-quadruplex. DNA can also form triple and quadruple helices, parallel helices, and cruciform structures. These structures are now beginning to be mapped in different cell types, a useful step in understanding any regulatory influences they may exert. It is difficult to imagine how some of the noncanonical DNA structures shown

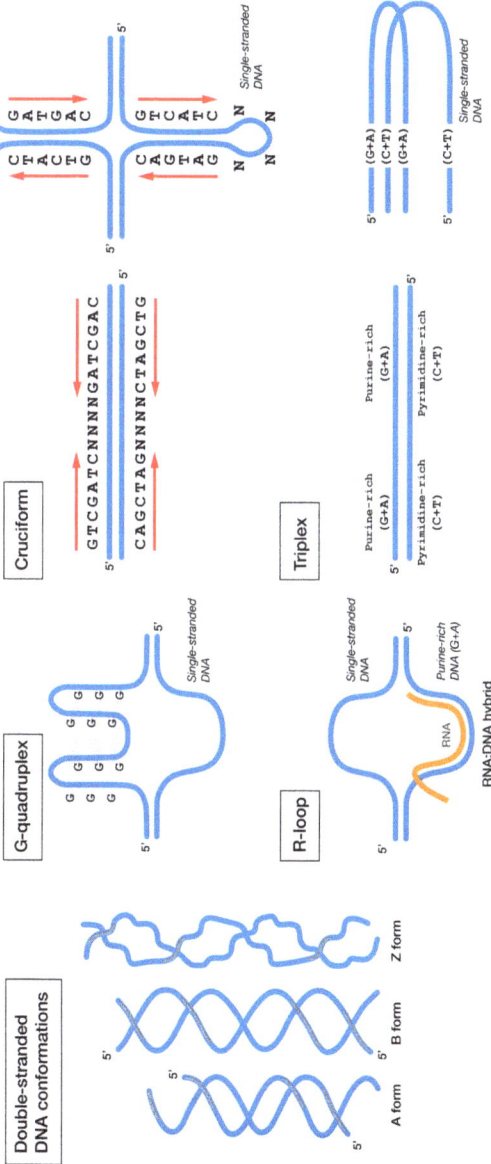

Figure 4.3. Some of the many ways that DNA can exist in vivo. Starting on the *left*, our default impression of DNA is as a right-handed double helix in the B form illustrated. However, the same number of nucleotides can exist in a more linearly compact form, the right-handed A form. If you can imagine untwisting the double helix so that it becomes left-handed instead, the zigzag or Z form of the DNA is an intuitive result. When something happens to cause one strand to find its own binding partner, the other is left floating as single-stranded DNA. The intramolecular association of four guanines at a time can create a stable G-quadruplex on one strand, and the binding of single-stranded RNA to a complementary DNA sequence (RNA:DNA hybrid) likewise frees up the strand with a sequence similar to that of the RNA to exist as single-stranded DNA. The presence of inverted repeat sequences can cause another type of intramolecular binding of DNA to create cruciform (cross-like) structures, whereas triplex DNA can exist in many forms, one of which is depicted, again with the result that a strand of DNA is left without a binding partner. Our ability to map these alternative DNA structures in living cells remains at an early stage but could be another source of variable use of the same DNA sequences between cells, with potential regulatory consequences.

in Figure 4.3 could be easily wrapped around histone octamers into nucleosome. The formation of single-stranded DNA is a common feature of many of these structures, raising the possibility that these loci may be occupied by proteins known to bind to single-stranded DNA or RNA.

DNA-binding proteins can bind in the major groove of double-helical DNA, in which the underlying DNA sequence composition is more apparent, or in the minor groove, where factors like the overall richness in adenines and thymines rather than the specific sequence of those nucleotides can promote binding of specific proteins. The shape of the DNA can influence protein binding, and in turn the binding of proteins can reshape the DNA. The methyl group of 5-methylcytosine protrudes into the major groove, where it appears to influence binding of certain transcription factors (TFs) and other DNA-binding proteins. It has also been noted that the methylation of cytosine influences the local shape of the DNA molecule (Rao et al. 2018), providing another means whereby DNA methylation may influence local protein binding.

▶ **TARGETING AND STABILITY**

Targeting: None of the enzymes that produce DNA modifications has sequence specificity, apart from the CXXC domains of DNA methyltransferases that bind to unmethylated CG dinucleotides, which, if anything, should help to reverse a preexisting pattern of an absence of methylation by bringing the enzymes to these sites that can add methyl groups to the DNA. The lack of sequence specificity means that the enzymes by themselves are therefore not making the decisions where to target DNA methylation in different cell types or in response to perturbations.

Stability: 5-methylcytosine is the paradigm for the property of stability through cell division, as shown in Figure 3.2. The other cytosine modifications (5hmC, 5fC, 5caC), being the products of 5mC, are not by themselves heritable from parent to daughter cell. There is at present no known capability for 6mA to be propagated to daughter cells, again with the caveat that this modification may not exist in mammalian cells.

If a DNA conformation is the result of local base composition, it should remain the same in daughter cells, as a genetic (DNA sequence–based) property. If the conformation is, at least in part, due to local binding of a protein or RNA molecule, then the stability (and sequence-specific targeting) of the local conformation change will be reproduced in daughter cells by the action of that effector molecule.

VARIABILITY OF RNA

Covalent Modifications of RNA

We have known about modifications of RNA for a long time. In 1956, 5-ribosyluracil was identified (Grosjean 2005), with the identification of other modified RNA nucleosides in intervening decades, including modifications such as methylation, deamination to produce inosine, and transglycosylation as major variants.

The best-characterized mRNA modification in mammals are methylation marks in general and 6-methyladenosine (6mA) in particular. Within the RNA transcript, the 6mA modifications cluster preferentially in the 3′ untranslated regions (UTRs) of genes and near stop codons of mammalian genes (Meyer et al. 2012; Yue et al. 2015). The m^1A modification, which is common in tRNAs, has been mapped in human mRNAs to 5′ UTRs (Li et al. 2017). There are thus distinct patterns of deposition of these modifications in numerous mammalian genes.

Enzymatically, we know that the addition of 6mA to RNA requires a methyltransferase complex that includes METTL3, METTL14, and WTAP and is removed by the FTO or ALKBH5 demethylases (Yue et al. 2015), whereas m^1A depends on the TRMT6/61A methyltransferase (Li et al. 2017), with this methyl group removed from mRNAs by ALKBH3 (Li et al. 2017). Deposition of m6A is enriched at the RRACH motif (R = A/G, H = A/C/U), but m6A is only deposited at a subset of these motifs. Whether these enzymes have the independent capability to target adenines within specific regions of the RNA molecule remains unknown at present. 6mA is bound by YTHDF2 (Yue et al. 2015), which mediates the decay of the bound RNA.

The functional effects of mRNA methylation have been inferred from a few sources (for an excellent review, see Gilbert and Nachtergaele 2023). One is the positioning of the methylation mark within the mRNA, prompting speculation that 6mA affects microRNA (miRNA, see below) binding or stop codon use (Meyer et al. 2012). A second inferred function comes from genetic manipulation of the methyltransferases or demethylases. When mouse *Mettl3* and *Mettl14* were depleted in embryonic stem cells, the RNAs became more stable and enriched in the cells, which lost their self-renewing capacity (Wang et al. 2014). The group of Chuan He has speculated that m6A is needed "to facilitate rapid transcriptome turnover during cell differentiation" (Roundtree et al. 2017).

The other source of information about potential function of these RNA methylation marks is from disease associations. The locus encoding the FTO (fat mass and obesity-associated) demethylase has been consistently linked to a risk for obesity (Frayling et al. 2007), but a description implicating a role in obesity may be inaccurate, as the regulatory elements within the *FTO* gene implicated in the obesity phenotype have been shown to interact with and regulate the expression of the *IRX3* gene located approximately a half million base pairs away (Smemo et al. 2014). Mouse models of *Irx3* deficiency show it to be involved in the regulation of body mass and adipose tissue composition (Smemo et al. 2014). A more recent study returned attention to the *FTO* gene as causative of the obesity phenotype, generating mice with the T>C variant at the orthologous intronic location in the *Fto* gene (Zhang et al. 2023). This revealed effects on *Fto* expression, but phenotypically the animals appeared to have phenotypic changes that should superficially be protective against obesity, including "increased brown fat thermogenic capacity and resistance to high-fat diet-induced adiposity" (Zhang et al. 2023). The potential link between RNA methylation and obesity or type 2 diabetes phenotypes therefore remains uncertain. A second interesting disease association came from studies of human immunodeficiency virus (HIV-1) infection of human $CD4^+$ T lymphocytes. Infection by HIV-1 was found to be associated with a substantial increase in m6A in both host and viral RNAs, benefiting the replication of the virus in the cell (Lichinchi et al. 2016).

The novelty of this area of research and its potential to shed light on some important human diseases is making the study of covalent modifications of RNA enormously exciting. To give this new field a succinct name, taking a cue from "... modified DNA bases, which form part of the 'epi'-genome (epi, on top) ...," the word epitranscriptomics was proposed (Saletore et al. 2012). This name has become widespread in its use and was proposed to describe "... an additional regulatory layer ..." (Sibbritt et al. 2013), the same idea driving the broader use of the epi+genetics back-translation. However, we should note that there is yet no evidence that the RNA modifications described above can mediate long-term cellular memories. In fact, the addition of a methyl group to the RNA molecule appears to be associated with a greater turnover of these molecules, which in effect is removing any memories mediated by RNA molecules. It is therefore important to avoid assuming that something called epitranscriptomics acts as a molecular mediator of cellular memory, if that is the presumed link with the broader field of epigenetics.

RNA Production Timing, Splicing, and Localization

The entire field of RNA biology is fascinating and much too large to address here, but a few intriguing observations should be noted, with the goal of linking them to other aspects of genomic regulation.

As will be described below, most of the genome is organized by being wrapped around nucleosomes, which themselves consist of two pairs of four different histones: H2A, H2B, H3, and H4. In a resting cell that is not undergoing division, the genes expressing these proteins don't need to be doing anything until the time comes to replicate the DNA and make histones for new daughter chromatids, following which the mRNAs encoding the histone proteins need to be destroyed quickly to avoid making proteins that are no longer needed. The mRNAs for these canonical histones are unique for not have polyadenylation at their 3′ ends. Normally mRNAs have a string of adenines (polyadenylation) added to the last part of the RNA to be transcribed (the 3′ end), which helps to stabilize them in the cell. The four core histone genes (plus a linker histone called H1) exist in a small cluster within the genome and are rapidly induced to express themselves as the cell undergoes the process of division. The short-lived mRNAs are quickly translated to proteins and then degraded actively when cell division is complete. The other distinguishing feature of these nonpolyadenylated histone genes is that they lack introns. When an intron was added to a histone gene, it led to alteration in the formation of the RNA's specialized 3′ structures (Pandey et al. 1990), which may be a reason why they are excluded.

There is another relationship between the molecular genomic regulatory processes described as epigenetic and the splicing of RNA. Splicing occurs when the primary RNA transcript is processed to remove introns, leading to the production of a shorter RNA molecule that contains only exons. This mRNA produced by this mechanism includes a protein-coding sequence that is carried to the cytoplasm for translation. When genome-wide DNA methylation studies started to be performed, there was an initial observation that there was more DNA methylation at exons than introns (Lister et al. 2009), but this was later found to be a spurious association, due to there being more guanines and cytosines (G+C) in exons than introns and thus more places that DNA methylation could target (Schwartz et al. 2009). However, the simultaneous emergence of genome-wide data mapping histones with post-translational modifications revealed that trimethylation of lysine 36 of histone H3 (H3K36me3) was not only a feature of transcribed loci, but it was distinctly increased in exons of genes. Trimethylation was not merely due to these sequences having an increased density of nucleosomes, which is in turn related to increased (G+C) content (Schwartz et al. 2009). H3K36me3 is deposited by the SETD2 enzyme, which has a domain that interacts with heterogeneous ribonucleoproteins (hnRNPs) which bind to RNA to influence splicing and transcription (Bhattacharya et al. 2021). This links the increased deposition of H3K36me3 in expressed sequences with the recruitment of splicing factors.

What may not be intuitive is why events in the transcribed chromatin have any effect on splicing of RNA, which is typically (in textbooks and on public resources like Wikipedia) represented in terms of the RNA molecule in isolation looping to juxtapose the exons. This is, however, misleading, as splicing almost always occurs as the RNA is being transcribed, described as co-transcriptional splicing. This being the case, the local properties of chromatin can indeed influence how the splicing machinery targets to specific parts of the primary RNA transcript and influence where it is normally spliced. Of major interest is how alterations in the locations of these molecular regulatory processes alter the splicing pattern as a consequence, leading to alternative splicing that modifies the protein produced by the gene and could have phenotypic or disease consequences.

It has been known for decades that chromatin contains a substantial amount of chromatin-associated RNAs (Paul and Duerksen 1975). In recent years, the use of genomic sequencing has allowed this fraction to be characterized more fully. The chromatin-associated RNA sequencing (ChAR-seq) assay successfully mapped the roX1 and roX2 RNAs in the male X chromosome of *Drosophila*, as expected for mediators of X-chromosome regulation. Additionally, there were a number of small nuclear RNAs (snRNAs) that showed patterns of specific enrichment in different genomic contexts (Bell et al. 2018). A similar assay in female mammalian cells would be expected to show the *Xist* functional RNA associated with an inactive X chromosome, as will be described below. In short, although a lot of the focus of epigenetics research is on the selection of specific genes for expression so that they can produce proteins, research areas within the broad field of epigenetics also overlap with how RNA molecules get spliced, or can associate with chromatin, and ultimately become a contributor to a complex system of cellular regulation rather than just a mediator of messages from the nucleus to the translational machinery.

> **TARGETING AND STABILITY**
>
> *Targeting:* As discussed above, there is evidence for specific targeting of nucleosome density and histone modifications to transcribed loci to increase the chances of local splicing and exon formation, related to some extent to underlying (G+C) content. Furthermore, specific chromatin-associated RNAs appear to find their own patterns of genomic targets, although we lack insight into how they find these loci.
>
> *Stability:* Neither RNA modifications nor the RNAs themselves are stable. In plants, the presence of RNA-dependent RNA polymerases allows small RNAs inherited from a parent cell to be regenerated in daughter cells, but such an enzymatic system does not appear to have an animal counterpart. RNA modifications do not appear to template or promote the production of similar modifications in other RNA molecules, another potential way that a message could be passed on.

FUNCTIONAL RNAs

One of the conclusions of the 2012 report of the ENCODE (ENCyclopedia of DNA Elements) Project was the controversial assertion that the majority of the human genome had "function," based on their evidence for >80% of the genome undergoing transcription (ENCODE Project Consortium 2012). The ENCODE project used short oligonucleotide microarrays to detect transcription, which had technical limitations, but an updated study in 2018 used the more sensitive and quantitative RNA sequencing approach in 31 tissues from hundreds of individuals and confirmed that there is indeed widespread transcription throughout the human genome (Pertea et al. 2018).

If a locus is transcribed, is it functional? If we find an RNA to be expressed in a tissue, should we infer that it is doing something? If the RNA encodes an amino acid sequence and is translated into a protein, then we have clear evidence for its function, but there are many more noncoding than coding RNAs expressed in the human genome (Pertea et al. 2018).

Perhaps inevitably, scientists diverge in their opinions when considering these noncoding transcripts (Doolittle 2018). Some prefer to assume that until evidence for function is demonstrated, we should consider these noncoding RNAs nonfunctional and reflective of transcriptional noise, whereas the more optimistic and enthusiastic among us like to speculate about the potential function of these RNAs before formal proof is established. Both views clearly have a place in productive scientific discourse. In this section, the characteristics of functional RNAs are described, with a focus on their regulatory and epigenetic properties, but recognizing that we can't necessarily attribute function to every transcript expressed in a cell.

Small RNAs

miRNAs: In the 1980s, the field of prokaryotic biology was beginning to find evidence for noncoding RNA molecules directly affecting the expression of other genes, based on experiments testing the *micF* RNA from *Escherichia coli* (Andersen et al. 1987). Concurrently, the world of plant biology was noticing odd phenomena involving transgenes or endogenous genes becoming silenced for reasons that were not obvious. In 1993, a breakthrough occurred with the discovery of the first miRNA. The story of how the *lin-4* miRNA was discovered is well worth a read, involving careful experimentation, intuition, and a bit of luck (Lee et al. 2004). After cloning and sequencing the locus, the researchers could not find a protein-coding sequence that caused the phenotype and eventually realized that the band of RNA running at the bottom of their RNase protection gel was, in fact, the mediator of the phenotype, a ~21-nt RNA (Lee et al. 2004). The 21-nt miRNA was found to be generated from a 61-nt precursor that formed a stem-loop structure and had complementarity to the 3′ UTR of the RNA it regulated, *lin-14*. The effect of binding of the *lin-4* miRNA was to reduce the translation of protein from the *lin-14* mRNA, effectively downregulating the gene.

siRNAs: The above action of miRNAs to repress translation represents one of the two mechanisms of RNA interference (RNAi). The second mechanism is through short interfering RNAs (siRNAs), which differ from miRNAs in several ways. They tend to be double-stranded RNAs, they need to match their target sequences exactly (whereas miRNAs tolerate a small degree of mismatch, broadening the range of genes they can regulate), and typically they are generated from DNA that is not native to the host cell (viruses being a good example) and end up regulating their own transcripts, whereas host miRNAs regulate other host genes. A major difference is the induction of cleavage of the target RNA by siRNAs and their apparent lack of conservation in mammals.

RNA-directed DNA methylation: In 1994, a separate RNA-mediated regulatory process was discovered. A group of researchers in Germany introduced potato spindle tuber viroid (PSTVd) cDNA into cells of tobacco plants. They wanted to know why transgenes inserted in multiple copies in the genome seemed to be prone to undergoing DNA methylation. They demonstrated that the RNA produced by the viroid mediated the methylation of the DNA from which it was expressed (Wassenegger et al. 1994). The phenomenon was called RNA-directed DNA methylation (RdDM). Although RdDM in plants requires RNA polymerases IV and V, which are not conserved in mammals, there is a comparable process in mammals for the Piwi-interacting RNAs (piRNAs, below), a mechanism that is not conserved in plants or fungi.

piRNAs: The original identification of piRNAs was in *Drosophila*, where they were discovered to exist at the *flam* locus, encoding sequences shared with transposable elements (TEs; see Fig. 4.4). The expression of these piRNAs was initially associated with cleavage of the RNAs produced by the targeted TEs (Goriaux et al. 2014), but subsequent studies showed that in fission yeast and in *Drosophila* the piRNAs had the ability to target heterochromatin formation to DNA with complementary sequences (Brower-Toland et al. 2007). In mice, piRNAs were shown to mediate genomic imprinting at the *Rasgrf1* gene on Chromosome 9. As a reminder, genomic imprinting involves the selective expression of a gene only when it has been inherited from a mother, with silencing of the paternally inherited copy or only when inherited paternally. Loci near imprinted genes have been found that have only DNA methylation only on the chromosome inherited from the father or only on the maternally inherited chromosome, described as differentially methylated regions (DMRs). What Hiro Sasaki and his team found was that mutations in mouse Piwi proteins were associated with the failure to perform normal DNA methylation at the *Rasgrf1* DMR. This was due to the failure to produce the normal amount of piRNAs from a Chromosome 7 location, sequences complementary to the transcript from a long terminal repeat (LTR)-type TE within the DMR (Watanabe et al. 2011). As a result, the DNA failed to become methylated. The results showed that the silencing effect of piRNAs in the male but not the female germline, targeting TEs for silencing, was causing an endogeneous gene to become imprinted as a bystander.

This study was focused on one mouse locus. With the question whether such effects were widespread in the genome, whole-genome bisulphite sequencing was performed on male germ cells in mice deficient in Piwi proteins. They found that the effects on LTR elements like that at the *Rasgrf1* locus were relatively limited, but that another class of transposable element, L1 LINEs, was more of a target for piRNA-mediated DNA methylation

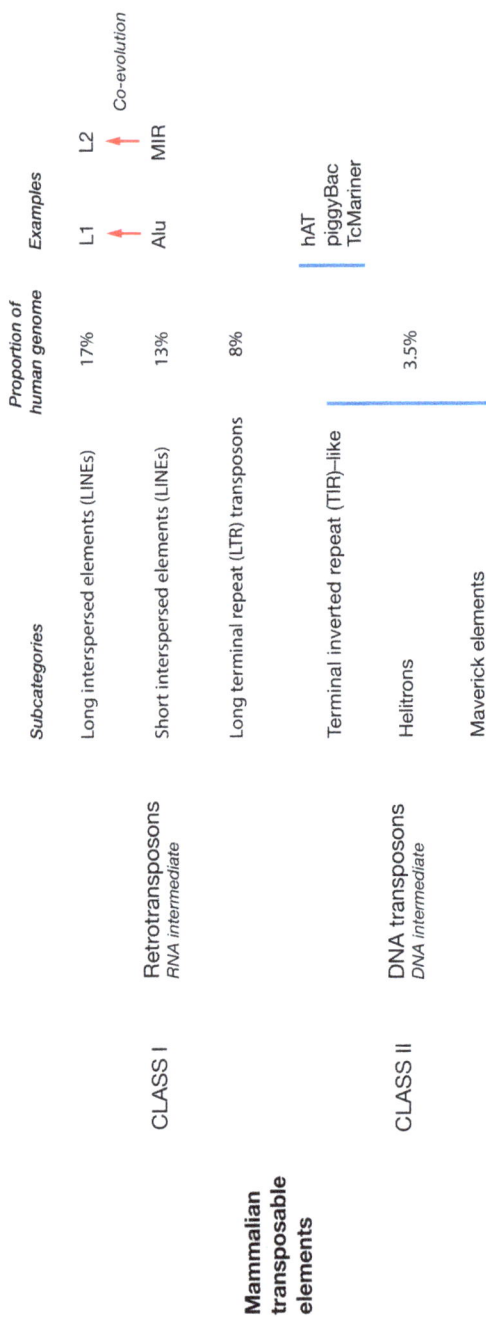

	Subcategories	Proportion of human genome	Examples	Co-evolution
Mammalian transposable elements				
CLASS I Retrotransposons *RNA intermediate*	Long interspersed elements (LINEs)	17%	L1	L2
	Short interspersed elements (LINEs)	13%	Alu	MIR
	Long terminal repeat (LTR) transposons	8%		
CLASS II DNA transposons *DNA intermediate*	Terminal inverted repeat (TIR)–like	3.5%	hAT piggyBac TcMariner	
	Helitrons			
	Maverick elements			

Figure 4.4. Mammalian transposable elements exist in two major classes—one that uses RNA as the intermediate for moving from one genomic location to another (retrotransposons), the other using a DNA intermediate (DNA transposons). There are multiple subcategories for each class, including paired LINEs and SINEs that co-evolved, as SINEs cannot replicate themselves without using proteins expressed from LINEs. Cumulatively these transposable elements make up >40% of the human genome. Although often considered unhelpful, parasitic sequences that need to be suppressed, transposable elements need to use host transcription factors (TFs) to be transcribed, each type associated using different types of TFs (Hermant and Torres-Padilla 2021), and thus allowing them to act on occasions as regulatory elements in the human genome (Ali et al. 2021).

and silencing, especially the younger and less-mutated L1 LINEs (Nagamori et al. 2015). These results implicated piRNAs in silencing of TEs in the male germline. The piRNA example is very interesting as a way that gene regulation can be targeted to specific loci in the genome. However, in mammals, these piRNAs only appear to be expressed in the male germline, limiting the ability to generalize this mechanism from the spermatogenic lineage.

FUNCTIONAL RNAs IN OTHER MODEL ORGANISMS

Tetrahymena: You hopefully recall from Chapter 3 the puzzle facing David Nanney in studying his *Tetrahymena*—how could genetically identical organisms have different mating types? And why did this seem to be mediated through the cytoplasm and not the cell nucleus? We will finish the discussion of short RNAs with the revelation of this mechanism, as it has been shown to involve small RNAs (Kataoka and Mochizuki 2011; Noto et al. 2015).

Tetrahymena, like other single-cell organisms, can reproduce asexually (by binary fission). The mating type becomes important when the organism responds to stress by reproducing sexually, requiring two cells of different mating types. The genome of the organism doubles up and creates two nuclei. One will be used to create a haploid copy for the next generation in sexual reproduction, but the other multiplies itself by ~50-fold to create numerous copies of each gene. The former nucleus remains transcriptionally silent and physically small and is described as the micronucleus, whereas the latter is bigger and becomes the transcriptional factory for the cell and is called the macronucleus.

Inheriting a genome that amplifies itself 50-fold each generation would not be a good idea, so this macronucleus degrades and is eliminated during sexual reproduction, regenerated anew from an unamplified micronucleus. The macronucleus can therefore be thought of as akin to the somatic cell information and the micronucleus as germline information in multicellular organisms.

As part of the endoreduplication of the DNA in forming the macronucleus, the DNA is first fragmented and ~15% is eliminated. Which DNA is eliminated defines the mating type of the *Tetrahymena*. These eliminated loci are referred to as internal eliminated sequences (IESs) and are mostly transposable elements living between the genes. When it was found that an Argonaut protein was needed for DNA elimination, attention focused on the role of small RNAs. What appears to be happening is that the micronucleus is transcribed to produce double-stranded RNAs that get processed to 28–29-nt small RNAs, called scan RNAs or scnRNAs, and that these travel to the macronucleus through the cytoplasm, finding their complementary IESs and initiating a program of degradation of these sequences.

How certain loci in the genome become IESs in one mating type and another in a different mating type is therefore presumably under the control of the repertoire of scnRNAs produced, although the mechanism for selecting

Continued

this specific scnRNA repertoire remains unclear (Kataoka and Mochizuki 2011). However, the fact that these scnRNAs must voyage through the cytoplasm to bind to the Argonaute protein and get from micronucleus to macronucleus gives insight into why Nanney found evidence that *Tetrahymena* transfers regulatory information through its cytoplasm, making Nanney a proponent of the plasmagenes hypothesis.

Arabidopsis: If the reader is interested in a similar story, but this time in multicellular plants, there is a compelling example of how small RNAs generated by the vegetative nucleus of *Arabidopsis* are used to target transposable elements in sperm cells. There are some obvious parallels with the *Tetrahymena* model but the small RNAs in *Arabidopsis* are guiding heterochromatin formation rather than elimination of DNA sequences. Again, this doesn't help us to understand mammalian processes because the molecular mediators don't have mammalian counterparts, but it's a great example of how a fascinating discovery was made (Calarco et al. 2012).

▶ **TARGETING AND STABILITY**

Targeting: An important epigenetic property of small functional RNAs is their ability to target specific genomic loci by having sequence complementarity.

Stability: Although small RNAs can be propagated by RNA-dependent RNA polymerases in plants, animals appear to lack these enzymes, so the duration of the effect of the small RNA depends entirely on the small RNA being present. There is, however, one potential mechanism for longer-term effects of small RNAs—if the small RNA induces heterochromatin formation, as do the piRNAs, and that heterochromatin formation can propagate itself (about which more below), then there can be long-term effects from the small RNA through this secondary mediator.

Long Noncoding RNAs

Just as there are short RNAs present in the cell, there are also long RNAs (long defined as >200 nt in length) that do not encode protein. In a catalog of transcripts from more than 30 tissues from hundreds of human cadavers, the total number of protein-coding genes identified appeared to be 20,352, but the noncoding genes exceeded this with 22,259, while transcripts that they couldn't confidently describe as genes were found at more than 650,000 genomic loci (Pertea et al. 2018). The authors concluded that transcription of the human genome is inherently noisy, and that many transcripts will lack any function.

There are, however, some very well-characterized long noncoding RNAs (lncRNAs) that serve as examples of how many lncRNAs may be acting functionally in cells. The earliest example of a functional lncRNA was the X inactive–specific transcript, *XIST*. Once again, the story starts with X-chromosome inactivation inspiring a young scientist, and, for some reason, the year is once again 1973. This was the year when Britain joined the European Economic Community, the Vietnam war ended and the Yom Kippur War took place, the Watergate hearings began, the U.S. Supreme Court handed down the Roe v. Wade decision legalizing pregnancy terminations, Pierre Trudeau was Prime Minister of Canada, the Sydney Opera House opened, and Steven Bergman was starting his residency in Boston, going on to recount the trauma of the experience under the pseudonym Samuel Shem in his classic book *The House of God*. On October 31, 1973, just over eight months after John Pugh had ridden his bicycle into the center of London to hear Ruth Sager speak, and the same year that Art Riggs was spending his sabbatical in the laboratory of Herbert Boyer, an undergraduate student called Huntington (Hunt) Willard was attending his Biology 113 class at Harvard. In the notes he preserved from this lecture (Willard 2010), he underlines "dosage compensation," the "Lyon hypothesis," and "facultative heterochromatin," foreshadowing his career in the research of X inactivation.

Hunt Willard gives credit to his then-trainee Carolyn Brown for the studies that led to the discovery of *XIST*: "It was she who had the original insight to recognize a gene that escaped X inactivation in a location that suggested that there could be many more such genes up and down the length of the chromosome, not just in the pseudoautosomal region or even in the ancient pseudoautosomal region" (Willard 2010). Their interest was in the region implicated in initiating X inactivation and "pursued a positional cloning strategy, using candidate DNA or cDNA clones from various regions of the X chromosome. These candidate clones had been sent to us by helpful X chromosome colleagues around the world" (Willard 2010).

As Carolyn Brown recounts it (2019, pers. comm.):

> In many ways it was straightforward—map the interval. In days without a complete genome sequence that wouldn't guarantee you found the players (and we still debate the ancillary players). So, it was serendipity that project 2 (identify escapees) came up with a gene in the interval ... and it was expressed not from both Xi and Xa [the inactive and active X chromosomes] but only Xi.

Having identified *XIST*, Willard and Brown "realized that a *cis*-acting functional RNA on the inactive X would be a more likely model than a protein product" for mediating X inactivation. They performed RNA in situ hybridization and showed that the *XIST* transcript did not leave the nucleus but instead formed a "cloud" located on the inactive X chromosome, the region within the nucleus recognizable as the Barr body (Brown et al. 1992).

How does *XIST* work? Why is this 17-kb noncoding RNA able to start the inactivation of an entire 153-Mb chromosome? Logically, you'd expect that this RNA should be binding to other molecules in some way, which may not be how we generally think about RNA. However, a clue should have been from the 1989 Nobel Prize in Chemistry, awarded to Sidney Altman and Thomas Cech for their finding that RNA could fold into three-dimensional structures, just as amino acids fold into proteins, and having formed the tertiary structure could then act as an RNA enzyme, or ribozyme. A simpler example of RNA forming a structure with functional properties is from the study of the viruses that infect bacteria, bacteriophage. Some RNA viruses have evolved sequences within the same strand that are complementary to each other, causing the RNA to fold back on itself to form a stem-loop structure. This stem-loop structure in three dimensions is able to bind to the coat protein of the phage that produced it (Horn et al. 2006). Such a nucleic acid structure is often referred to as an aptamer and can exist in nature like the bacteriophage examples or can be generated in vitro.

When attention became focused on human *XIST* and mouse *Xist*, it was apparent that some of the RNA produced was repetitive, existing in multiple copies within the ~17-kb transcript. In Figure 4.5 the properties of the *XIST* transcript are shown. The approach that first yielded insights into the function of different domains within *Xist* was from the laboratory of Rudolf Jaenisch, involving the incorporation of a copy of *Xist* (the mouse version) onto the X chromosome in male mice who only have a single X, so that silencing of genes on the only copy of the X chromosome present would be detectable by the failure of survival of these cells. When they showed that switching on the full length *Xist* caused silencing, they could then chop different pieces out of the RNA and see what mediated the silencing effect. They implicated the A-repeat region as the major component of the RNA mediating silencing (Wutz et al. 2002). Edith Heard discovered that the B and F repeat regions within *Xist* recruit the Jarid2 component of the

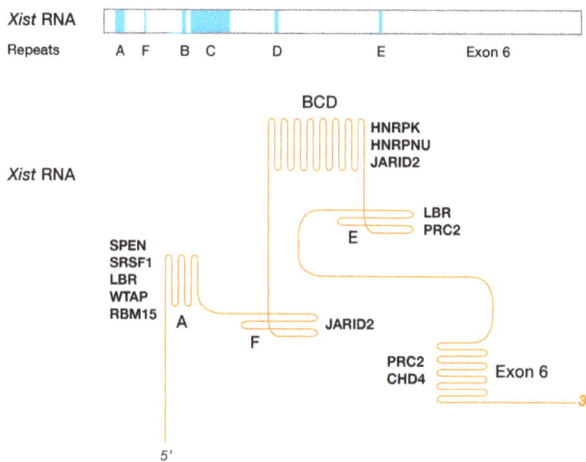

Figure 4.5. The mouse *Xist* transcript does not encode protein but contains several areas of repetitive sequences that can fold to form higher-dimensional structures. These structures are associated with the binding of proteins, some better-studied examples of which (Lu et al. 2017) are shown.

Polycomb repressive complex 2 (PRC2) (da Rocha et al. 2014), a protein complex with repressive properties. Jeannie Lee found evidence that repeat C binds to the YY1 transcription factor (Jeon and Lee 2011), the "YY" referring to the Yin and Yang properties of the transcription factor, which can both activate and suppress gene expression.

Other binding partners for *Xist* have been identified: RBM15 binding to A-repeats, several RNA-binding proteins to the E-repeat, and SAF-A to the molecule as a whole (Pintacuda et al. 2017). This confirms the logical assumption that this long noncoding RNA that spreads to cover the inactive X chromosome binds to regulatory molecules that help to silence the chromosome. Although *XIST* is a paradigm for functional noncoding RNAs, it turns out that the mechanisms of action of other noncoding RNAs are very diverse, making it difficult to generalize about how these transcripts work. For example, there is a noncoding RNA that binds to and stabilizes p53, another that interferes with the activated glucocorticoid receptor as it tries to generate a transcriptional response, others that affect enzyme activity, and others that promote the unusual DNA conformations described earlier in this chapter (reviewed by Marchese et al. 2017). Long noncoding RNAs will continue to surprise us with their range of unexpected properties as they continue to be discovered and characterized.

> ▶ TARGETING AND STABILITY

Targeting: Just as small RNAs can target specific loci, long noncoding RNAs are also capable of binding to complementary DNA sequences and direct their functional properties locally, so they have the necessary epigenetic property of being able to target specific loci.

Stability: However, without an active RNA-dependent RNA polymerase that can propagate the RNA molecule itself, the continuing effect of the functional RNA depends on its reexpression in daughter cells following cell division, so these RNAs cannot by themselves mediate long-term stability during cell growth. The heritability of the inactivated state needs to be mediated by secondary mechanisms.

VARIABILITY OF CHROMATIN

Chromatin organization is possibly the first thing that springs to mind when people talk about an epigenetic regulatory process in the genome. The degree of variability in many aspects of how DNA and proteins come together to form chromatin is immense, is functionally strongly tied to gene regulation, and can be associated with numerous human diseases. There are voluminous textbooks on this specific topic that are worth delving into for detailed insights into this field.

Harold (Hal) Weintraub (1945–1995) was an influential voice in the 1970s. His work included studies of red blood cell formation and the expression of globin genes. He studied how DNA in the region of globin genes was normally packaged by winding around an octamer (8 unit) of histones (two sets of H3, H4, H2A, and H2B), with the DNA and globular assembly of proteins collectively forming what is called the nucleosome. He used the DNase I nuclease to digest DNA preferentially where there were no nucleosomes, revealing an organization associated with the regulation of these genes. His provocative question in 1978 was "How is the active nucleosome conformation propagated from mother to daughter chromosomes?" (Weintraub et al. 1978), echoing the question about the stability of X-chromosome inactivation that had attracted the attention of John Pugh and Art Riggs a few years earlier. Weintraub proposed that the proteins associated with DNA were nonrandomly segregated during DNA replication to the daughter chromatids (DNA and associated proteins), stimulating research into the information carried by these proteins as mediators of heritability of information through cell division by means other than DNA sequence alone.

As it turns out, there are mechanisms for propagating chromatin states through cell division in ways that Weintraub might not have anticipated. In the discussions below, multiple facets of chromatin organization will be described, aiming to encompass the breadth of what is known in this field but leaving plenty of room for deeper diving into specific areas of research.

Post-Translational Modifications of Histones

When a string of amino acids is translated from a messenger RNA, this represents a starting point for a protein's potential complexity, as the peptide chain folded into a three-dimensional protein can be cleaved at specific sites, and there can be chemical groups added to individual amino acids. These post-translational modifications (PTMs) may be needed to allow the protein to become functional, and in the case of the addition of chemical groups, may be reversible, allowing the protein to switch between states.

Despite the octamer of histones in the nucleosome being the same in most of the genome (to be discussed further below), and the sheer number of nucleosomes being immense (with ~6 billion base pairs in a diploid human genome containing a nucleosome every 200 bp, we have roughly 30 million nucleosomes per cell), the presence of many potential post-translational modifications on each histone generates a lot of potential diversity. Lysine methylation of histones in calf thymus tissue was first identified in 1964 (Murray 1964), with recognition of addition of acetylation and of the small ubiquitin-related modifier (SUMO) onto lysines, the phosphorylation of serines, and ubiquitylation or ADP-ribosylation of other residues. Up until 2002, the way of identifying these modifications was by a process called Edman degradation, which tested the first 20–30 amino acids from the amino-terminal end of the protein, which for histones was the unstructured tail region. As a result, most early discoveries focused on modifications of histone tails, but the advent of mass spectrometry allowed the identification of further modifications in the globular part of the histone. We now know of hundreds of post-translational modifications of the core histones (H3, H4, H2A, and H2B) and of the histone variants described below.

There are excellent catalogs available for reference when exploring the specific PTMs of histones known at the time of assembly of the catalog (e.g., see Zhao and Garcia 2015). Rather than describe these exhaustively, it is probably more valuable to introduce here the way of thinking about the complexity and information contained in these modifications. In Table 4.1

Table 4.1. Examples of histone post-translational modifications

Modification	Molecule added to protein
Acetylation	Acetyl
Mono-ADP-ribosylation	ADP-ribose
Biotinylation	Biotin
Butyrylation	Butyryl
Citrullination	Citrulline
Crotonylation	Crotonyl
Deimination	Conversion of arginine into citrulline
Formylation	Formyl
Glutathionylation	Glutathione
2-Hydroxyisobutyrylation	2-hydroxybutyryl
Isomerization	None: transformation into different isomers
Malonylation	Malonyl
Monomethylation, dimethylation, trimethylation	Methyl (one, two, or three)
O-GlcNAcylation	O-GlcNAc (β-linked *N*-acetylglucosamine)
Hydroxylation	Hydroxyl
Oxidation	Oxygen
Phosphorylation	Phosphorus
SUMOylation	Small ubiquitin-related modifier proteins
Ubiquitination	Ubiquitin protein

we can see that there are many different modifications that can take place in histone proteins, representing a first axis of information contained by these modifications. The second axis is where the modification takes place in the protein. Adding three methyl groups to a lysine at position 4 of histone H3 (H3K4me3) is usually associated with the DNA nearby activating a gene's expression, whereas the same trimethylation (me3) of a lysine (K) at position 27 of H3 (H3K27me3) is found at transcriptionally inactive loci. A third axis to consider is the combination of modifications in the individual histone or nucleosome, and a fourth is temporal, the sequential order in which these modifications are acquired. The fifth axis of information is where in the genome these histone modifications occur. These fourth and fifth axes will be addressed in Chapter 6, in which the epigenome is discussed.

A leading figure in the development of the field of chromatin biology, David Allis (1951–2023), was puzzled by yet another issue, how the same modification appeared to have different properties. He was intrigued by the

H3Ser10ph (histone H3 serine 10 phosphorylation) mark which appeared to act as "both a mitosis mark (condensed chromatin) and immediate-early gene activation (decondensed chromatin) … there our work and that of others suggested that nearby acetyl marks with H3S10ph helped to distinguish the latter … hence an early concept of 'combinatorial readout' of histone marks. Additionally, the idea that all these marks did was change the charge on this histone tails (witness: phosphate and acetyl) didn't easily apply to methylation, and with the very recent discovery of bromodomains in 2000, we felt that various 'readers' might use various combinations of binding modules to various marks to bring about different downstream events" (D Allis, 2018, pers. comm.).

He wrote a review with Brian Strahl (Strahl and Allis 2000), who (B Strahl, 2018, pers. comm.) recalls: "I believe we had mentioned something like histone PTMs forming a language not unlike a 'Morse Code.' A reviewer pointed out that while technically correct for the idea we had, the Morse Code itself was no longer widely used and was considered antiquated. In addition, there was also a chromatin researcher in the field with the last name Morse, which might cause confusion."

David Allis (2018, pers. comm.) eventually learned who the reviewer was: "Later I came to learn that Tim Richmond was that reviewer only because he told me he was, and he gave me a hard time that only he and I were old enough to know what a 'Morse Code' was. Randy 'Morse' was the chromatin researcher Brian was referring to."

Allis and Strahl revised the review and described the concept of combinatorial information in histone PTMs as "the histone code," which immediately captured the imagination of the field. Just as DNA was found to contain information in its four nucleotides, histone PTMs seemed to have a similar capacity to encode information, neatly fitting into idea of epi+genetics concept of a layer of information that can vary for the same DNA sequence.

When Allis refers to "readers" of histone PTMs above, he is referring to another conceptual framework that helped engage people to think creatively about this field. In a 2001 review with Thomas Jenuwein (Jenuwein and Allis 2001), Allis started to describe "readers" and "erasers" of histone PTMs. By readers, they meant the proteins that could bind to histones only when they had a specific PTM. For example, proteins with bromodomains (so-called because the domain was first described in the *Drosophila* gene *brahma*) bind to lysines to which an acetyl group has been added, whereas lysine methylation is bound by chromodomains (from chromatin organization modifier).

An example of a protein with a chromodomain is HP1, heterochromatin protein 1, which binds to the trimethylated lysine at position 9 in histone H3 (H3K9me3) to form heterochromatin. In this way we can see why finding H3K9me3 at a locus is usually associated with its being transcriptionally silent, a general property of heterochromatin. The transcriptional activating properties of histone acetylation are in part attributable to the affinity for H3K14ac by the BRG1 protein, which is part of a protein complex that can remodel chromatin, meaning that it can remove a nucleosome from the DNA.

Allis and Jenuwein, in describing erasers, meant the enzymes that could remove the PTM from the histone, allowing the mark to be reversible. An acetyl group could be removed by a deacetylase enzyme, whereas a methyl group is, unsurprisingly, removed by a demethylase. Later the third mediator, the "writer" that put in place the PTM to start with, became part of the attractive combined conceptual framework of writers, readers, and erasers of chromatin marks.

With hundreds of PTMs now described for histones, you might imagine that the language that they combine to create is highly complex. However, as has been reviewed by Oliver Rando (Rando 2012), only a tiny proportion of the complexity is used in vivo, with the same combinations of chromatin marks occurring repeatedly in the genome. As Rando puts it (Rando 2012): "At present, an intellectual schism exists between biochemists on one hand, and geneticists and epigenomics researchers on the other. Genome-wide mapping of histone modifications invariably shows that histone modifications occur in groups of multiple highly correlated modifications, demonstrating that the huge potential space of modification combinations is not utilized in vivo." This lack of diversity argues against the idea that combinations of modifications mediate complexity in the regulation of gene expression, although Rando also cites evidence for some readers that bind preferentially when a histone bears a specific combination of PTMs. The results of genome-wide mapping will be a focus of Chapter 6, but the take-home message is that although histone PTMs generate enormous potential for complexity, only a small fraction of that complexity appears to be used in living cells. Although the histone code is an attractive idea and has stimulated extraordinarily productive research, the degree to which it appears to be underutilized is surprising. Such unexpected findings generally prompt clever experiments, so we should expect that approaches like "epigenetic editing" (Chapter 10) can be used to perturb the normal patterns of clustering of histone PTMs to see what sort of effects they have.

▶ **TARGETING AND STABILITY**

It is intriguing that there are some enzymes that catalyze the addition of a specific PTM that also contain a domain that binds to the same PTM. In other words, the enzyme "reads" the modification that it "writes" (Zhang et al. 2015). One example is the presence of a chromodomain in Polycomb proteins that adds the trimethyl modification to histones, so that the H3K27me3 PTM is bound by the chromodomain, tethering it to the locus where it can act in the immediate vicinity. This is interesting, because it suggests that during cell division and after the formation of daughter chromatids, a re-association of nucleosomes locally would leave the area "diluted" with the unmodified histones incorporated in each chromatid, but that these areas could attract Polycomb proteins through chromodomain binding, allowing reconstitution of the H3K27me3 PTM to the extent that existed in the parental chromatid. Supporting this idea are observations like that in the nematode *Caenorhabditis elegans*, showing that when you remove Polycomb function, H3K27me3 can still persist for several generations before being lost, whereas the restoration of Polycomb function supports a model of de novo reestablishment of H3K27me3-modified nucleosomes in each generation (Gaydos et al. 2014). Thus, there is evidence for some degree of self-propagation through cell division of some histone PTMs at specific genomic loci and, therefore, support for this epigenetic property.

By the same reasoning, the same histone PTMs are capable of targeting to specific loci, another epigenetic property. However, it has to be remembered that this targeting allows maintenance of histone PTMs, but not de novo establishment of histone modifications at new loci. In *Drosophila*, the Polycomb-response element (PRE) is a long-established paradigm for targeting of Polycomb, as the name suggests. The PRE is characterized by the binding of sequence-specific TFs, which then recruit Polycomb complexes. In mammalian cells, there is some evidence for TFs having the same properties, but there is also evidence that other preexisting histone PTMs and long noncoding RNAs can have roles recruiting Polycomb de novo (Aranda et al. 2015), whereas CpG islands also have properties that allow them to recruit Polycomb (Riising et al. 2014). The property of targeting of histone PTMs is therefore complex—for some, there is evidence that they can retarget loci already enriched for that PTM, but that de novo targeting requires the sequence-specific actions of other mediators.

Nucleosomal Positioning

If DNA methylation can be regarded as a default state throughout the mammalian genome, so, too, can the organization of DNA into nucleosomes. Where the DNA is not wrapped around histone octamers is the minority of the genome and has been used as a marker of the locations of regulatory sequences for decades. By exposing chromatin to nucleases (such as DNase I), researchers have been able to map sites where the DNA is more easily digested, which reflects the absence of nucleosomes at those sites. Mapping DNase hypersensitive sites, initially using gel electrophoresis and radioactive probes, revealed that these sites could be located long distances from genes but have the ability to regulate their expression.

A great example of this was the domain of β-globin genes on the short arm of human Chromosome 11. This is a fascinating region of the genome, very (A+T)-rich, and a place where many olfactory receptor genes reside. Olfactory receptor genes are odd for being expressed one at a time in olfactory sensory neurons—not just choosing one of the many genes for expression but also choosing just one of the two alleles of the gene at that locus. Not quite as dramatic but still reflective of developmental regulation of gene expression is the sequential activation of different β-globin genes during red cell development. It is as if these (A+T)-rich regions are like a desert climate in which the genomic flora differs compared with the rest of the terrain.

The regulatory properties of these DNase-hypersensitive sites distant from the β-globin genes was vividly demonstrated by the human disease of β-thalassaemia, in which people living near the Mediterranean Sea ('Thalassa' being Greek for 'Sea') failed to produce one of the β-globins, leading to red cells having more α- than β-globins, resulting in anemia. The increased incidence of β-thalassaemia for Mediterranean people is probably the result of evolutionary selection for protection of carriers against *Plasmodium falciparum* (malaria) infection in times gone by. What the β-globin domain revealed was that these distant DNase-hypersensitive sites could exert regulation of gene expression over tens of thousands of base pairs, referred to collectively as the locus control region.

The question of how nucleosomes organize themselves can therefore take two avenues: (1) what causes a small minority of loci to be depleted in nucleosomes and to have gene regulatory properties and (2) what influences the positioning of the nucleosomes elsewhere in the genome.

Nucleosomes get removed from chromatin by nucleosome-remodeling complexes, which use energy (ATP) and share the common feature of having SWI/SNF enzyme family members. The SWI/SNF enzymes were first discovered in yeast as being required for sucrose fermentation (switch/sucrose nonfermentable). One such complex also has histone deacetylation activity (the nucleosome-remodeling and deacetylation [NuRD] complex), linking histone PTMs and nucleosomal positioning. The related imitation SWI (ISWI) complex has been observed to bind to different genomic regions in what appears to be a nonspecific way most of the time with little to any effect on nucleosome occupancy, but at certain loci the binding is stabilized, and remodeling occurs more efficiently (Erdel et al. 2010). When mapped in the genome using chromatin immunoprecipitation (Chapter 6), chromatin-remodeling complexes are found at all open chromatin, suggesting a model of generalized affinity for naked DNA, but without explaining how these loci were chosen to become nucleosome-depleted in the first place (Bornelöv et al. 2018). In the section on Transcription Factors below, the model linking targeting by transcription factors with recruitment of nucleosome-remodeling complexes will be described.

The positioning of nucleosomes has two major influences. One is the position where the nucleosome-free region stops. The first nucleosome is typically positioned at a specific location, followed by a spacer sequence and the second nucleosome, and so on, with increasing variability as you move from the nucleosome-free region. We see exactly this sort of phenomenon in cars parked on the side of the street in New York City—to avoid a ticket, the first car parked at a fire hydrant is located almost exactly the regulation 15 feet (4.6 m) away, the next car pulls in leaving enough space to manuever out, and so on, the precision of positioning decreasing with distance from the fire hydrant.

DNA sequence also appears to have some influence. The B-form double helix undergoes a complete 360° twist every ~10.5 base pairs and has to bend significantly to wrap itself around the nucleosome, a bending favored by oscillating patterns of specific dinucleotides. Regions with strong poly(dA:dT) homopolymers (all As on one strand, all Ts on the other) are organized as nucleosomes with great difficulty (Struhl and Segal 2013). These are general rules, made to be broken by the demands that nucleosomes reposition in single-cell organisms when new patterns of genes are switched on and off or when cells in multicellular organisms differentiate. DNA sequence cannot be an absolute determinant of nucleosome positioning when the same DNA has to be organized into different nucleosomal patterns in different circumstances. DNA sequence is an influence; it facilitates certain patterns of nucleosomal organization, but it does not dictate these patterns.

And then we have the influence of linker histones. In case you were wondering why the histones in nucleosomes start off by being named H2, followed by H3 and H4, now we can introduce the missing histone H1. Although known for a long time to bind to DNA somewhere in between nucleosomes, recent structural studies have shown the linker histone to bind to the DNA as it enters and as it exits the nucleosome. The combination of the nucleosome (DNA and histone octamer) and the linker histone create a structure called a "chromatosome" (Fyodorov et al. 2018), a rarely used term.

Like other histones, H1 linker histones can undergo post-translational modifications and exist with numerous linker histone variants as alternatives. It is thought that H1 helps to bring together groups of up to eight nucleosomes at a time—"clutches" of nucleosomes that represent the higher-order structure of chromatin beyond individual nucleosomes. Genomic studies indicate that histone H1 is not uniformly distributed in the genome, raising the question whether its enrichment in certain regions leads to distinctive functional properties. Biochemical studies associate the presence of histone H1 with decreased local actions of histone acetyltransferases and methyltransferases associated with transcriptional activation and the facilitation of repressive chromatin states such as that resulting from Polycomb action. Despite these associations, the partial depletion of histone H1 is associated with few transcriptional changes, although it is intriguing that these few changes include loci undergoing genomic imprinting, where DNA methylation was found to be altered (Yang et al. 2013). Although histone H1 and the methylcytosine-binding protein MeCP2 have been found in vitro to compete with each other (Ghosh et al. 2010), genomic studies indicate that this may not occur to any substantial extent in vivo (Ito-Ishida et al. 2018). There remains a lot to learn about linker histones and their roles in transcriptional regulation and higher-order chromatin organization.

> ▶ **TARGETING AND STABILITY**
>
> There is no evidence that linker histones and nucleosome positioning can independently self-propagate to daughter cells through cell division, nor do we know of any mechanism that causes them to become preferentially deposited at specific loci or broader regions. The case that they can mediate epigenetic properties of cells remains to be made.

Histone Variants

Mentioned briefly above was that there can be variants of histone H1. Variation in histones can occur within an organism (e.g., there being multiple genes for histone H2A-like proteins that can take the place of histone H2A in a nucleosome) or between organisms (the H2A-like protein produced by different organisms may differ in their amino acid sequence).

Sorting through this variability is challenging and was the focus of an entire EMBO Workshop on Histone Variants and Genome Regulation held in Strasbourg, France, in 2011 (Talbert et al. 2012). Paul Talbert and Steven (Steve) Henikoff have reviewed this topic in detail (Talbert and Henikoff 2021). In this section, the focus will be on a few, relatively better studied examples of histone variants that reveal some interesting properties of genomic regulation (Fig. 4.6).

Let's start with a clear example of a major difference in the use of histone H3. At centromeres, H3 is replaced in nucleosomes by a centromere-specific histone, which in mammals is the CENPA protein. Of interest is the possibility that centromeric nucleosomes contain only four (not eight) histone proteins when they contain the centromere-specific H3 alternative (Steiner and Henikoff 2015). In mammals, centromeres typically form at repetitive sequences (e.g., the 171-bp α-satellite sequence in primates) and can extend over hundreds of thousands of base pairs. However, centromeres can sometimes be formed ectopically in regions lacking centromeric repeats

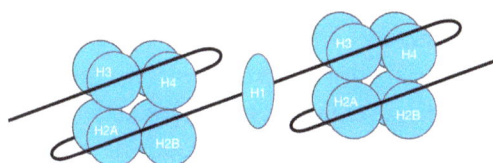

	Canonical histone(s)	Histone variants: somatic	Histone variants: germline
H1	H1.1 H1.2 H1.3 H1.4 H1.5		H1.6 H1.7 H1.8 H1.9
H2A	H2A	H2A.1 H2A.B, H2A.L, H2A.P, H2A.Q H2A.X H2A.Z macroH2A	H2B.1 H2B.W
H2B	H2B		H2B.1 H2B.W H2B.L
H3	H3.1 H3.2	H3.3 CENP-A H3.X H3.Y	H3.4 (H3T) H3.5
H4	H4	H4G	

Figure 4.6. Examples of variants that can replace the canonical histones in humans. The histone octamer, DNA, and H1 linker histone are depicted *above*, with histone variants listed *below*, distinguishing those that are specifically expressed in the germline—in particular, during spermatogenesis.

("neocentromeres"), whereas evolutionarily new centromeres have been seen in the family of *Equidae* (horses, donkeys, and zebras), in regions lacking both genes and repetitive DNA within which centromeres can position themselves variably, a sliding model of centromere formation (Nergadze et al. 2018). New long-read sequencing technologies indicate that similar variability of centromere positioning occurs in human chromosomes (Logsdon et al. 2024). The influence of DNA sequence on nucleosome positioning was discussed earlier; the evolution of centromere-specific sequences favoring targeting and stability of this structure is a striking example of co-evolution of histone protein properties with DNA sequence.

An intriguing variant is histone H3.3. It differs from histones H3.1 and H3.2, the usual versions of H3 in nucleosomes, but only at a handful of amino acids, all located within the non-tail component of the protein, the globular, histone fold domain. Also, unlike H3.1/H3.2 and the other canonical histones, the H3.3 gene has introns, generates a polyadenylated transcript, and is expressed throughout the cell cycle. When mapped in mammalian genomes, H3.3 was found to be located in very specific locations—active genes, transcription factor binding sites, and telomeres—localization that was dependent on protein "chaperones" that escort the histone to these locations (Goldberg et al. 2010). What this pattern appears to represent is loci in which nucleosomes are turning over in chromatin not undergoing replication. With H3.3 the histone variant preferentially available in these other times of the cell cycle, that is likely to be why it ends up being the H3 variant sent to loci that need incorporation of new nucleosomes. The amino acid changes in H3.3 serve to make the nucleosome in which it is incorporated less stable and more prone to dissociate into component histones, which may further aid with turnover of nucleosomes at these loci.

Not all histone variants are so subtly different from the core histone that they replace. The H2A variant H2A.Z is only ~60% identical to the H2A of the same species and has been found to have multiple potential functions, including the regulation of transcription, repair of DNA, formation of heterochromatin, and segregation of chromosomes during cell division (Bönisch and Hake 2012). Interesting models have been proposed about its selective targeting in the genome, each with a degree of support. One model is based on its introduction into chromatin by specific chromatin remodelers acting at those loci, although the obvious question arises how the chromatin remodeler found its way there in the first place. A second model is that the histone variant integrates randomly and is selectively removed from most sites, whereas a third model is that the variant confers preferential stability when replacing both H2A molecules in a nucleosome. This creates a so-called homotypic state, in which the two histones that can fill the role of H2A are identical, compared with a heterotypic state, in which they are different, usually the core H2A and an H2A variant.

Even more strikingly different is macroH2A, which not only deviates from the amino acid sequence of H2A but then adds onto the histone fold domain an H1 linker histone-like sequence followed by a "macro" domain unlike anything seen in other histones. The resulting histone is about three times the size of other histones, leading to the macro description. This extra macro domain has several intriguing properties, including binding to nicotinamide adenine dinucleotide (NAD) metabolites, which are involved in both energy production and cell signaling processes. MacroH2A has also been found to be involved in heterochromatin formation and DNA repair and at the cellular level has been implicated in the inhibition of reprogramming following nuclear transfer (Bönisch and Hake 2012). Structural experiments from the group of Karolin Luger indicate that a homotypic macroH2A:macroH2A state in a nucleosome is less stable than a heterotypic H2A:macroH2A state (Chakravarthy and Luger 2006). In our group, we found macroH2A to be incorporated in G_1, after the formation of new daughter chromatids, at a time when it would need to actively displace the H2A incorporated during the G_2-S phase of the cell cycle, and that it occupies large domains, especially heterochromatic regions on the human genome, in a heterotypic configuration, in approximately one-quarter of all nucleosomes in the genome (Sato et al. 2018). The scale of the domains and the timing of incorporation for macroH2A have parallels with the better-studied CENPA model for centromeres, suggesting that similar processes may be involved in the regulation of macroH2A deposition.

The most extreme example of unusual histone variants is, however, in sperm, in which the haploid genome is tightly packaged for delivery to the ovum. What happens during the condensing spermatid stage of mammalian (and *Drosophila*) spermatogenesis is that the core histones are replaced by testis-specific H3, H2A, and H2B variants and subsequently by the aptly named transition proteins and the linker H1 histone by testis-specific H1 variants (Bao and Bedford 2016). The protamines contain a lot of arginines, facilitating tight binding to DNA, and cysteines, which allow disulphide bonds to form within and between protamines, helping to compact the sperm chromatin (Rathke et al. 2014). The replacement of histones is not complete, however. In the human genome, it is estimated that 10%–15% of DNA remains organized by histones and not protamines, but only 1% in mice. These histone patterns are predominantly retained at centromeres and telomeres but also appear to occur at other loci in the genome, with conflicting evidence that these loci retaining histones are enriched at certain gene promoters or at nongenic regions. The higher proportion of DNA organized as histones in human sperm may reflect the inclusion of sperm in which histone replacement is incomplete (Yoshida et al. 2018), indicating that the question remains open whether here are specific patterns of retention of histones in mammalian spermatogenesis.

Finally, and without getting into what would be an extensive description, histone variants of all kinds above are being found to undergo post-translational modifications, adding a further layer of complexity to the information potentially carried by chromatin states. In the Further Reading section at the end of this chapter an excellent review will be recommended on this topic (Millán-Zambrano et al. 2022).

▶ **TARGETING AND STABILITY**

Targeting: The targeting of certain histone variants has been discussed above. The example of CENPA is interesting—this centromeric histone is retargeted following cell division to the loci where CENPA was located in parent cells, but only after the daughter chromatids introduce histones H3.3.and H3.1 as placeholders during the G_2-S phases of the cell cycle. In G_1, after cell division, it appears that the chaperone HJURP (Holliday junction recognition protein, named after Robin Holliday) acts as a chaperone, bringing new CENPA to these sites in a process that also requires transcription of the centromeric region. This allows efficient retargeting of centromeres and highlights how histone chaperones play a role in directing variants to specific loci. In Chapter 8 we will consider the overexpression of CENPA in cancer, leading to its ectopic location and predisposition to chromosome breakage.

Stability: Although there remains a lot to be learned about histone variants, at least the CENPA example indicates that they may be able to retarget the same sequences in daughter cells after cell division. How de novo targeting occurs during cell differentiation or reprogramming is less clear.

Chromatin Insulation

The partitioning of the genome into discrete functional domains was always an attractive idea. Studies of the X chromosome showed clusters of genes escaping the inactivation process, imprinted genes were found to be located in physical groups, developmentally regulated genes like the α- and β-globin genes formed their own domains, and there was much to suggest that regulation occurred for specific regions and did not affect immediately adjacent genes.

It was in *Drosophila* that the first chromatin insulators were discovered. Insights were being sought into the higher-order organization of DNA, going beyond nucleosomal packaging: "It seems probably that interphase chromosomes may have a defined structure and that nuclei are not just bags full of chromatin" (Benyajati and Worcel 1976). It was assumed that chromosomes contained a series of looped domains or chromomeres, anchored to a structure like the presumed nuclear matrix. A great candidate region to study was the heat shock locus on the right arm of *Drosophila* Chromosome 3, in the 87A7 band. In normal conditions, these *Hsp70* genes are transcriptionally silent, but upon heat induction they switch on quickly, accompanied by a microscopically apparent physical expansion of the chromatin in *Drosophila* polytene chromosome preparations (in which multiple chromosomes align along their length into a composite structure). This appeared to involve a discrete domain of only a few kilobases responding to an environmental stimulus, an excellent model for study. What Paul Schedl found was that this domain was flanked by what he called specialized chromatin structures (scs and scs'), unusually nuclease-sensitive chromatin that flanked the region responding to increased heat (Udvardy et al. 1985). This seemed like a perfect model in which to address a vexing question (Kellum and Schedl 1992):

> The ability of enhancer and silencer elements to control the transcriptional activity of one or even several different promoters located at positions many kilobases away raises a seemingly important question about the regulation of gene expression in higher eukaryotes. How are enhancers or silencers directed to act on the appropriate target gene(s) but prevented from acting on other genes? Without some way of limiting the activity of these auxiliary elements to their specific targets, one might imagine that it would be difficult, if not impossible, to generate distinct patterns of transcriptional activity for different genes and that in the extreme, the consequence of such unrestricted activity would be regulatory havoc.

Kellum and Schedl developed what they called an enhancer-blocking assay to test whether the candidate scs and scs' sequences could prevent spread of regulatory signals beyond a defined domain. They confirmed that both scs and scs' were able to block enhancers from acting on promoters and described this as an "insulation" of the promoter and gene from position effects and from enhancer and silencer elements (Kellum and Schedl 1992). The enhancer blocking assay then became a valuable tool for testing candidate chromatin insulators, revealing one to be encoded by a fly transposable element called gypsy (Geyer and Corces 1992). An alternative assay was transgenic, testing whether a gene could be expressed in the same way in multiple genomic locations, some of which would be unfavorable for expression, thus avoiding "position effects." However, one way that a transgene could shrug off the negative effects of surrounding repressive chromatin would be by including a very strong enhancer sequence, like the locus control region of the mammalian beta globin domain, so the two assays were not equivalent.

In Chapter 3 the phenomenon of genomic imprinting was introduced. It now returns to provide a striking example of a chromatin insulator that explained an odd behavior of an imprinted domain responsible for the human Beckwith–Wiedemann and Russell–Silver syndromes. Shirley Tilghman was looking for the regulators of the α-fetoprotein gene, focusing on genes that shared its expression pattern of elevated levels in livers of the BALB/cJ strain of mice compared with other strains.

The cDNA library that they were testing had a clone in a multiwell plate at column H and row 19, leading it to be referred to within the laboratory as the *H19* gene (Tilghman 2014). Her group showed that *H19* was noncoding and a neighbor of the imprinted *Igf2 (Insulin-like growth factor 2)* gene, and also imprinted, but expressed from the maternal chromosome, whereas *Igf2* was expressed from the paternal chromosome. Through mutations of the nonexpressed sequences around *H19*, they also showed that the same enhancers worked on *H19* and *Igf2* and that there was a region critical for imprinting of both genes just on the *Igf2* side of *H19*.

It turned out that this imprinting control region (ICR) was a chromatin insulator, blocking the effects of the shared enhancers on *Igf2* when *H19* was active on the maternal allele, but allowing the enhancers to bypass the insulator when it was inactivated with *H19* on the paternal chromosome. They then showed that the insulator bound to the CTCF protein, which was being implicated as a mediator of chromatin insulation, but that this did not occur when the CTCF binding site was methylated or mutated (Fig. 4.7).

The field of chromatin insulator research has progressed substantially since these early paradigms (Phillips-Cremins and Corces 2013), including linking CTCF functionally with cohesins, a ring-like complex that encircles DNA when brought to its target site by DNA-binding proteins like CTCF, helping to build higher-order topological structures of the genome, as will be discussed next.

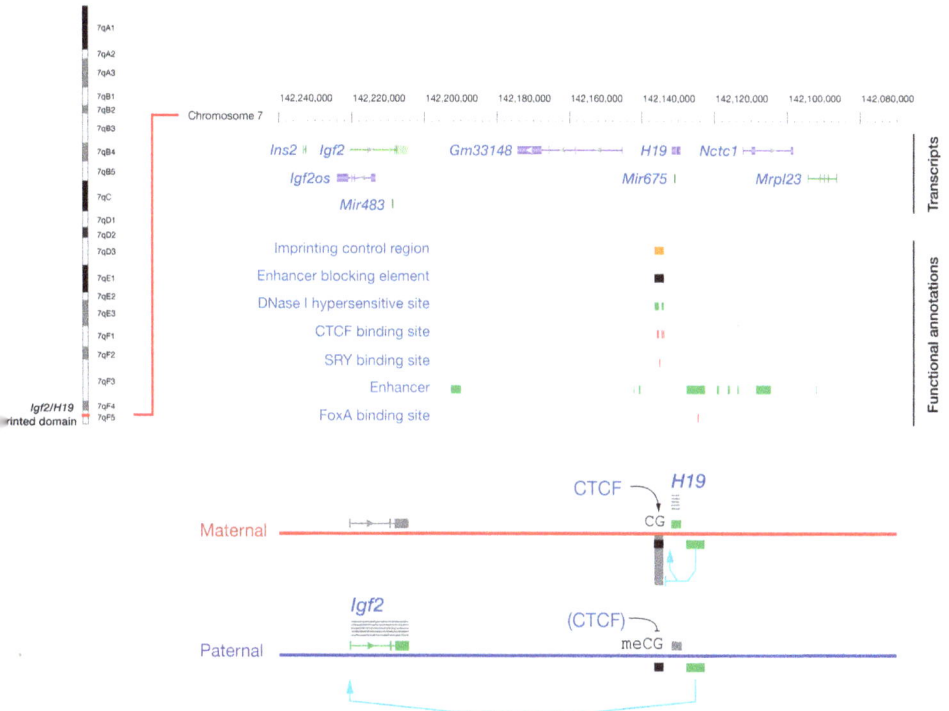

Figure 4.7. The mouse *Igf2/H19* imprinted domain. To keep the figure consistent with those of prior publications (Bartolomei et al. 1993), the domain is reversed to a telomeric to centromeric orientation. Current annotations of *cis*-regulatory elements are shown (from https://www.ncbi.nlm.nih.gov/gene/14955, NCBI RefSeq Annotation GCF_000001635.27-RS_2024_02). The "enhancer competition" model is based on a primary role for DNA methylation leaving the maternal CTCF binding, enhancer-blocking element unmethylated, allowing the local enhancers to drive *H19* expression, but when DNA methylation occurs on the paternal allele, CTCF cannot bind, allowing the enhancers to activate the *Igf2* gene instead.

▶ **TARGETING AND STABILITY**

Stability: As DNA methylation can be propagated through cell division and can help to define where CTCF binds in the genome, this aspect of chromatin insulation can have the epigenetic property of stability.

Targeting: Likewise, this is a situation in which DNA methylation appears to be instructive and defines whether a DNA-binding protein can bind to a specific locus. We will also encounter an example of abnormal global DNA methylation in cancer influencing CTCF binding and altering chromatin insulation in Chapter 8.

Chromatin Looping

In 1885, Carl Rabl (1853–1917) published his study of chromosomes from salamanders in his classic paper *Über Zelltheilung* (On cell division) (Rabl 1885). He made two proposals arising from this work, the first being that the number of chromosomes is constant in a given animal. The second proposal was based on the observation that chromosomes appeared aligned during mitosis, leading to his proposing that spatial order was maintained in interphase cells. In Chapter 2, the work of Theodor Boveri was described in terms of his observations of chromosomes during meiosis and the realization that there was a reduction division of the cell to create a haploid number of chromosomes in the cell. As excellently reviewed by Thomas and Christoph Cremer (Cremer and Cremer 2006), Boveri's other major contribution was to coin the term (in German) "chromosome territory," in accordance with Rabl's idea, but now driven by the model of chromosomal inheritance that was also being promoted by Walter Sutton at Columbia University in New York. The idea that chromosomes could carry specific combinations of genes was not reconcilable with the idea that the chromosome disintegrated in the interphase nucleus. Boveri's idea that chromosomes retained structure, therefore forming territories within the nucleus, was a necessary theoretical foundation for the Boveri–Sutton chromosome theory.

It took 80 years and the invention of in situ hybridization techniques, but Boveri's bold idea of chromosome territories was proven to be correct. Researchers had been using somatic cell hybrids which fused human with rodent cell lines, often leaving as few as a single human chromosome mixed with the rodent chromosomes. By biotinylating the DNA from these cell lines and hybridizing it to human metaphase preparations, they could confirm the presence of the human chromosome when looking at individual chromosomes in metaphase, but could also see that the interphase cells in the same preparations had patches of hybridization signal representing individual chromosome territories (Manuelidis 1985; Schardin et al. 1985). Subsequently it became possible to flow sort and amplify the DNA from individual human chromosomes and use that for hybridization, finding similar patterns of chromosome territories (Lichter et al. 1988). A few interesting observations followed. The nearly identically sized Chromosomes 18 and 19 had very different intranuclear positioning; the gene-rich Chromosome 19 was much more centrally located, with the gene-poor Chromosome 18 located at the nuclear periphery (Croft et al. 1999). Three-dimensional reconstruction of the active and inactive X chromosomes revealed that they had distinct shapes but similar volumes (Eils et al. 1996), which did not fit with expectations about a heterochromatic, inactive chromosome being more condensed and compact than its active homolog. The subset of genes escaping X-chromosome inactivation was found to exist of the surface of the inactive X-chromosome territory (Clemson et al. 2006), suggesting a link between subnuclear organizational characteristics and gene expression regulation. In 2008, Bas van Steensel and colleagues extended these studies by mapping the loci in the human genome located at the nuclear periphery, using proximity to lamin B1, part of the nuclear lamina (Guelen et al. 2008). They described sharply demarcated lamina-associated domains (LADs) averaging ~1 Mb in size where there were more histone modifications associated with heterochromatin formation and decreased gene expression, further supporting the model of peripheral location within the nucleus being a repressive environment for gene expression.

◆ ◆ ◆

Job Dekker did not set out to study chromosome topology (2019, pers. comm.):

> I was interested in meiosis, where the homologs find each other prior to recombination. I was a graduate student working on DNA replication when I first got to know about Nancy Kleckner's work. I heard her speak about biophysical and mechanical models for chromosome pairing she was developing. I was so inspired by her creativity and that is why I went to her lab for my postdoctoral studies. Best decision/opportunity of my career.

He developed the technique later known as chromosome conformation capture (3C) (Dekker et al. 2002) with the goal of mapping the points of contact between homologous chromosomes (J Dekker, 2019, pers. comm.):

> I set out to develop 3C to specifically to look at interactions between homologous chromosomes in meiosis (I was hoping that by mapping such contact points I would be able to learn how homology is recognized). However, in the process I learned that chromosome folding and chromosomal interactions in general are way more variable than I had thought and worse: my PCR-based approach could not distinguish between homologs unless I introduced RFLPs [restriction fragment length polymorphisms]. Doable, but not very efficient. But I did learn that 3C was great to look at chromosome structure in general! So, the original plan for why I developed 3C did not work, but it led me to study chromosome organization in general, and it led me to set a new goal for myself in 2003: solve the structure of the mitotic chromosome.

We made a lot of progress there, with the help of great collaborators (Leonid Mirny, Bill Earnshaw). And, also, now with hybrid crossed there are now systems with high SNP densities, and now interhomolog contact can be studies using 3C-based assays! That others are now doing that is great to see for me!

What was 3C, and how did it usher in a new era of understanding chromatin's topological organization and relation to gene regulation? The idea was simple enough and started by using a chemical fixation step to retain the DNA and associated proteins in their in vivo configuration. The DNA was then digested by an enzyme that leaves ends dangling and available to religate to nearby digested DNA. Although the original DNA strand that was digested has a good chance of reannealing and being ligated together, if there is another DNA sequence close by, it can also be ligated. By quantifying the amount of PCR product for different potential ligation events, regions of DNA that exist in three-dimensional proximity to each other can be identified by occurring more frequently than by chance.

3C began a series of progressively more informative technical advances that are described in Chapter 6, culminating in the Micro-C assay that tests for all interactions genome-wide (Fig. 4.8). On the scale of the entire chromosome, high-order patterns of interaction extending over tens of millions of base pairs that are reminiscent of chromosomal banding patterns are apparent, but strikingly few interchromosomal interactions are. On the scale of megabases, and within these larger blocks, interactions within discrete blocks become evident. These blocks are described as topologically associating domains or TADs, equivalent to the bands and interbands of *Drosophila* polytene chromosomes. TADs are proposed by Leonid Mirny and others to be formed by a combination of extrusion of loops of DNA by SMC complexes (cohesins) (Fudenberg et al. 2016), which then end up forming the boundaries of each TAD.

This sounds like a scale of organization that is difficult to reconcile with the scs/scs′ example described earlier, which is only kilobases in size, whereas TADs extend over megabases. The picture is indeed more complex: Within individual TADs appear to exist intra-TAD loops anchored by CTCF with insulator properties (Matthews and Waxman 2018). This smaller-scale regulation is more likely to reflect the examples of chromatin insulators described in the previous section. On a global regulatory scale, when DNA methylation and Polycomb repressive marks are substantially altered during mouse embryogenesis, three-dimensional chromatin organization is substantially modified, but as Richard Meehan and Wendy Bickmore caution, this may not be responsible for transcriptional changes observed (McLaughlin et al.

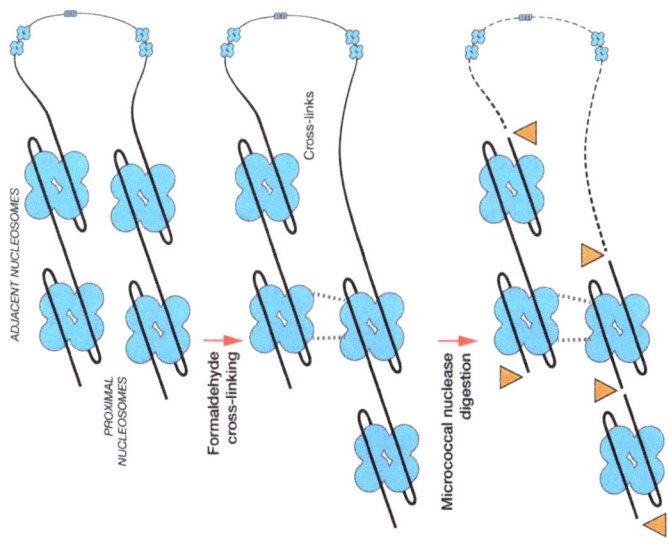

Figure 4.8. Of the many assays that reveal three-dimensional organization of chromatin in the cell nucleus, the example of the high-resolution Micro-C assay is shown (Hsieh et al. 2015). The assay begins with cross-linking using formaldehyde to stabilize spatial relations between nucleosomes in proximity to each other. Micrococcal nuclease is used to optimize the recovery of dinucleosomes, which may be adjacent in the genome or in proximity to each other in three dimensions. Enzymatic steps lead to a situation in which the DNA linking proximal nucleosomes is specifically marked by biotin incorporation, allowing specific recovery of these DNA fragments, from which a library for sequencing can be generated. By identifying the locations of the DNA at each end of the sequenced molecule, the positions in the genome that are more proximal than would be expected by chance can be identified.

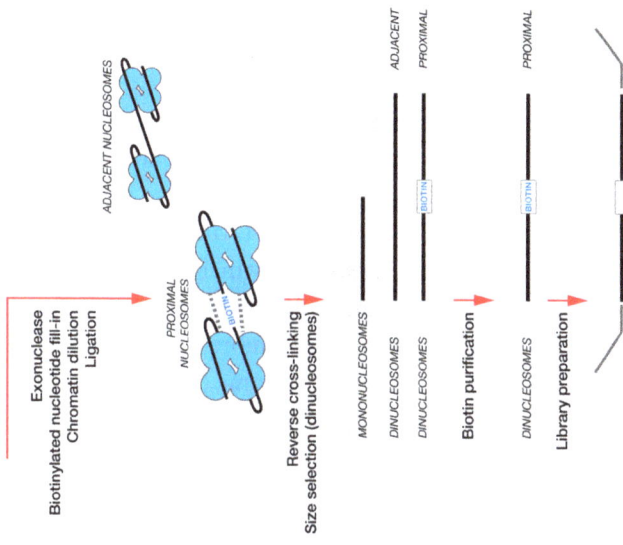

2019). In terms of variability between cell types, it has been appreciated for decades that cytogenetic bands are the same in all cell types, and so it is not surprising that the most variation in topological characteristics appears to be within and not between the TADs (Matthews and Waxman 2018). Adding to the intriguing nature of higher-level chromatin organization is the finding by Christine Disteche and colleagues that whereas the active X chromosome looks like autosomes in having multiple TADs, the inactive X forms just two huge individual domains, separated by a repetitive DNA sequence called *DXZ4* (Deng et al. 2015). This may be the molecular reason for the structural differences seen by the Cremer brothers using imaging (Bischoff et al. 1993) and demonstrates the role of a single large repetitive sequence, *DXZ4*, in the organization of an entire chromosome.

Do TADs help to regulate gene expression? It would seem logical that they should, having gone to the trouble of creating them in the first place, and an example of dysregulation of a CTCF binding site in cancer will be discussed in Chapter 8 that certainly supports this likelihood. There is, however, an intriguing study in *Drosophila* in which chromosomes and TADs are rearranged extensively, without any obvious effects on gene expression (Ghavi-Helm et al. 2019). There is clearly more to discover in this fascinating field in which structure meets chromosome function.

> ▶ **TARGETING AND STABILITY**
>
> The nature of the extrusion process indicates that TADs are formed dynamically de novo every time a new chromatid is formed following cell division and is therefore not heritable. The relative similarity of TADs across cell types indicates that they are strongly determined by DNA sequence properties, attributable to DNA-binding proteins such as CTCF, YY1, and others.

Heterochromatin and Euchromatin

Finally, we get to the molecular underpinnings of chromatin organization on the cytological scale, the oldest recognized heterogeneity of functional organization within the genome. The history of the terms euchromatin and heterochromatin has been nicely described (Jost et al. 2012) and reflected the advances in microscopy and staining techniques being used in the early twentieth century. Heterochromatin ("other" chromatin) stained more darkly than euchromatin ("good" chromatin) because of its denser packaging, reflecting the general silencing of DNA contained within it.

The relatively heterochromatic appearance of the inactive X (Barr body) was one example of a cytological correlation with gene activity on a chromosome-wide scale. Proximity to heterochromatin within the cell nucleus was also an indicator of gene repression. An evocative early example was the Ikaros protein in B lymphocytes, which was found to associate with genes being silenced in that cell type and to migrate into close proximity to heterochromatic parts of the cell nuclei (Brown et al. 1997). It is not clear why the name Ikaros was chosen for this protein (Georgopoulos et al. 1992), but it was well-chosen in retrospect, bringing to mind the vision of a transcription factor flying too close to heterochromatin in the nucleus and becoming transcriptionally inactivated.

The second description of heterochromatin is in terms of whether it is present at that genomic location all the time, in which case it is described as constitutive, or whether it is only present on occasions, such as an X chromosome in female cells in mammals. In the cytological era, specific banding technologies of chromosomes were used to reveal heterochromatic DNA bands in chromosomes.

The link between heterochromatin and gene silencing came from work on *Drosophila* in the 1930s by Hermann Muller, as will be described in Chapter 5, the observation that mutations that caused a gene to relocate beside the heterochromatin at the centromere led to the silencing of the gene, so-called position effect variegation. When a *Drosophila* mutation causes a clear phenotype like this, it becomes possible to do screens to test for proteins that influence the phenotype. A gene mutation that reduces (suppresses) position effect variegation is likely to be contributing to forming heterochromatin. The suppressor of variegation (Su(var)) genes were therefore candidates for forming heterochromatin, whereas the enhancer of variegation (E(var)) genes have the opposite effect. In this way the molecular properties of heterochromatin and euchromatin could be inferred, allowing a transition of the definition of these kinds of chromatin from cytological to molecular characteristics.

Heterochromatic regions of the genome are not merely silent and uninteresting. An excellent review of heterochromatin by Allshire and Madhani is a good starting point for those who want to learn more about this curious genome-within-a-genome (Allshire and Madhani 2018). Heterochromatic regions can be functionally essential, such as centromeric and telomeric areas of the genome, whereas a major function for heterochromatin appears to be the silencing of certain types of transposable elements in the genome. Molecular characteristics include specific patterns of PTMs, such as trimethylation of histone H3 at lysines 9 and 27 (H3K9me3, H3K27me3). We recognize readers of these PTMs, with H3K9me3 PTM bound by heterochromatin protein 1 (HP1), and writers of the PTMs, the Polycomb repressive complexes adding the trimethyl group to H3K27. In Chapter 6 we will focus on the patterns of distribution of heterochromatic and euchromatic molecular characteristics in the genome.

How is heterochromatin targeted to specific loci? In *Drosophila*, there exist Polycomb-response elements (PREs) that are bound by specific DNA-binding proteins and recruit Polycomb locally to regulate developmental patterns of gene expression. In mammals there is some evidence for TFs recruiting Polycomb complexes (Aranda et al. 2015) but also a preference for targeting to CpG islands or loci where specific histone PTMs already exist. Intriguingly enough, it is now appreciated that noncoding RNAs can bind to specific loci in the genome and to Polycomb proteins (Achour and Aguilo 2018), allowing a sequence-specific targeting of heterochromatin.

Something not often appreciated about the cell nucleus is the very high concentration of proteins contained within it. If a mammalian nucleus is ~6 μm in diameter, its volume should be ~113 μm^3 (~10^{-12}) mL, containing ~250 pg (10^{-12} g) of nuclear protein, for an approximate concentration of 2.2 g/mL, which is strikingly high and doesn't include the other molecules (like nucleic acids) that are also present. Furthermore, some of these proteins bind to and can cause bridging between different areas of chromatin. Somehow the intranuclear organization that keeps nucleoli, heterochromatin, and euchromatin in distinct partitions occurs despite molecules being able to diffuse across each compartment. A biophysical concept of "phase separation" has been invoked to explain how these compartments maintain themselves without expending energy and appears to be an emerging insight into one of the influences on subnuclear organization (Erdel and Rippe 2018).

▶ **TARGETING AND STABILITY**

Some of the writers of heterochromatin also have readers in the same molecule, as discussed earlier, providing at least some stability to these patterns in the genome through cell division. The targeting property of heterochromatin is complex—sequence-specific targeting through TFs and noncoding RNAs does indeed occur, but some of the targeting preference of Polycomb involves base compositional properties of DNA (CpG islands), whereas the targeting to loci with preexisting histone PTMs raises the question how those PTMs got targeted in the first place.

GEDANKENEXPERIMENT

I think we can all agree that compound German nouns are fascinating, especially when attributable to Albert Einstein, so we instead of performing a mere thought experiment, we'll take it to the next level and perform *ein Gedankenexperiment*. You'll need a piece of paper and a pen. Or a pencil, which will look more like a serious, old-school scientist. You may also want to consider smoking a pipe at the same time to look more like Waddington.[1]

This is an exercise to help you to understand a central problem in epigenetics. On your piece of paper, draw a big circle covering most of the paper and within it a smaller circle. Within the smaller circle, which represents the nucleus of the cell, write two gene names in different parts of the nucleus, *ABC1* and *XYZ2*. (For the nitpickers (*die Tadelsuchten* if you're still following along in German), assume it's a haploid locus on a sex chromosome, and that's why each gene is only in one place within the nucleus) (also, draw the names so that they only occupy a small proportion of the nucleus, don't ruin the game).

Now close your eyes and think of your favorite epigenetic mark, and whatever acronym is used to describe it, like 5mC or H3K27ac. Keeping your eyes closed, spin the piece of paper around a few times and then write that mark on *XYZ2* but not *ABC1*.

It is highly likely that you failed to put the epigenetic mark on your gene of interest. This is a problem with the assumed mechanisms of epigenetic regulators—that they somehow know where in the genome to target their activities. All these epigenetic regulators, whether histone acetyltransferases, DNA methyltransferases, or others, are incapable of directing themselves to a specific sequence of interest. Which is exactly how you want them to be; as you will need them to target different loci in different cell types, they shouldn't be going to the same places every time. So, what is it that partners with the blind epigenetic regulator, takes it by the hand, and leads it to its destination in a specific cell type?

---

[1] Don't smoke. It causes mutations, and DNA methylation biomarkers in your peripheral blood make it impossible to lie when you claim you have never smoked.

THE CURIOUS CASE OF TRANSCRIPTION FACTORS

In Arthur Conan-Doyle's story *The Adventure of Silver Blaze*, Dr. Watson asks Sherlock Holmes a question:

"Is there any point to which you would wish to draw my attention?"
"To the curious incident of the dog in the night-time."
"The dog did nothing in the night-time."
"That was the curious incident," remarked Sherlock Holmes.

A plant epigenetics researcher recounted a story in 2013 about the time she had given a talk at the Gordon Research Conference on Epigenetics, during which she mentioned the role of a transcription factor. After the talk, she was taken aside by one of the more seasoned researchers attending the meeting, who explained to her that "This is an epigenetics meeting. We don't talk about transcription factors." This anecdote illustrates how we traditionally dissociated the study of epigenetics from that of transcription factor biology. The silence about TFs in the epigenetics literature is our nonbarking dog. This omission is deliberate, not accidental. Why might this have happened?

We have learned in Chapter 3 from the experiences of researchers in the 1970s like Art Riggs and John Pugh that they regarded the lac operon paradigm as insufficient to explain long-term memories in development like X-chromosome inactivation. The concept of DNA methylation being maintained through cell division and influencing gene expression was much more attractive as a concept. Furthermore, when paradigms like X-chromosome inactivation and genomic imprinting were considered, the role of TFs could only be regarded as secondary and subordinate to some other regulatory process that was silencing one of a pair of alleles in each cell nucleus: Despite the TFs being present and capable of switching on genes on one allele, they were unable to activate the other allele. Adding to this were the results of experiments like Su(var) and E(var) screens—when one of these "epigenetic" regulators was mutated, that's when TFs started to function in different ways. The example of DNA methylation leading to a failure of CTCF to bind at the *H19* imprinting control region also looked like evidence for TFs being instructed by epigenetic regulators.

One of the discoverers of the lac operon was Mark Ptashne, who began in 2007 to push back against the pervasive use of the word epigenetic to describe all aspects of transcriptional regulation (Ptashne 2007). He argued that if DNA methylation is not present in insects, the view that this molecular mark is critically important for normal development was much too mammal-centric a perspective. Although a provocative viewpoint at the time, his perspective caused some degree of soul-searching on the part of the epigenetics community.

TFs have been referred to in passing up to now but merit more attention at this stage. They are defined as proteins that bind to specific DNA sequences and regulate transcription (Lambert et al. 2018). To have these linked functions, the TF proteins tend to have two domains: one to bind to DNA, the other to bind to other proteins (the *trans*-activation or effector domain). DNA binding on its own has no likely consequence, but when the DNA sequence with the TF binding site is bound, this anchors sites for other interacting proteins to bind to and influence the chromatin locally.

When a locus is bound in the genome by a TF, it tends to have several characteristics. One is that the TF is rarely working on its own, but it frequently accompanied by other TFs binding locally that form a module of cooperating regulators. The DNA where the group of TFs is binding undergoes active displacement of nucleosomes, with very precisely located nucleosomes flanking the resulting nucleosome-free region (NFR). The flanking nucleosomes tend to have specific patterns of PTMs that are typical of loci with *cis*-regulatory properties, such as H3K4me3 flanking gene promoters or H3K4me1 flanking enhancers. The promoter is also the site at which RNA polymerase can access the DNA and initiate transcription, with enhancer RNAs expressed at distal regulatory elements, so the removal of the nucleosomes and creation of a local environment of specific histone PTMs is important in facilitating the RNA polymerase activity.

With the emphasis above on the core epigenetic properties of stability over time and through cell division, and the ability to target specific loci, TFs clearly do the latter through their DNA-binding properties. What is the evidence that they can mediate long-term memories, and that these can be propagated though cell division?

The answer is complicated by the fact that different TFs do different things. Some TFs are involved in cell fate decisions—in other words, causing the cell as a whole to change its properties, including and based on its transcriptional regulation. In embryogenesis, the initial signal could be a morphogen, a molecule diffused within the developing embryo, which binds to a transmembrane receptor and triggers a cell-signaling pathway, which in turn activates one or more TFs, often by adding a PTM-like phosphorylation that causes the TF to enter the nucleus and act on chromatin.

We then have the problem that the TF must bind to sites in the genome that are probably wrapped up by nucleosomes. Ken Zaret has created a nice conceptual framework for how to think about this process. The TF may be targeting loci that are already nucleosome-free, because they were opened by other TFs previously. They may be targeted chromatin that is nucleosomally organized but otherwise not distinctive, which he called low-signal (L) chromatin. His idea was that certain kinds of TFs have the inherent ability to find their target binding sites and to displace the nucleosomes locally, which he called "pioneer" TFs (Iwafuchi-Doi and Zaret 2014). This may in part involve allosteric (binding-induced) interactions between these TFs and local nucleosomes (Tan and Takada 2020). The subsequent description of other classes of TFs as "settlers" and "migrants" (Sherwood et al. 2014) adds further terms (that may grate on people from the many countries worldwide with a history of being colonized). Instead, the idea that some TFs can open chromatin and allow the subsequent binding of other TFs locally is valuable conceptually and could be used in describing TF properties more accurately. Zaret makes a distinction between low signal and repressed (R) chromatin, the latter defined by having additional repressive marks, such as the H3K9me3 and H3K27me3 PTMs described above as being typical of heterochromatin. These genomic regions are proposed to be refractory even to pioneer TF binding (Fig. 4.9). The pioneer TF model was tested by the group of Barak Cohen, selectively inducing the expression of the pioneer TF FOXA1 and a nonpioneer TF HNF4A in K562 lymphoblast cells, in which neither TF is normally expressed (Hansen et al. 2022). This comparison found no differences between the chromatin and transcriptional responses to each TF, undermining the pioneer TF hypothesis, while suggesting to these researchers that the ability

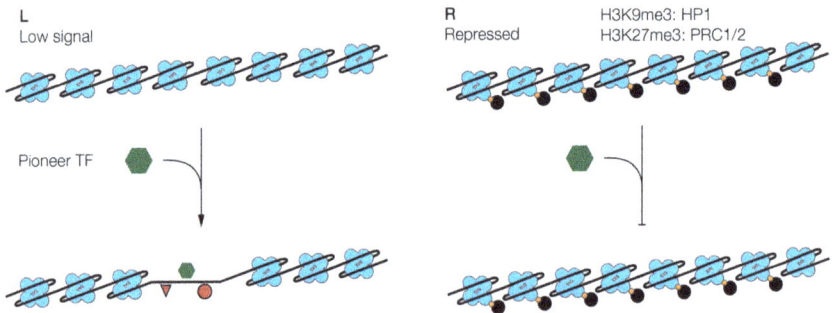

Figure 4.9. The pioneer transcription factor (TF) model, defined as a property of a subclass of TFs that can bind to nucleosomally packaged DNA, with the co-recruitment of partner TFs and the remodeling of the chromatin to displace local nucleosomes. This can only happen at L/Low signal chromatin, but R/Repressed chromatin, which has repressive histone modifications like H3K9me3 or H3K27me3, which in turn recruit repressive proteins and complexes like HP1 and PRC2, respectively, is refractory to binding by pioneer TFs. This model can explain why imprinted loci exist if one allele is in the R/Repressed chromatin state.

for a TF to induce a loci to become activated is instead dependent upon the "free energy balance between TFs, nucleosomes, and DNA" (Hansen et al. 2022), a more complex model.

How the TF activation and binding induces a new cell state is proposed to be mediated by network models such as feed-forward loops (Sasse and Gerber 2015). One simple example of a feed-forward loop is of a TF activating two genes, one of which is another TF, and the other a target gene for activation by both TFs. Without getting into this fascinating area of biology too deeply, the lesson is that TFs involved with cell differentiation drive a series of transcriptional regulatory changes that newly activate a cascade of genes while leading to repression of many that had been expressed prior to the TF-mediated change.

What is the evidence that TFs can maintain binding to DNA through cell division? In 1997, David Levens exposed cells to potassium permanganate ($KMnO_4$), typically used to modify thymine residues in single-stranded but not in double-stranded DNA, allowing nucleotide resolution analysis of the conformation of DNA. His group studied promoters of several genes, including *MYC* and *HSPA1A* (*Hsp70*, the mammalian ortholog of the *Drosophila* heat shock gene studied by Paul Schedl), showing that the conformational changes typical of the active core promoter TATA box persisted during mitosis. They suggested the term "mitotic bookmark" to describe how a TF continues to bind to its target sequence as DNA is replicated during mitosis, allowing the information carried by the TF to be passed to daughter chromatids and to daughter cells (Michelotti et al. 1997). Pablo Navarro has assembled a list of numerous TFs that appear to have the property of acting as mitotic bookmarks (Festuccia et al. 2017). This is an emerging and understudied field but there is certainly evidence to support TFs being capable of propagating their targeted binding through cell division, making it reasonable to include at least some TFs in the category of genuinely epigenetic regulators, capable of targeting specific loci and maintaining their effects following cell division.

How do we reconcile this view of TFs with the issue raised earlier—that the silencing of one allele in genomic imprinting or X inactivation overrides TF activity? The answer probably resides with Zaret's definition of R chromatin being especially refractory to TFs. Genomic imprinting and X inactivation are both characterized by strongly repressive heterochromatic organization of the silenced allele, involving the activity of Polycomb repressive complexes, and would certainly qualify as R chromatin. It should also be recalled that there are targeting events involved in X inactivation and genomic imprinting; X inactivation involves an interplay between *XIST* and the YY1 transcription factor (Jeon and Lee 2011), whereas the ZFP57 and ZNF445 TFs act to select loci for genomic imprinting (Takahashi et al. 2019), demonstrating that sequence-targeted events precede allelic inactivation and involve TFs in these paradigms of long-term transcriptional regulation effects.

A focus on TFs allows one further model for propagation of states through cell division. Many TFs are regulated by cell signaling pathways. When a cell divides, the symmetrical distribution of organelles into daughter cells is accompanied by a similar propagation of cytoplasmic proteins, which is likely to maintain a cell signaling state of the parent cell in the daughter cells. This makes it likely that the same TF regulation will persist in the daughter cells, a possibility that has some experimental support. A study looking at the effects of alterations of cell stress or mitogenic activity in parent cells revealed an influence on cell cycle commitment in the daughter cells. The researchers found this effect to be mediated by transmission of the p53 protein and the cyclin D1 (*CCND1*) mRNA through mitosis (Yang et al. 2017).

Waddington would have been fascinated by the role TFs play to make cell fate decisions at the bifurcations of his creodes. As will be discussed in Chapter 5, another way that TFs can create a long-term memory of a past event is by changing cell fate decisions, creating a distinctive repertoire of cells within a tissue that can persist and contribute to phenotypes.

▶ **TARGETING AND STABILITY**

The discussion above makes the case that TFs can target loci and perpetuate their activities through cell division, thus making them excellent candidates for mediating epigenetic properties.

An observation by David Knowles during our co-authorship of a review on environmental epigenomics (Wattacheril et al. 2023) provides an indirect mechanism for TFs to mediate stability of regulation through cell division. The foundational observation is that TFs are mostly regulated by cell signaling pathways, and that cell signaling states can be transmitted to daughter cells after cell division (Yang et al. 2017). It therefore follows that propagation of cell signaling states may, in certain situations, be the primary means of passing on a memory of a parental cell state to daughter cells, mediated by effects on TF activities.

TRANSCRIPTION

The outcome of all these regulatory events is, ultimately, the appropriate expression pattern of specific genes in the genome. Gene expression happens when an RNA polymerase complex copies the DNA message into an RNA molecule. There are three eukaryotic RNA polymerases: RNA polymerase I, which acts on ribosomal RNA genes in the nucleolus; RNA polymerase III, which transcribes tRNA and 5S rRNA genes; and RNA polymerase II, which transcribes mRNA and several types of small RNAs (miRNA, snRNA, and snoRNA). Noncoding RNAs are transcribed by all three types of RNA polymerases (Bunch 2018).

Focusing on RNA polymerase II, the first step in its starting to transcribe a gene is have a protein complex called TFIID bind to a specific sequence (TATAA) known at the TATA box, with or without a second sequence called the Initiator (Inr) element. TFIIB is then recruited, in turn binding RNA polymerase II, and then three further components are bound, TFIIF, TFIIE, and TFIIH. The polymerase complex is paused at this promoter site until the serine amino acid at position 2 in the carboxy-terminal domain of RNA polymerase II becomes phosphorylated, at which point it starts to move along the DNA and generates the RNA transcript.

What has been observed in organisms from prokaryotes to mammals is that the transcription of a gene within an individual cell is not a constant steady state but is instead more like a discontinuous series of pulses or bursts of expression. This is difficult to appreciate when studying a population of cells in bulk, and was first revealed by Rob Singer using imaging techniques of individual live cells (Chubb et al. 2006). This observation raises the intriguing idea of a transcriptionally active gene in fact oscillating between active and inactive states, which is not our typical way of thinking about gene expression regulation. By altering these patterns of transcriptional bursting, the cell can be more adaptive in its responses to stimuli (Corrigan et al. 2016).

Although we have considered chromatin and DNA modification states in the discussions above as regulators of the activity of a gene, the act of transcription through a locus itself leads to changes of both chromatin and DNA modifications. These will be described further in Chapter 6 in which we will look at the patterns of transcriptional regulators decorating the genome.

A COOPERATIVE SYSTEM FOR CELLULAR MEMORY

DNA methylation and nucleosomal organization are quite similar in many ways—as we will learn in Chapter 6, they are both distributed to most of the genome and tend to colocalize. In other words, in the small subset of the genome in which there are no nucleosomes, there tends not to be DNA methylation either. Both nucleosomal organization and DNA methylation are quickly reestablished following DNA replication, so although we think of DNA methylation as being a prime example of a heritable molecular regulator, much the same could be claimed for nucleosomal organization.

What is most interesting in understanding transcriptional regulation are the deviations from these default states, and how these occur in distinctive patterns in different cell types and in response to specific stimuli. These patterns need mediators that can recognize distinct sequences, for which TFs and noncoding RNAs that can bind to DNA are our major candidate mediators. When Polycomb and heterochromatin formation are targeted to a region, this appears to be enough to maintain a silent state even when the pioneer TFs normally capable of accessing chromatin are present in that cell. Once a TF or group of TFs binds at a locus and recruits complexes that modify chromatin and DNA methylation locally, this appears to help make the transcriptional regulation more efficient and stable, locking in the program for gene regulation. Although simplistic, this looks like being a reasonable framework for thinking about the interactions of the multiple regulatory processes described above, a cooperative system for cellular memory.

In terms of the competing definitions for epigenetics from Chapter 3, we can paint these biochemical insights onto each definition with varying degrees of success. Waddington would have recognized current insights about the properties of TFs as central to the mechanism of epigenesis, or cell differentiation. Nanney would have focused on the capability of some biochemical mechanisms to self-propagate to the same loci in daughter chromatids as a mediator of persistent homeostasis, or cellular memory. We know that Holliday was drifting away from thinking about DNA methylation's epigenetic properties in terms of mutations that didn't involve nucleotides and instead toward a role in regulating genes, so he would welcome the association of DNA methylation with gene expression but would have been perturbed by the suggestion that it merely footprints other primary events like TF binding. The epi+genetics definition was made for all the observations above, defining epigenetic properties as anything that regulates DNA sequence, whereas those interested in multigenerational memory would welcome the insights into the potential mediators of such memory, with a particular focus on anything that is maintained robustly from parent to daughter cells.

This is how the multiplicity of definitions survives today. We have seen how the definitions of epigenetics have evolved accidentally and without much regard to prior knowledge about or use of the term. To address this frustrating situation, the next chapter is focused on rethinking the definition of epigenetics in a way that integrates its various roots as much as possible and gives us a framework for thinking about how to use the concept of epigenetics to understand phenotypic variability. Understanding phenotypic variability was the reason the term was used and modified in the first place and still represents a major reason why we study these molecular processes today.

RECOMMENDED FURTHER READING

Gilbert WV, Nachtergaele S. 2023. mRNA regulation by RNA modifications. *Ann Rev Biochem* **92**: 175–198. doi:10.1146/annurev-biochem-052521-035949

This comprehensive review is valuable not only for describing what we currently understand about the effects of modifying mRNA but also highlights where current gaps in knowledge exist.

Loda A, Collombet S, Heard E. 2022. Gene regulation in time and space during X-chromosome inactivation. *Nat Rev Mol Cell Biol* **23**: 231–249. doi:10.1038/s41580-021-00438-7

The current state of knowledge about X-chromosome inactivation in mammals.

Millán-Zambrano G, Burton A, Bannister AJ, Schneider R. 2022. Histone post-translational modifications—cause and consequence of genome function. *Nat Rev Genet* **23**: 563–580. doi:10.1038/s41576-022-00468-7

Everything you need to know about the functional effects of post-translational modifications of histones.

Popay TM, Dixon JR. 2022. Coming full circle: on the origin and evolution of the looping model for enhancer–promoter communication. *J Biol Chem* **298**: 102177. doi:10.1016/jbc.2022.102117

A productively critical, lucidly presented review of chromatin looping and gene regulation.

Talbert PB, Henikoff S. 2021. Histone variants at a glance. *J Cell Sci* **134**: jcs244749. doi:10.1242/jcs.244749

Not just a review of a fascinating topic but also comes with a poster you can print to decorate your wall.

CHAPTER 5

Epigenetic Properties as Those of Cells: Fates and States

DNA METHYLATION AND CELL SUBTYPE PROPORTIONS

Karl Kelsey, his longtime friend and collaborator John Wiencke, and his Brown University colleague Eugene Andres (Andy) Houseman were getting frustrated. Their paper, which combined mathematics and biology, had been submitted to a biostatistics journal and rejected because the mathematical approach was deemed too routine. It was now under review at a journal whose title and contents dealt exclusively with bioinformatics; the paper was getting bounced back repeatedly by reviewers, who were asking for more and more molecular experiments to justify the authors' challenging conclusions. Finally, after four rounds of submission, the paper was rejected by the bioinformatics journal. The authors resubmitted the paper to *BMC Bioinformatics* where it was published in 2012 (Houseman et al. 2012) and has, at the time of writing, been cited more than 3200 times.

What was this controversial study that elicited such a hostile response from the reviewers? Houseman, Kelsey, Wiencke, and their co-authors, Brock Christensen and Carmen Marsit, pointed out that:

> The composition of leukocyte populations is well known to reflect disease states and toxicant exposures and can be altered by signaling cascades that prompt migration of whole classes of cells into or out of tissues,... [and that] ...it is generally the overall balance of leukocyte subclasses in circulation or tissue that most prominently influences pathogenesis.

At that stage, Karl Kelsey had been interested for more than 20 years in the question of how the properties of peripheral blood leukocytes reflected exposures and diseases (Cullen et al. 1992). This insight helped to prompt the recognition that:

...changes in the distributions of blood cell types alone could account for disease associated DNA methylation,... [causing the authors to warn that] ...it is crucial to the development of this new avenue of biomarker research to delineate effects due to the immune cell distribution itself from other "non cell type" alterations in DNA methylation. We term the differences among human populations attributed to cell distributions to be "immunologically mediated".

The key moment in the genesis of the whole idea was classic across-discipline collaboration. Kelsey and Wiencke were working diligently (in a reductionist fashion) with colleagues Christensen and Marsit to define individual loci where DNA methylation defined a single cell lineage. Houseman, quietly evaluating the effort, realized that this could much more efficiently be accomplished using deconvolution methods. The collaboration meant that ideas could be exchanged and everyone's data sets could be used to test Houseman's approach.

In the end, this approach was challenging the interpretation of every DNA methylation study of disease at the time. The field was making the assumption that a change in DNA methylation seen in an assay of a bulk population of cells reflected changes of DNA methylation within those cells. What Houseman and colleagues were demonstrating was that a systematic change in the proportion of a cell subtype in a population will cause the same effect, and that there may be no individual cell types changing their DNA methylation at all. The logic behind their conclusion was based on the fact that there are loci where DNA methylation is specific to a cell subtype. When the proportion of that cell subtype is different between the groups compared, these loci are identified as differentially methylated (Fig. 5.1).

This is probably the point at which the field of people studying DNA methylation changes in disease started to pay substantial attention to the fact that we were actually studying cells and not just nucleic acids. Despite the thoughtful discussion in the Houseman and Kelsey paper about how peripheral blood cell subtype changes could reflect disease states, the response by the field as a whole was to treat cell subtype proportional changes as a confounding effect on DNA methylation variation, leading to the development of strategies to eliminate rather than understand the effect of variable cell subtype proportions.

While these insights and technical developments were being applied for DNA methylation, the same was happening for gene expression studies (Shen-Orr et al. 2010; Gong et al. 2011; Gaujoux and Seoighe 2012). In the years since, with the development of single-cell genome-wide assays, we have become even more acutely aware of the heterogeneity of cell subtypes

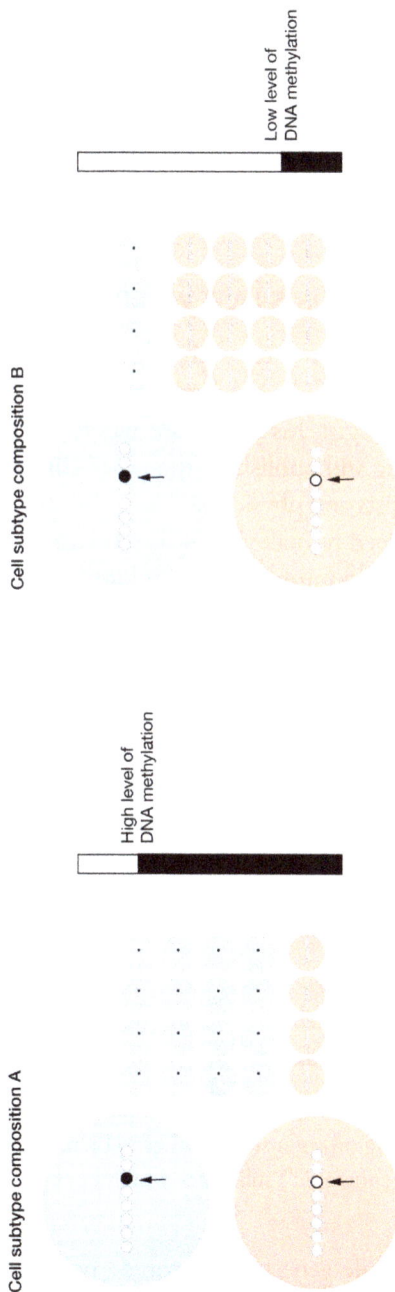

Figure 5.1. Cell subtype effects on DNA methylation. Two cell types are represented: (*top*) one having a methylated cytosine at the black arrow; (*bottom*) one unmethylated at that locus. In the mixture on the *left*, in which the *top* cell comprises 80% of the cells, the DNA methylation at that locus should be 80%, whereas on the *right*, in which the *top* cell type is now only 20% of the mixture, the DNA methylation value is now a corresponding 20%. In a situation in which DNA methylation differences are being tested for between samples, the above possibility can occur, without any cells having undergone a "reprogramming" of DNA methylation.

even beyond what we can define by sophisticated flow cytometry (Jaitin et al. 2014; Oetjen et al. 2018). It appears that we may have been preoccupied with the difficulties involved in developing and analyzing the results of genome-wide assays and neglected to remember that the unit of organization of all of the molecular events we study in these assays is the cell. In this chapter, we are going to focus on how we can rethink epigenetic properties as those of cells, mediated by, but not merely equivalent to, the biochemical molecular properties discussed in the last chapter.

WADDINGTON'S LAST WORD

Conrad Hal Waddington remained a very active figure in science until his untimely death in 1975 at the age of 69. Although he could always have been described as a theoretical biologist, his later work was focused even more strongly on theory, organizing and publishing the proceedings of symposia that brought together biologists and physicists.

Perhaps because of the need to make his ideas more readily communicable to new colleagues lacking a foundation in biology, and because some of the fundamental rules of genetics were becoming clear, his later writings read much more lucidly and simply about what he meant by the epigenetic landscape, creodes (which he had started to spell as chreodes), and the relationship between genotype and phenotype.

In his book *Towards a Theoretical Biology* (Waddington 1968), we find Waddington re-introducing epigenetics in a prologue chapter on "The Basic Ideas of Biology":

> Epigenetics has two main aspects: changes in cellular composition (cellular differentiation, or histogenesis) and changes in geometrical form (morphogenesis).

He appeared daunted by the challenge of understanding morphogenesis, but felt that it was possible to talk about the mechanism of differentiation of cells:

> It is in fact conventional to say that the basic elementary process is the derepression (or possibly switching on) of a structural gene by means of a cytoplasmic gene-recognizing ('genotropic') substance which has been produced by some other genetic locus.

He described this model of single-gene activation as derived from studies in bacteria and cautioned that genes in cells from higher organisms instead tend to switch on groups rather than as single genes. The cytoplasmic genotropic substance from a separate locus acting on a gene is a good description of the transcription factors that we now appreciate to be central to cell differentiation decisions.

Waddington's 1968 musings can be summarized as defining epigenetic properties as those of cells and tissues, with speculation about the underlying mechanisms of these cellular epigenetic properties. The resurrection of the use of the word epigenetics a few years later by Pugh and Holliday would be focused solely on molecular mechanisms, prompted by the difficulty of understanding how a *lac* operon/transcription factor–like model could generate long-term stability of gene repression in biological processes like X-chromosome inactivation. The attractiveness of DNA methylation as a repressive mechanism over time and through cell division subsequently became an intriguing preoccupation for Pugh, Holliday, Riggs, and others and reversed the way of thinking espoused by Waddington, with the new primary focus on molecular mechanisms, and the assumption that this molecular focus would lead to insights into cellular properties secondarily.

If we now revisit the idea of epigenetic properties being primarily those of cells, mediated by but not defined by molecular processes, we have a conceptual model that is much closer to Waddington's evolving ideas about epigenetics in 1968. In this chapter, the focus is on the value of a cellular definition of epigenetics. The case is made that we gain substantially by developing ideas about phenotypes and their variability when we think of epigenetics in a cell-centric way, while still allowing us to delve into the mechanisms of phenotypic perturbation in terms of the genomic regulators described in the last chapter.

OUR EPIGENETIC PARADIGMS WERE ALWAYS ABOUT INDIVIDUAL CELL DECISIONS

It is worth revisiting some of the biological observations that were foundational for the field of epigenetics, as they give us some insights into how we ended up focusing on molecular mechanisms when there were some interesting cellular events going on that we ignored.

Position-Effect Variegation

In 1930, Hermann Joseph Muller (1890–1967) described an unusual eye phenotype in irradiated *Drosophila*. Although it was already known that mutation of the *white* gene caused complete loss of the normal red pigment in the eyes of *Drosophila*, Muller's flies were doing something strange—only some of the cells in the eye were white, and others remained red (Muller 1930). This should not be occurring if the gene was completely inactivated, as would be the typical outcome of mutation, and therefore represented a mechanistic puzzle. Subsequent studies of these flies revealed that the mutation had not directly damaged the protein-coding sequence of the *white* gene but had instead caused an inversion of the chromosome containing the gene, juxtaposing the gene beside the pericentric heterochromatin of the chromosome. The mechanism for silencing was of the repressive heterochromatin spreading onto the *white* gene in some cells, causing the gene to inactivate in those cells. However, in cells in which the heterochromatin did not spread and inactivate the gene, the cells remained red. The problem was therefore not within the gene itself but due to its ectopic position in the genome, and the effects were reflected by a variegation of colors in the eye, causing the effect of the mutation to be referred to as position-effect variegation.

These flies turned out to be very valuable for genetic screens to identify heterochromatin components, as described in Chapter 4, by looking for effects of mutation of genes throughout the genome on the degree of variegation. The position-effect variegation model in *Drosophila* is a paradigm of an epigenetic process. With a focus on biochemical mechanism, the value of the model is clear in understanding mediators of domains of repressive signals in the genome. However, if we think instead in terms of cellular properties, it is equally intriguing, because the model suggests that cell decisions are probabilistically influenced but not absolutely determined by molecular regulators of gene expression, as cells are making individual decisions whether to be pigmented or not. Waddington's epigenetic landscape model does not predefine that the differentiating cell will end up in a specific lineage. Instead, the landscape represents a range of options with probabilistic decisions being made at the bifurcation points. The heterochromatin effects on an individual eye cell have the same nondeterminative outcome.

Metastable Epialleles in Mice

Position-effect variegation was subsequently demonstrated in mice. The groups of Liane Russell (1923–2019) and Bruce Cattanach (1932–2020) showed that the translocation of an autosomal coat color gene onto the X chromosome in female mice led to variable repression of that gene due to X inactivation (Cattanach 1961; Russell and Bangham 1961). More than

40 years later, using modern molecular approaches, Emma Whitelaw described a comparable pattern of expression of a transgene inserted into a
random genomic position expressing green fluorescent protein (GFP), regulated by α-globin gene promoter and enhancer sequences. These animals
expressed the transgene in only around half the red cells in blood, which
Whitelaw called a "metastability" of expression of the transgene. Her group
exploited this system to identify regulators of variegated expression of this
transgene, named modifiers of murine metastable epialleles, or *Momme*s for
short (Blewitt et al. 2005; Blewitt and Whitelaw 2013). In each of these
mouse examples, the cell probabilistically ends up adopting one of two
states, a probability that can be influenced by the balance of regulators of
activation and repression in those cells.

The best-known example of a "murine metastable epiallele" is probably
the viable yellow mouse, introduced in Chapter 1. In May 1960, the Jackson
Laboratory was maintaining its pedigreed expansion stock of the C3H/HeJ
line and noticed that one female offspring had been born with an unexpected yellow coat color, instead of the usual attractive mousey brown of the
parents. These parents did not go on to have another yellow offspring, indicating that this mouse had undergone a de novo mutation (Dickies 1962).
Interestingly, when this mouse was crossed with a littermate, and in later
generations, the presence of the mutation did not always segregate with a
strikingly yellow phenotype—some mice were very yellow, others looked
like they had the fur color of wild-type animals, and there was an extensive proportion of mice with intermediate phenotypes, with patches and
streaks of yellow mixed with brown fur, reflecting regulatory decisions made
by individual clones of cells manifesting in the animal's fur. Like Muller's
Drosophila, this was not how a gene mutation was expected to act. The viable yellow mutation and how it exerts its effects at the *a (nonagouti)* locus
were discovered in 1994 by Greg Barsh, revealing the de novo mutation to
be the insertion of a mouse endogenous retrovirus, an intracisternal A parti-
cle (IAP) element, ~100 kb upstream of the gene, as depicted in Figure 5.2.
Although there is much more to discuss later (Chapter 7) about the viable
yellow mice, the point relevant to the current discussion is that this again
represents a set of choices being made by cells in these animals whether
to develop the unusual yellow fur color. Unlike the *Momme* transgene example, in which about half the cells consistently silence the transgene, the
viable yellow animals are very variable in the proportion of cells silencing
the *a (nonagouti)* locus. This could be related to the inactivation decision

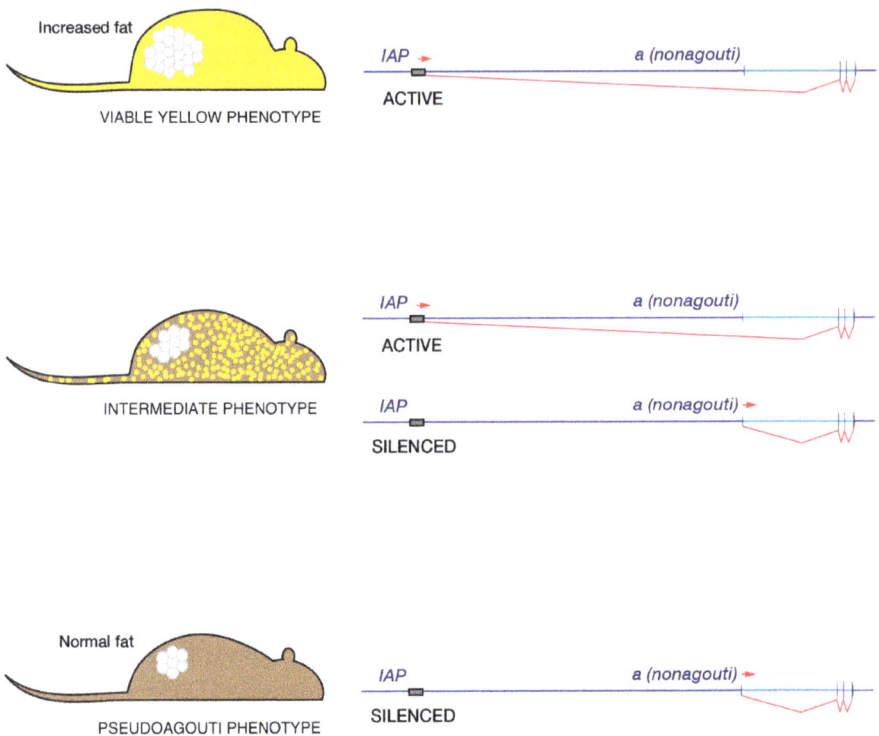

Figure 5.2. The viable yellow mouse. The yellow fur color is due to the inappropriate expression of the *a (nonagouti)* gene by the upstream IAP transposable element. In some animals (*top*), the IAP element drives expression in all cells to cause the viable yellow phenotype, whereas other animals express the gene in some (*middle*) or no (*bottom*) cells. The obesity phenotype is also caused by IAP-driven ectopic expression of the *a (nonagouti)* gene.

occurring earlier in viable yellow mice, which would be supported by the finding that early in vitro culture of zygotes and blastocysts influences the silencing at this locus (Morgan et al. 2008).

X-Chromosome Inactivation

As a final example of a long-standing epigenetic paradigm, we return to the phenomenon that had fascinated both John Pugh and Art Riggs, X-chromosome inactivation. The molecular mechanism of this process is described further in Chapter 6. In essence, it involves the cell counting how many X chromosomes it contains and then inactivating all but one of them. For the vast majority of individuals, this involves no inactivation in those with only one X chromosome and inactivation of a single chromosome in those two. The choice of which of the two X chromosomes to inactivate is generally random and occurs early in development. This means that, for each cell in a female, either the paternally inherited X is silenced or the maternally inherited X is instead.

The vivid example of X inactivation typically used for illustration is fur color in cats. There is a gene on the cat X chromosome called *O* (*Orange*)[1] that has one type of allele that causes eumelanin (black pigment) to be deposited in the hair shaft, and a second type of allele that leads to the orange-colored pheomelanin to be deposited instead. When the cat is female (or, extremely rarely, when the cat is XXY male, the human Klinefelter syndrome genotype) and is heterozygous for the black and orange pigment alleles, then the cell will inactivate one allele in each cell, which then causes hair to grow in patches of black and orange. The mottled fur color is described as a "tortoiseshell" pattern. When the cat also has a mutation in the separate, autosomal *S* (*white spotting*) gene, the pigment-producing cells (melanoblasts) appear to migrate to the skin in fewer numbers and cause two things to happen—patches of white fur also develop, and the individual patches of black and orange fur are bigger, possibly representing greater spatial expansion of each clone of melanocytes (Schmidt-Küntzel et al. 2009). In the United States, these are commonly referred to as calico cats (tortoiseshell elsewhere) and represent a striking demonstration of how X-chromosome inactivation causes variegated expression of genes within individual clones of cells in a mammal, with the unique pattern of fur color for each cat a very visible manifestation of these cell decisions.

MISSING HERITABILITY, THE LIMITS OF GENETIC DETERMINISM, AND EPIGENETIC FACE CREAM

Position-effect variegation, gene expression metastability, X-chromosome inactivation, and other examples of gene regulation became of interest to biologists because of the insights they offered into how the transcription of genes is regulated, even at the scale of large parts of entire chromosomes when considering position-effect variegation or X-chromosome inactivation. In

---

[1] *O* (*Orange*) was recently, simultaneously, and independently discovered to be a 5-kb deletion upstream of the *Arhgap36* gene by the teams of Gregory (Greg) Barsh (Kaelin et al. 2024) and Hiroyuki Sasaki (Toh et al. 2024).

Chapter 6 the fascinating mechanistic insights from the study of these processes will be described. However, we may be ignoring another, equally important insight from these studies, exemplified by the results of cloning of a calico cat in 2002. Although the cloned cat was genetically identical (probably give or take a few somatic mutations in the cloned cell) to her mother, as the authors noted, "This kitten's coat-coloration pattern is not a carbon copy of its genome donor's" (Shin et al. 2002). In other words, identical genotypes do not confer identical phenotypes.

This should be no surprise, if you have ever met identical (monozygotic) twins and noted that although they can look very similar, they will have enough subtle differences in appearance that they can be distinguished, if only by their family members. Furthermore, as Kevin Mitchell points out (Mitchell 2018), the same genetic program runs twice in the same individual, creating the left and right sides of the body. The asymmetry that is inherent to normal development is another manifestation of the way that genetic information is not fully deterministic in morphogenesis.

Monozygotic twins are frequently discordant in developing heritable diseases, despite what are assumed to be identical DNA sequences. This has raised the question whether factors other than genetic predisposition govern the risk of disease. It is often assumed that external, environmental factors help to mediate disease risk. This is a perfectly reasonable assumption for many diseases; although type 2 diabetes mellitus has a component of genetic predisposition, diet and physical activity are also strong contributors to risk. One way to identify diseases with environmental influences on susceptibility is to track their incidence over time. As it is not possible for the genetic makeup of a population to change substantially over the course of years or decades, a major trend in incidence in the disease is likely to reflect alteration in these environmental exposures; type 2 diabetes mellitus is a great example of this.

There are two logical leaps from this kind of example of environment influencing genetic susceptibility to disease. The first seeks a mechanism for interactions of environmental factors and the genome and has focused on the epi+genetic regulatory mechanisms as leading candidates for mediating these environmental influences. The second takes a further leap of faith to speculate that we can override genetic susceptibility to disease by modifying epi+genetic regulators through some sort of positive environmental exposures. Distressingly for the purist scientists toiling in their laboratories, what has emerged based on this enthusiastic overoptimism are consumer products like epigenetic yoga, epigenetic face cream, and epigenetic diets.[2] There is almost no evidence to support these kinds of interventions as having any positive effects. Did the emergence of these pseudoscientific health products reflect something that we did wrong as a scientific field?

It's probably impossible to prevent intriguing scientific ideas from leading to well-intentioned but underinformed adoption by enthusiastic nonspecialists. However, we scientists have probably been distracted by the lure of understanding molecular biological processes and may have lost sight of why Waddington proposed the epigenetic landscape model in the first place—he wanted to have a means of understanding how genes give rise to phenotypes. His focus was on cellular differentiation (epigenesis) and Nanney's interest was in cellular memory, whereas the later paradigms of epigenetics that drove discovery of molecular mechanisms (like X-chromosome inactivation, position-effect variegation, and metastable epialleles) represented cell fate decisions made by individual cells that yielded distinctive organismal phenotypes. The study of epigenetics has therefore from its origins been about cellular properties. The essence of the epigenetic landscape model is that the ball rolling down the creodes in the hillside is not predestined to end up in one of the terminal creodes, it is governed by the branching patterns and depth of the creodes to create a probabilistic rather than a deterministic outcome. Running the same program twice—whether rolling balls down the epigenetic landscape, developing as monozygotic twins, or as the two sides of a body—is never going to give you identical outcomes. This probabilistic model can explain by itself some of the nondeterminism involved when two people with the same genetic susceptibility to disease are discordant for the phenotype. Environmental influences represent another factor in susceptibility, but we don't generally think of these influences in a way that involves cell decisions. Is there a value to thinking beyond molecular biology and instead in terms of cellular properties?

One area of human disease research is referred to as the developmental origins of health and disease, or DOHaD. This field of research assumes that perturbations in early life, especially during development, can give rise to susceptibility to disease in later life. Given that this latency between perturbation and disease development sounds like a kind of a memory of past events, and some of the definitions of epigenetics have emphasized this as

---

[2] Seriously. Do an internet search, you'll find these to be the tip of the epigenetic pseudoscience iceberg.

a property of molecular mechanisms like DNA methylation, a substantial focus of the DOHaD field has been on potential molecular mediators of these past perturbations. There is, however, a very interesting paradigm that should influence us to think more broadly about how a perturbation during development can cause an adult disease phenotype through cell development decisions—as discussed in the next section, a study performed by a developmental biologist who was not trying to perform a DOHaD study at all but instead wanted to understand the role of retinoic acid signaling in mouse lung development.

FIVE DAYS WITHOUT VITAMIN A

Wellington Cardoso had been dissecting out the tiny foregut from mouse embryos for years. He was fascinated by how the pouch that pushed itself out of this miniscule tube grew and branched to form the entire complexity of the lungs, with its many cell subtypes organized meticulously into an interface between air and blood for the body. He was especially interested in the role of retinoic acid in lung development. We get retinoic acid from vitamin A in our diet, and it was already known that when the developing human embryo was being carried by a mother deficient in vitamin A, the embryo was at greater risk for problems of lung formation.

Cardoso decided to make the diet of pregnant mice completely deficient in vitamin A for the 5 days in which lung formation starts, embryonic days 9.5 to 14.5, and then return the mothers to a normal amount of vitamin A in their diets. When he sacrificed the pregnant mothers at embryonic day 14.5 and studied the fetuses, he found the lungs to look generally normal, but with one striking difference—an increase in the amount of smooth muscle encircling the distal airways of the fetuses who had been deficient in vitamin A for the past 5 days. When he looked at the smooth muscle around blood vessels in the lung, there was no difference between the groups of mice on different maternal diets; the smooth muscle effect was confined to the airways (Chen et al. 2014).

Was this a big deal? Fetuses don't breathe, so some extra smooth muscle around distal airways isn't going to have any effect as the fetus floats in the amniotic fluid. Cardoso repeated the experiment, but this time allowed the pups to be born from each group of mothers. The vitamin A–deficient pups looked healthy and grew to adulthood without apparent problems. At that time Cardoso decided to sacrifice the animals and look at their lungs in detail. What he found was that the increased amount of smooth muscle around the distal airways was still there, having persisted from the middle third of pregnancy.

When someone takes an inhaler containing a bronchodilator for their asthma, what they are doing is delivering a medication to their distal airways to relax the contracted, thickened smooth muscle that is in spasm and constricting the flow of air. We have usually thought of this smooth muscle accumulation as being secondary to the inflammation of asthma, a consequence of the immune system abnormality of the disease. Cardoso was now wondering whether his vitamin A–deficient mice had the pulmonary characteristics of human asthma. Although this is relatively easy to test in humans (asking them to blow into a tube as hard as they can), mice needed to be anesthetized and put on a ventilator to breathe through a tube, into which was sprayed a medication, methacholine, that causes smooth muscle to contract. The more smooth muscle that was present, the greater the airway resistance in response to the drug. When Cardoso performed the lung function measurements, he found significantly increased airway resistance in the offspring of mothers fed the vitamin A–deficient diet for 5 days in mid-pregnancy.

This is exactly the kind of outcome being sought in the typical DOHaD study—a perturbation (vitamin A deficiency) at a critical period of development (during the middle third of pregnancy) that leads to a disease phenotype later in life (the airway resistance of asthma). Waddington was daunted by the mechanisms of morphogenesis, and this is a good example why—only by performing detailed histological studies of the lung was it possible to identify the accumulation of one cell type to a greater extent than normal in one critical region of the lung. As described earlier in this chapter, typically the alteration of a cell subtype proportion in a tissue is seen as a confounding variable in epigenetic studies of disease. If Cardoso's experiment had been performed in a different way—taking the lungs out of mice with vitamin A deficiency and increased airway resistance and comparing them with control samples by grinding them up, extracting RNA and DNA, and comparing gene expression and DNA methylation differences—the effect of an increased proportion of smooth muscle cells in the vitamin A–deficient mice would have been discarded as a confounding variable, using current analytical approaches.

As with the epigenetic paradigms involving cellular variegation, what is happening in the lungs in the vitamin A–deficient embryos is a systematic set of cell choices—in this case the decision of progenitor cells to differentiate to smooth muscle. Retinoic acid binds to the retinoic acid receptor, which joins with another nuclear receptor to find a DNA sequence motif and bind as a transcription factor. There is therefore plenty to intrigue the molecular epi+geneticist also, as we can study the biochemical, transcriptional regulatory events involved in making these cell fate decisions. But, at the end of the day, what was important in giving these mice a partial replica of the human asthma phenotype was not dependent on a molecular change imprinted into a specific cell type but instead the production of more of a certain cell type in a specific part of the lung. This is going to require a new way of thinking about how we link what we call epigenetic processes with phenotypes, zooming out from molecules to cells, and thinking of epigenetic properties as those of cells.

CELLULAR MODELS OF PHENOTYPES AND DISEASES

We need to be repetitive, once again revisiting the definitions of epigenetics over time, noting the drift from cell-based to molecular definitions. Waddington was focused on cell differentiation and how genetic mutations could influence lineage commitments. Nanney was trying to understand mating type decisions and how cells maintained a memory of past events. Pugh and Holliday and Art Riggs were all prompted by the paradigm of X-chromosome inactivation to explore mechanisms that could maintain long-term cellular memories of decisions made early in development. As interest in molecular processes displaced these initial questions, there was an attempt to restrict the focus on molecular processes acting to regulate genes that were heritable from parent to daughter cells, which was not very successful but reflected the idea that epigenetic properties involved some sort of memory of past events or perturbations involving progenitors of the cell.

This is the logic of many studies seeking to link disease or other phenotypes with epigenetic processes—in this case meaning molecular regulators of transcription like DNA methylation. The model, often unstated, is one of a specific cell type undergoing a kind of reprogramming of its transcriptional regulatory properties. In other words, the cell looks like a regular myocyte, a lymphocyte, or a neuron, but it bears a memory of a past event that is encoded in its transcriptional regulatory program, a program that is different enough from other cells of the same type to make it distinctive but not enough to cause the cell to become a different cell type.

Tuuli Lappalainen and I have called this the cellular reprogramming model of epigenetic properties of cells (Lappalainen and Greally 2017). This is the cell model being tested when cells from individuals differing in their past exposures are compared in epigenetic association studies, even when not stated explicitly by the researchers involved.

There is, however, another cellular model to consider. This is prompted in part by the Cardoso example of intrauterine vitamin A deficiency and in part by some musings by Waddington himself. Having introduced the idea of the epigenetic landscape, Waddington was interested in how it might vary either between individuals or within an individual following an environmental exposure. He depicted this as an alteration in the relative depths of the creodes in otherwise similar epigenetic landscapes (Fig. 2.4). The outcome would be a difference in the relative proportions of cells committing to one or another lineage. To depict the Cardoso example for cells in the distal airway, you could depict the epigenetic landscape for cells differentiating locally as having a deepened creode for smooth muscle after prenatal vitamin A deficiency. However, there is no current reason to believe that the hyperplastic cells of the distal airway in vitamin A–deficient mice are reprogrammed in any way; there are just more of them, and that causes the asthma-like phenotype. We need a way of describing this, something less cumbersome than "polymorphism of creode use." The word polymorphism ("many forms") is used today to refer to differences in nucleotide sequences, an equivalent to "variability," itself a corruption of the word's original meaning. We coined the term *polycreodism* as a mild provocation, to make the idea of variability of cell fate decisions explicitly neo-Waddingtonian (Lappalainen and Greally 2017) and to avoid further compounding the misuse of the word polymorphism.

The idea of polycreodism nicely captures the property of variability of cell subtype proportions in a tissue as a result of developmental events. If cellular reprogramming is a model for memory of past events on the scale of individual cells, polycreodism is the memory of past events on the scale of the whole tissue. In reality, there are further influences on cell subtype composition within tissues in human diseases, including cellular infiltration (often involving inflammatory cells) and new patterns of differentiation of cells within the tissue as part of the disease process, such as the generation of fibrogenic myofibroblasts by transdifferentiation from hepatic stellate cells in chronic liver disease (Tsuchida and Friedman 2017). Polycreodism therefore represents one of several causes for systematic cell subtype compositional differences in human diseases.

Both cellular reprogramming (altered cell states) and polycreodism (altered cell fates) are likely to contribute to phenotypic variability and disease. Cellular reprogramming describes the altered properties of individual cells, whereas the idea of polycreodism makes us think in terms of whole tissues and how changing their cell subtype composition can predispose to disease. Although these are valuable ways of forming hypotheses for epigenetic association studies, these are not new concepts at all and have been studied by pathologists for a long time.

REDISCOVERING HISTOPATHOLOGY

There are more established ways of understanding how cellular properties influence phenotypes and diseases, and they have been around for more than a century. Studying the tissues affected by or mediating a disease using microscopy has not only been a common approach, but histopathology is also still a definitive test for many diagnoses. From a 2017 editorial discussing advances in microscopy and the application of machine learning to histology (Histopathology is ripe for automation 2017):

> For many diseases, histopathology is the clinical gold-standard technique used for diagnosis, and in research laboratories it is typically used to benchmark or confirm findings from basic and translational biomedical research.

To stay consistent with how topics have been introduced in this book, we can start by adding in some historical context. The techniques still used in modern histopathology began to be developed in the second half of the nineteenth century. The use of formaldehyde to fix tissues had been introduced. The fixed tissue was then dehydrated and hardened and embedded in paraffin, allowing the chunk of tissue to be sliced on a microtome to allow a very thin tissue section to be placed on a microscope slide. The first dyes to be used on these sections included cochineal and picric acid (Cook 1997) and a natural compound called hematoxylin, extracted from a tree native to Central America called *Haematoxylum campechianum*. The German pathologist Heinrich Wilhelm von Waldeyer-Hartz (who coined the words "chromosome" and "neuron") popularized the use of this dye (Scheuerlein et al. 2017), but it was the addition of alum to the hematoxylin by Böhmer that allowed this chemical to be used effectively to stain the nucleus (Cook 1997).

Around this time, chemists were beginning to develop techniques for manufacturing new, inexpensive dyes for the clothing industry; previously dyes had been extracted from natural products. The key compound was aniline, which gave the color to the natural product indigo. By treating aniline in various ways, the color of the resulting compound could be altered. An academic chemist called Adolf von Baeyer, working in Strasbourg, was developing a series of derivatives of aniline, including the triphenylmethane compounds (green in color) and the phthaleins, one of which was fluorescein. He gave a sample of this fluorescein to Heinrich Caro, who was working at the company destined to be called Badische Anilin- und Sodafabrik (Baden Aniline and Soda Factory [BASF]). Caro added four bromines to fluorescein and created a new yellow–red dye. He wanted to name this after a girl he had admired growing up, who was known by the nickname Eos (after the goddess of the dawn in Greek mythology), so the smitten Caro named his new dye eosin (Reinhardt and Travis 2000). By the 1870s, hematoxylin and eosin were being used in a combination that persists to today, commonly abbreviated as H&E.

Today, when making a diagnosis of a disease, diagnostic histopathology remains a gold standard for many diseases, with H&E surviving as a first-line staining reagent. We can infer from imaging, blood chemistry analyses, or other indirect tests that there is fat accumulation in the liver, inflammation of the large intestine, damage to the structure of the kidney, or a neoplastic mass in the lung, but only when we get a sample of the affected tissue can we be certain of our diagnosis; treatment decisions often rely on the direct histopathological evidence rather than the inferred evidence from other diagnostic approaches.

What is a pathologist like my father looking for when scrutinizing the sample under the microscope for a disease diagnosis? Fundamentally, a few patterns are important: the change in morphological characteristics of a cell type, the presence of cells in the tissue that are not normally there, and structural alteration of the tissue (e.g., the appearance of scarring, bleeding, or other abnormalities). The morphological studies are not limited to individual cells but also the physical relationships of cells within a tissue. We have rediscovered all these principles of pathology in modern molecular biology: describing morphological properties of the cells as cellular reprogramming; the alteration of cell subtypes within a tissue as a consequence of polycreodism, infiltration, or transdifferentiation; and the physical relationships of cells to each other in terms of influences of the microenvironment on a cell type of interest. It is not uncommon for new technological approaches to rediscover findings already known through existing approaches, and our reinterpretation of the results of genome-wide assays in terms of cellular properties aligns us back to the principles that have guided diagnostic pathologists for more than 100 years.

GENETIC VARIATION AND CELLULAR MODELS OF DISEASE

It is worth remembering something about Waddington's epigenetic landscape at this point. As discussed in Chapter 2, he was trying to reconcile the viewpoints of the embryologists studying epigenesis with those of the geneticists studying the morphological effects of DNA mutations. His proposal was that genetic mutations could change cell fate. He was not excluding genetic mutation as a mechanism for phenotypic variability at all, whereas the modern definition of epigenetics appears to be solely focused on nongenetic events. It was only with Nanney's resurrection of the term that it began to encompass how cells could be affected by nongenetic influences, and this persisted in the resurrection of the idea of epigenetics in the 1970s.

It is worth noting that the cellular models of reprogramming and polycreodism are also applicable when studying the outcomes of genetic variants and mutations. In fact, the idea that a damaging genetic variant changes the property of the cell is probably unquestioned as the default cellular model. There are plenty of examples in which this is clearly the case. Sickle cell disease has a specific mutational mechanism involving globin gene production, leading to the functional consequence of altered oxygen transport by the erythrocyte, and the morphological characteristic of sickle-shaped red blood cells (which are called drepanocytes, from the Greek δρέπανον [*drepanon*], meaning "sickle") due to polymerization of the hemoglobin following hypoxic stress. Sometimes the morphological change is the mechanism of the disease; the range of conditions termed ciliopathies involve mutations of genes involved in the formation of cilia, projections from the cell that are involved in how the cell moves and senses its surroundings. The mutations in ciliopathies are defined by how they lead to the failure to form normal cilia, with consequent motility and sensing problems for the cell. Some mutations lead to intrinsic problems for the cell that are not reflected by morphological changes, such as the cardiac channelopathies that can cause a fatal arrhythmia without any histological changes (Webster and Berul 2013).

Not all mutations cause intrinsic properties of cells to change; some alter the differentiation of cells. Some severe congenital immune deficiencies are due to blocks in differentiation to generate mature cells capable of fighting infection, an extreme phenotype. Benign ethnic neutropenia (Atallah-Yunes et al. 2019), on the other hand, describes how some human populations normally have lower proportions of neutrophils in their peripheral blood: In the case of some people with African ancestry this reflects a variant in the Duffy blood group antigen (Reich et al. 2009) that was under strong evolutionary selection for protection against malaria (Ntumngia et al. 2016). Variants in genes can thus be involved in changing the proportion of cell types in a tissue.

Variants outside genes in the noncoding 98% of the human genome are increasingly recognized to have disease-causing consequences. In fact, this is where the world of genome-wide association studies (GWASs) has led us— these studies revealed associations between individual variants within haplotypes (DNA sequences usually inherited as blocks) and human phenotypes. The most strongly associated variants were typically located between and not within genes, prompting the question whether they were having effects on the expression of genes nearby. This in turn drove the field of research to identify functional variants in the genome in specific cell types, where DNA sequence variation led to an alteration in a functional property of the genome, such as gene expression, DNA methylation, or chromatin structure.

The implicit assumption is that the variant influences an intrinsic property of a specific cell type. This is perfectly reasonable, but there is room to be misled experimentally. As described for DNA methylation at the beginning of this chapter, changes in something like gene expression can be caused by a systematic alteration in a proportion of a cell subtype in the tissue tested. To be more explicit, consider the following situation. There is a gene called *WXYZ1* that is only expressed in monocytes in peripheral blood. In one group of individuals, the average expression level for *WXYZ1* in peripheral blood is 3, but in the second group the average expression level is 9 (arbitrary units). However, the first group has an average proportion of monocytes in the white cell population of 2%, whereas the second has an average proportion of 6%. The threefold increase in expression of *WXYZ1* is therefore likely to reflect only the threefold increase in the proportion of monocytes, and there may be no cells in the second group that have intrinsic changes in the expression level of the gene (Fig. 5.3).

Now consider the possibility that you're looking for a variant that is associated with the change in expression of *WXYZ1*. Typically, you look for a variant located somewhere nearby on the same chromosome, a so-called *cis* expression quantitative trait locus (eQTL). If there is a variant that is in this

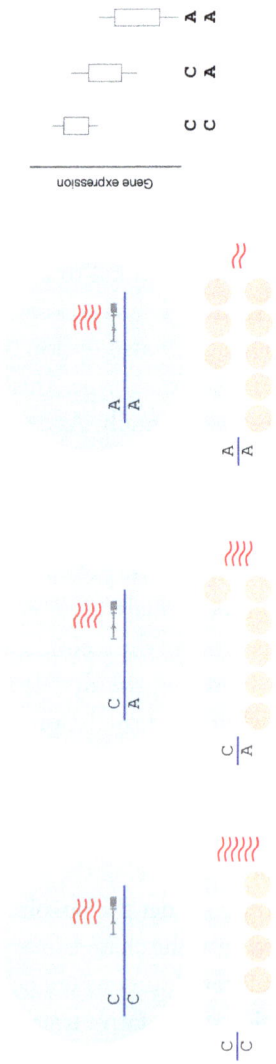

Figure 5.3. How a genetic variant near a gene expressed specifically in a cell subtype can act to influence the level of expression of the gene as an expression quantitative trait locus (eQTL, *top*). In the situation depicted in the *bottom* part of the figure, a variant that influences the proportion of that cell subtype located in proximity to a gene expressed specifically in that cell subtype will generate a very comparable association, without influencing the level of expression of the gene in an individual cell. It is therefore worth considering that some variants currently described to act as eQTLs may actually be influencing cell subtype proportions in a tissue.

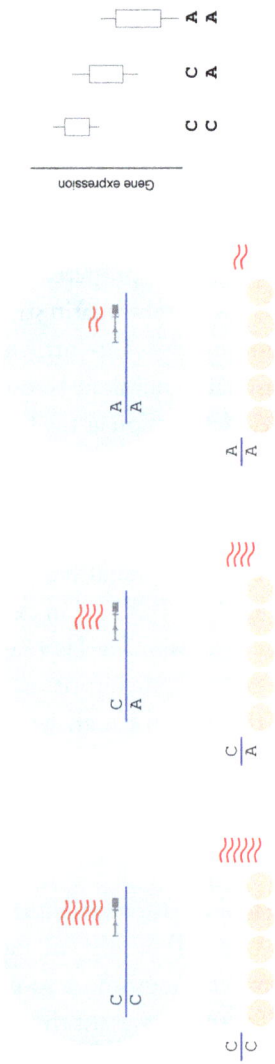

region (limited to a megabase around the gene's transcription start site) that is significantly overrepresented in one of the compared groups, this would get called a *cis* eQTL as an innocent bystander. The variation with which it is really associated is, however, the cell subtype proportion change. Can DNA sequence variants between genes influence cell subtype proportions?

Obviously, there is no point in asking rhetorical questions like that without the answer being positive. GWASs testing blood cell subtype proportions as the phenotype have revealed noncoding variants in the genome to be associated with altered cell subtype proportions (Orrù et al. 2013; Roederer et al. 2015; Aguirre-Gamboa et al. 2016). The possibility that what is being called a *cis* eQTL is in fact mediating its effect by altering cell differentiation is therefore open but will be unlikely. The chances are relatively lower that this "cell count QTL," or ccQTL as it has been described (Aguirre-Gamboa et al. 2016), resides within 1 Mb of a gene expressed solely in the cell type changing proportion in the tissue. More of the time the ccQTL will instead look like a *trans* eQTL, associated with expression of many genes expressed solely in the regulated cell type. *Trans* eQTLs are very challenging to identify, as they are defined as being located >5 Mb from a gene's transcription start site or on another chromosome, creating a problem of a huge number of genome-wide associations that will include many spurious associations due to noise.

DNA sequence variability in noncoding regions can therefore also cause a polycreodism effect. Waddington's illustration of variable creode use (Fig. 2.6) was depicting changes in cell lineage choices that he believed could be due to either genetic or environmental influences. We can therefore think of phenotypes in terms of the cellular models of reprogramming and polycreodism, mediated by coding and noncoding germline genetic variation, or mediated by transcriptional regulatory decisions. By focusing on the cellular models as epigenetic, which is justifiable in term of the historical roots of the term, we bring both transcriptional regulatory and DNA sequence influences under this umbrella.

As mentioned earlier, the tendency of the field of epigenetics has been to regard the cell subtype proportion influence as a confounding factor that should be eliminated. Less attention has been paid to DNA sequence variability, which has been estimated to have strikingly strong influences on DNA methylation, estimated to account for between 18% (Bell et al. 2012) and 66% (Cheung et al. 2017) of DNA methylation variability between individuals, depending on the cell type. Because the typical current viewpoint is that epigenetic properties exclude anything to do with DNA sequence variation, it would be reasonable to treat sequence polymorphism as another confounding influence.

What is the advantage of being rigorous about removing cell subtype and DNA sequence variation as confounding influences? The implicit message is that we are, as a field, focused on the cellular reprogramming hypothesis, solely because of transcriptional regulatory changes, and consider any genetic influences to be separate and not informative. This is a constraint, and suggests that we should rethink our overall goal, which is to understand phenotypes.

CELLULAR GENOMICS

In Chapter 7 the focus will be on how to identify epigenetic mechanisms of disease. Merely by using the word "epigenetic," we can see that we are loading preconceptions onto how these studies should be performed, and the typical approach is going to deliberately exclude mechanisms that include variation in cell subtype composition and genotypic contributions.

When we get a sample of tissue from a human suffering from some sort of disease, there has been an investment made. Even a blood draw involves some degree of risk and discomfort, personnel time, consumables, and care with the transfer of the sample to its appropriate destination. The goal of genomics studies should be to maximize the insights we can get into the pathogenesis of the disease when provided with these samples. This is the updated, human disease–focused version of Waddington's goal when coming up with the model of the epigenetic landscape of trying to understand phenotypes.

When presented with the human tissue sample, we can think in terms of the two cellular models discussed above. Have individual, canonical cell types in the sample changed their innate characteristics compared with what is normally expected of them (as in the cellular reprogramming model)? Or are there differences in the composition of the tissue in disease, reflecting new differentiation decisions in forming the repertoire of cells in the tissue (polycreodism), new infiltrative cells, or transdifferentiation of cells already within the tissue (as in the cell subtype composition model)?

Once both these cellular models are being considered, the question then becomes how to find evidence for each. Histopathology should probably be standard when performing these studies, as a strongly established way of making diagnoses and gaining at least some insights into both cellular models. In the field of cancer genomics, for example, it is accepted practice that histopathological studies be performed on a sample to be sequenced for somatic mutations to make sure that it contains a high proportion of tumor cells.

Genomic studies can then be selected. When looking for cell subtype proportion changes, single-cell transcriptional studies represent an exciting new way to reveal changes that might not otherwise be apparent. Gene expression and DNA methylation studies of cells in bulk from the tissue can be influenced by cell subtype proportion changes, as described earlier. There exist ways to adjust for such changes that can be applied if the cell subtype composition is already measured or by using reference data and analytical deconvolution techniques. These assays, once adjusted for cell subtype composition, give insights into cellular reprogramming. The same approaches to account for cell subtype effects are not yet developed for chromatin assays, which produce a different kind of data, as will be discussed in the next chapter. Chromatin changes provide another reflection of cellular reprogramming.

The next question would be whether the evidence for one or other type of cellular model is due to genetic influences or instead to an independent effect on transcriptional regulation. Genotyping of the individual's germline DNA can reveal functional variants like eQTLs or ccQTLs, allowing them to be identified as an influence. Not discussed up to now is the possibility that one or more somatic mutations could have occurred in the affected tissue at some stage and could be causing one of the cellular models. This would require sequencing of the affected tissue, which is technically feasible.

This hypothetical approach is designed to maximize the chances of revealing insights into cellular events occurring as part of disease pathogenesis. It is open to the idea that disease is caused by transcriptional regulatory or genetic variation and by cellular reprogramming or cell subtype compositional changes. This is sufficiently different from typical epigenetic studies that it may warrant a more specific description. From now on in this book this comprehensive approach will be described as the cellular genomic approach to understanding human disease pathogenesis. Studying germline DNA from blood or saliva has been a useful way of understanding genetic variation and its relationship to human phenotypes, but we need to have approaches that can be used on tissues affected by or mediating disease and get as much information as possible.

In the previous chapter we learned about the many regulatory mechanisms influencing the function of the genome. There are genome-wide assays that test most of these mechanisms. When applying them to gain insights into cellular properties, these assays have certain strengths and weaknesses. The next chapter will deal with these "epigenomic" assays in detail.

RECOMMENDED FURTHER READING

Cuomo ASE, Nathan A, Raychaudhuri S, MacArthur DG, Powell JE. 2023. Single-cell genomics meets human genetics. *Nat Rev Genet* **24**: 535–549. doi:10.1038/s41576-023-00599-5

The emergence of single-cell molecular assays is now highlighting both cell subtype composition and molecular reprogramming within cell subtypes in tissue samples. This review describes how these cellular genomic insights show promise for human disease studies.

Wattacheril JJ, Raj S, Knowles DA, Greally JM. 2023. Using epigenomics to understand cellular responses to environmental influences in diseases. *PLoS Genet* **19**: e1010567. doi:10.1371/journal.pgen.1010567

Our review of the field of "environmental epigenomics," designed to be a practical guide to the design and execution of these kinds of studies, with a focus on using molecular assays to understand cellular properties.

Zechner C, Nerli E, Norden C. 2020. Stochasticity and determinism in cell fate decisions. *Development* **147**: dev181495. doi:10.1242/dev.181495

A review that manages to communicate mathematical modeling and cell biology simultaneously and effectively, creating a framework for how the reader should think about stochasticity and determinism in the understanding of cell fate choices.

CHAPTER 6

Epigenomics: Testing Whole Genomes

In Chapter 4 we looked at a number of molecular mechanisms influencing genomic function. Some of the insights provided, such as the chromatin looping information, were only possible because of the development of genome-wide assays. When a genome-wide assay is testing something that has regulatory activity for the genome, following the epi- (above/upon) -genetics (DNA sequence) definition, the assay is described as epigenomic—another definition that is often used to describe these as functional genomics assays.

In this chapter, there will be a discussion of the evolution of epigenomic assays as well as descriptions of how the data they generate are analyzed, because another way we think of "the epigenome" is as the organization of transcriptional regulatory events along the DNA sequence. Because this is a massive field and can't really be described comprehensively in one chapter, some choices have been made about what to include. How assays generate data and how analyses turn those data into information is the technical foundation for the chapter. The distribution of regulatory events in terms of annotations of the genome is a major focus of this chapter, but the overall aim is to understand how to interpret the information from these assays—not just at the molecular and genomic level but also at the cellular level—getting back to the emphasis of Chapter 5. The goal is to create the foundation for understanding why some assays and analyses are better than others for the applied topics of the next three chapters, with their focus on understanding human disease.

FISHING EXPEDITIONS AND STARTING HYPOTHESES IN EPIGENOME-WIDE STUDIES

Proposing to use genome-wide assays to study samples from individuals with a disease compared with samples from unaffected individuals is not always met with enthusiasm from scientific colleagues. An easy, dismissive response is that these studies represent "fishing expeditions," implying that the researcher has not developed a starting hypothesis and will instead be guided by the results of complex assays after having generated their data. This approach makes it difficult to predict a priori whether the study is likely to be successful, unless there is already a substantial amount of preliminary data generated that suggests a likely outcome.

Epigenomic assays are typically genome-wide and generate an extraordinarily large amount of data. Anyone proposing to perform epigenomic assays is therefore potentially vulnerable to the criticism that they are merely embarking on one of these fishing expeditions. For those who have performed such studies, this concern fails to recognize that sometimes you just have to do the study before you know what you're going to find, which does indeed carry an element of risk and strays from the model of hypothesis-driven research.

The solution may be to update the idea that science must always work from a foundation based on a starting hypothesis, even though this is explicitly what major funding agencies often define as being required in grant proposals. David Glass has argued that the advancement of a prior hypothesis is just not appropriate for projects that generate massively multidimensional data, and that proposing an hypothesis may be, in fact, a source of bias, driving the scientist to favor results that support their starting hypothesis (Glass and Hall 2008; Glass 2010). Glass proposes that the approach should instead be to pose a question—not a directed hypothesis—and to allow the answers generated from experimental testing of the question to prompt a model to explain the results. This solution reverse engineers what we already do in practice—for example, a genome-wide association study does not start with a hypothesis about the causation of a disease but merely asks the question whether certain alleles are overrepresented in individuals with specific phenotypes. The answer to the question posed can be positive or negative, and, if positive, allows the creation of a model for why the phenotype has a genetic contribution based on the genes found at the loci with overrepresented alleles.

In this chapter, the question/model paradigm is taken a step further for epigenetic studies of human disease, with the question framed in terms of the cellular models described in the last chapter, allowing the assays to serve as ways of testing questions based on cellular events, an inherently lower-dimensional set of outcomes. It becomes difficult to dismiss the use of epigenomic assays as fishing expeditions if they instead represent tools to understand cell biology and pathophysiology. By maintaining this focus in the current chapter, the hope is that more productive projects can be designed that satisfy even the most curmudgeonly of reviewers.

EARLY EPIGENOMIC ASSAYS

The technique of restriction landmark genomic scanning (RLGS) was developed to scan for differences in digestion patterns of infrequently cutting restriction enzymes, allowing identification of polymorphisms at the megabase-scale range (Hatada et al. 1991). The assay involved two-dimensional electrophoresis: initially digesting with enzymes cutting rarely in the genome, running that digestion in a first dimension, then cutting out the strip of DNA and labeling those digested sites by filling in the overhanging DNA fragments with radionucleotide, digesting the DNA with more frequently cutting restriction enzymes, and laying it sideways at the origin of a second gel. Exposure of the resulting gel to film allowed the individual restriction fragments to be visualized as spots, with polymorphism between samples identified by the difference in patterns of spots. The assay was further developed to use the same methylation-sensitive restriction enzyme in two tissues to scan for sites where the enzyme cuts in one but not the other, defining a tissue differentially methylated region (tDMR) (Hayashizaki et al. 1993). The father of Japanese epigenetics, Kunio Shiota, has described to me how he assigned RLGS-M (methylation) projects among his University of Tokyo trainees to those who appeared to have enough upper body strength to lift the >30-kg gel electrophoresis rig, possibly representing the only assay in the history of genomics for which brawn was advantageous.

Almost all epigenetics research in the 1990s was focused on small regions of the genome, such as imprinted domains, the X-inactivation center, or the β-globin locus control region (LCR). Although the RLGS-M assay was extraordinarily difficult to perform, it caused great excitement as a means of looking beyond one locus or region, hinting at the promise of discovery in the genome as a whole. The most used restriction enzyme for RLGS-M was NotI, which cuts the sequence GCGGCCGC. This motif contains two CG dinucleotides in which methylation of either CG blocks digestion. Such a CG dinucleotide and (G+C) mononucleotide-rich NotI cleavage sequence was much more likely to occur in CpG islands, helping to map these loci and focus attention on their DNA methylation patterns. The focus on CpG islands and their regulatory characteristics became a recurring theme in the early days of epigenomic studies, when we didn't even have a full human reference genome from which to work.

The next set of widely used assays that scanned throughout the genome for differential DNA methylation could be described as fingerprinting approaches. DNA digested by methylation-sensitive restriction enzymes was PCR-amplified and separated by gel electrophoresis, and the banding pattern was compared to find differential methylation between samples. The bands were then cut out of gels and cloned to identify the specific loci undergoing differential DNA methylation (Frigola et al. 2002; Liang et al. 2002). Once again, the results were heavily skewed toward identification of events in CpG islands.

It was therefore unsurprising that the earliest microarrays used for detection of DNA methylation represented CpG islands, the first a 276 CpG island array spotted onto a membrane (Huang et al. 1999). Microarrays of bacterial artificial chromosomes (BACs) were next used (Weber et al. 2005), aiming for 80-kb resolution in the human genome. Promoter (Keshet et al. 2006) and CpG island (Rauch and Pfeifer 2005; Gebhard et al. 2006) microarrays soon followed, in parallel with the transition to the oligonucleotide microarrays (Adorján et al. 2002; Gitan et al. 2002) that remain highly used today. DNA methylation lost its unique position as the focus as the target for epigenomic assays when yeast microarrays were developed by Rick Young to map the binding sites of transcription factors (Ren et al. 2000), with the first mammalian equivalent developed by Michael (Mike) Snyder and his Yale colleagues, who generated 74 PCR fragments, each of about 1 kb in length, tiling through the entire human β-globin LCR and the developmentally regulated β-like gene domain (Horak et al. 2002). They performed chromatin immunoprecipitation (ChIP) for the GATA1 transcription factor in K562 human erythroleukemia cells and hybridized the enriched DNA to the PCR fragments that had been spotted onto a glass slide as a microarray "chip." What was then called the "ChIP-chip" assay revealed GATA1 binding sites in these cells. Microarrays allowed any source of enriched DNA to be mapped and quantified in the genome, permitting their use for an expanding range of assays until direct DNA sequencing offered a more recent alternative, as will be described below.

THE CpG ISLAND

The CpG island has already made its presence known in the preceding text and is worth highlighting as it will continue to be a focus of discussions below. But what is a CpG island? In Chapter 4 we were introduced to the CG dinucleotide (archaically but for some reason currently called a CpG to refer to the cytosine-phosphate-guanine 5' to 3' order, not that anyone ever reads the genome from 3' to 5'), the preferred target of DNA methyltransferases. The presence of a methyl group at position 5 of the cytosine ring does not prevent the nucleotide from being recognized as cytosine. However, the amine group at the adjacent position can be lost and replaced by oxygen, a deamination process that occurs spontaneously in vivo. If the cytosine is unmethylated, the deaminated cytosine is now chemically a uracil, paired to the guanine on the opposite DNA strand that had previously been paired with the cytosine. Uracil is clearly not a native part of the DNA sequence and is easily recognized by DNA repair mechanisms that excise the damaged nucleotide and replace it with the cytosine templated by the guanine on the opposite DNA strand. If, however, the cytosine had already added a methyl group at position 5, and then lost its amine group, the resulting nucleotide would be a thymine. Thymine, unlike uracil, is native to DNA, so when the DNA repair machinery gets to the site it finds a mismatch between a thymine (T) on one strand and a guanine (G) on the other. Picture the DNA repair molecules scratching their figurative heads wondering which nucleotide to replace (Fig. 6.1). There is a specialist protein called MBD4 that takes charge in these situations, binding to the mismatched T and G and using its glycosylase activity to remove the thymine selectively. This makes the choice for the DNA repair machinery, which then replaces the abasic site with a cytosine, just as it would have done with the repair of an unmethylated cytosine.

The problem is that this process doesn't seem to work perfectly efficiently. The reason we know this is because organisms that methylate their DNA do not have as many CG dinucleotides as would be expected based on their numbers of Cs and Gs individually. In the human genome, we have Cs followed by Gs at only ~24% of the rate expected, whereas in the mouse genome, the proportion falls to 19%, and in marsupials (*Monodelphis domestica*), the number falls further to 13%. When we pass on our DNA to the next generation as mammals, we have been passing on mutations in which CG dinucleotides have been transformed to TGs because of deamination of 5-methylcytosine to thymine, causing the overall CG dinucleotide proportion to drop markedly.

The decay of CGs does not happen uniformly in the genome, however. The use of enzymes like NotI to map promoters in RLGS reflected the

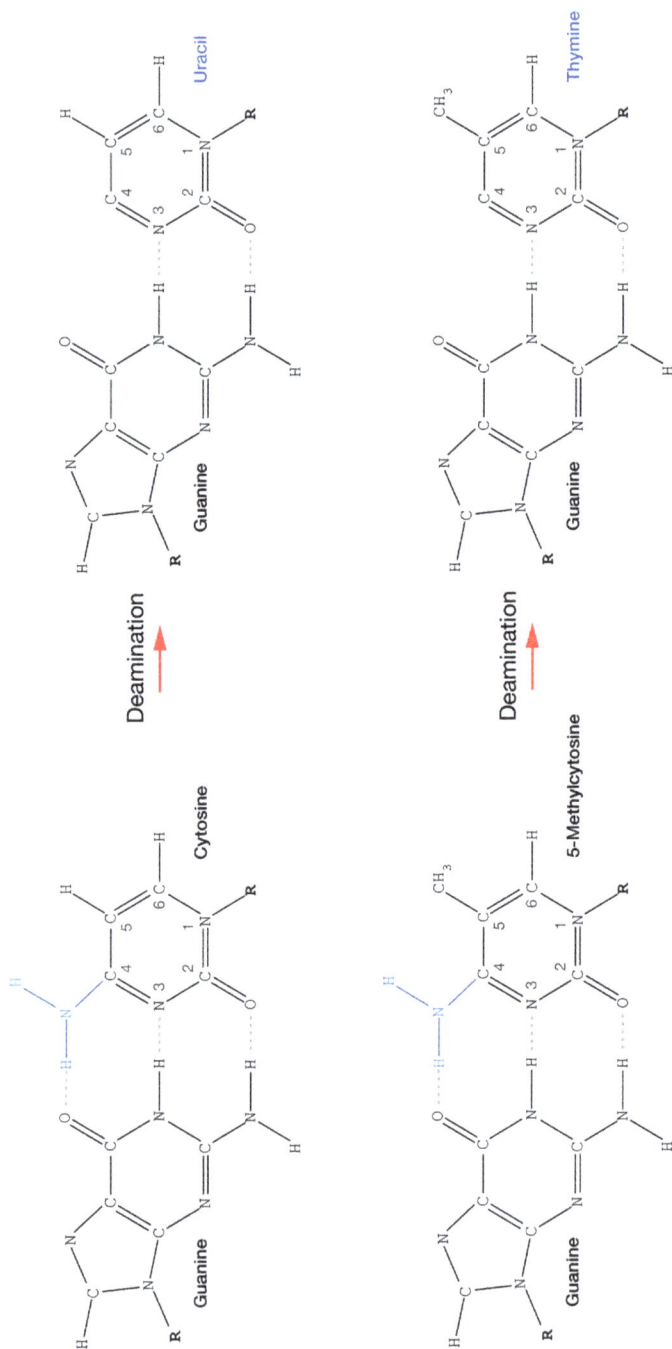

Figure 6.1. Deamination of cytosine and 5-methylcytosine. When the amine group at position 4 of the cytosine ring is spontaneously lost in vivo, cytosine becomes uracil (*top*), readily detected as foreign within the DNA sequence and removed by DNA repair enzymes to create a gap in the sequence. The repair machinery then sees the guanine on the other strand and uses that to guide the insertion of a new cytosine at the locus, repairing the damage. In the *bottom* part of the figure, deamination of 5-methylcytosine leads to the production of a thymine. This could be a genuine part of the DNA sequence, creating a decision for the DNA repair machinery: Should it remove the thymine or the guanine on the other strand? The fact that CG dinucleotides are so underrepresented in species methylating their genomes indicates that the decision is frequently incorrect.

recognition as early as 1961 that there exist some discrete sites in genomes where CG dinucleotides are more abundant (Josse et al. 1961). Marianne Frommer and her graduate student Margaret Gardiner-Garden explored this using the GenBank data available at the end of 1985, when the entire database had 5.2 Mb of sequences banked, only a subset of which was mammalian. They selected "sequences which extended more than 200 bp upstream from the translation start site...for analysis, with the exception of sequences of genes such as immunoglobulins that require rearrangement of genomic DNA for transcription." They described their analysis as a survey: "For the purpose of this survey, regions of DNA with a moving average of %G+C over 50 and Obs/Exp CpG over 0.6 have been classed as CpG-rich regions. CpG-rich regions over 200 bp in length are unlikely to have occurred by chance alone, so, as a working definition, have been labelled as CpG islands" (Gardiner-Garden and Frommer 1987). Despite the limitations of the data available to them, the authors created a valuable working definition of CpG islands based on DNA sequence characteristics that would go on to be valuable as a means of annotation of genomes with the deluge of sequence data that subsequently became available.

There are limitations inherent to this survey-based definition of the CpG island. Using the same definition in different species in which the degree of CG dinucleotide depletion differs leads to some loci apparently looking like they are not conserved as CpG islands between species, something even noted by Gardiner-Garden and Frommer with the limited DNA sequence data available to them (Gardiner-Garden and Frommer 1987). Although CpG islands are typically unmethylated, the sensitivity of the annotation in genomes with widely diverging base composition was found to be poor for the purpose of identifying unmethylated sequences (Long et al. 2013). We found some interesting loci in the human genome, like satellite repeat sequences, to be very CG dinucleotide–rich but also very rich in (A+T) mononucleotides, and that young retrotransposons are also very dense in CGs (Glass et al. 2007). When browsing the CpG island track in a genome browser, what is not obvious is that the track is created only after removing any sequences defined as repetitive, and that the inclusion of these repetitive sequences, largely derived from transposable elements, increases the number of "CpG islands" greatly (Glass et al. 2007). With a definition purely founded on base composition, why the same base composition in transposable elements is deemed nonfunctional appears unnecessarily arbitrary and has probably led to a selective focus on DNA methylation at CpG islands in unique sequences. Pervasive assumptions dating from an era when we had little genome-wide information are that transposable elements are universally highly methylated and that this is required for their suppression (Yoder et al. 1997). However, DNA can be less methylated in young transposable elements (Reiss et al. 2010) in which CG dinucleotide content is relatively higher (Babenko et al. 2017), suggesting that ignoring what we call CpG islands just because they occur in repetitive DNA is probably unwarranted.

In this book there will be many references to CpG islands in the discussions of mapping of regulatory processes in mammalian genomes. What should be borne in mind while reading about these phenomena is that the CpG island annotation is neither sensitive (if you want to use them for identification of unmethylated loci) nor specific (sites with comparable base composition in repetitive DNA vastly outnumber those in unique sequences). Today, with our ability to map different types of transcriptional regulatory events in specific tissue and cell types, the utility of a "fixed" annotation defined by base composition is fading. The CpG island is an historically useful annotation of the genome, and the question remains why these sites retain CG density compared with the rest of the genome. What is emerging as an area of current exploration is the unusual physical structural properties of the DNA in CpG islands and how this is influenced by DNA methylation (Pérez et al. 2012), whereas the relationship of CpG islands to the recruitment of Polycomb and the consequent formation of bivalent chromatin domains (see below) is also an area of ongoing studies.

THE EPIGENOMICS ASSAY TOOLSHED

Today we are still using microarrays as cost-effective reagents for some assays, especially when studying large number of samples in cohort studies, but increasingly the field is relying on sequencing to generate epigenomic data. Microarrays can only offer a sampling of loci in the genome, so although technically this can be described as genome-wide (i.e., not focused on one specific region), we have used the term "survey" to describe how such sampling assays perform (Ulahannan and Greally 2015). Some sequencing-based assays are also surveying limited samples of the genome, as will be highlighted below, allowing cost-effectiveness at the expense of comprehensiveness.

As in Chapter 4, in which transcriptional regulatory mechanisms were described in overview, the goal of the section below is to illustrate major examples of how we identify where these mechanisms are occurring in the genome, using different types of assays. The way that we currently use raw data generated by a sequencer and convert them into data about the regulatory process is an inherent part of any assay and is described in parallel with the sequencing-based approaches below. The descriptions are kept relatively simple and high-level rather than detailed to make this section a bit more accessible to less specialized colleagues.

Nucleic Acid Modifications

5-Methylcytosine and Related Modifications

5-methylcytosine (5mC) and its derived modifications 5-hydroxymethylcytosine (5hmC), 5-formylcytosine (5fC), and 5-carboxylcytosine (5caC) are all present in the mammalian genome. Both 5fC and 5caC are present in very low amounts, making their mapping difficult. 5hmC is also present as a very small proportion of 5mC overall in the genome but has been more tractable for mapping.

When RLGS-M was supplanted by microarray-based techniques, they were based initially on the use of methylation-sensitive restriction enzymes. The basic idea was not dissimilar to the fingerprinting approach mentioned above, which digested DNA from two different sources and compared the bands produced electrophoretically on a gel. Instead, the digested material was hybridized to a microarray to look for differences in signal intensities at certain loci represented on the array. In the plant community, where methylated sequences represent the minority of the genome and usually involve transposable elements, the optimal enzyme-based approach was to enrich methylated sequences for hybridization, but for mammalian genomes the opposite proved to be a better approach, enriching the unmethylated minority of the genome and hybridizing that to the microarray (Schumacher et al. 2006). Although the early focus was on intergenomic comparisons (in which DNA methylation differed between samples), the focus began to include intragenomic comparisons, identifying where DNA was more or less methylated at different genomic locations within the genome. Our HELP assay was one of many such assays, involving the comparison of representation at a locus by HpaII digestion (methylation-sensitive) and MspI (methylation-insensitive), helping to control for PCR and hybridization inefficiencies at different loci in the genome (Khulan et al. 2006).

Restriction enzymes had the disadvantage that you could only test where they cut. In an attempt to create a genuinely comprehensive genome-wide assay, affinity techniques were developed, commonly using an antibody binding selectively to 5mC but also using natural ligands of 5mC such as methyl-binding domains from mammalian proteins. The antibody-based approach, methylated DNA immunoprecipitation (MeDIP), got the most widespread use and generated some valuable insights. A problem that ended up requiring some analytical ingenuity to solve was the very divergent distributions of CG dinucleotide densities in the genome and the effect on signal that resulted (Fig. 6.2). This compromised the accuracy of intragenomic comparisons but was less of a problem for intergenomic comparisons within a species. An alternative affinity-based approach mentioned earlier involved using protein domains that bound specifically to unmethylated CGs (Long et al. 2013). Given time, the experience of the community performing affinity-based approaches may have converged with that of the researchers using restriction enzyme–based approaches, seeing advantages for enriching the unmethylated minority of the genome; however, by the time

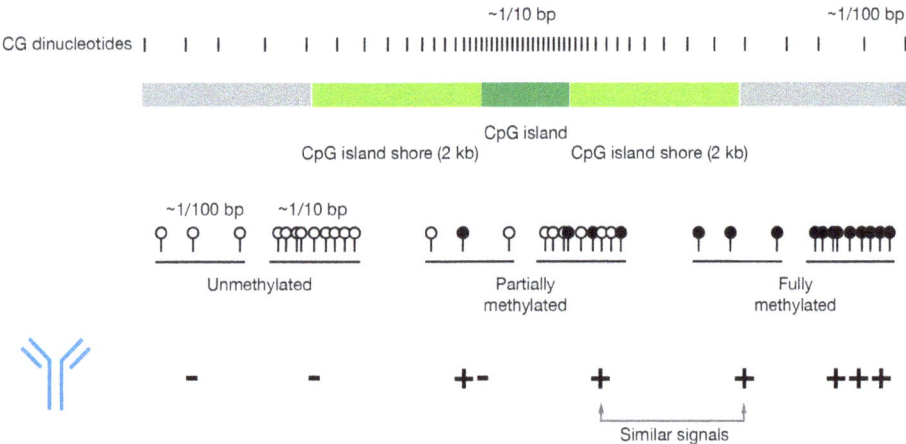

The effect of CG dinucleotide density on affinity-based DNA methylation assays

Figure 6.2. The density of CG dinucleotides is strikingly different within CpG islands (~1/10 bp) compared with the genome-wide average (~1/100 bp). When DNA is sheared to ~300-bp fragments, most molecules will have ~3 CGs, but those from CpG island will have ~30 CGs. When an affinity approach was used to enrich DNA with 5mC, CG-dense regions that were fully methylated were strongly enriched, but a partially methylated fragment from a CG-dense region gave the same enrichment as a fully methylated fragment from a CG-sparse region. The assay was very good at identifying methylated CpG islands but was less informative for the remaining majority of the genome.

such techniques like CAP-seq (Illingworth et al. 2010) were published, the DNA methylation microarray field had moved to an approach that allowed single CG resolution (Bibikova et al. 2006).

The dominant application for microarrays was and still is their use for genotyping, immobilizing an oligonucleotide on the glass surface and hybridizing DNA with a complementary sequence to that oligonucleotide. The focus will be on the Illumina microarray system, not as an endorsement but because it remains by far the most commonly used system today. The foundation for the DNA methylation microarray was called the BeadArray, using what was called GoldenGate genotyping. The genotyping system was conceptually straightforward—following hybridization of the cellular DNA, the oligonucleotide would end just upstream of the single-nucleotide polymorphism (SNP) being tested, allowing the next nucleotide to be filled in using a single base extension step, incorporating a nucleotide with a fluorescent label that differed depending on which allele was present in the cellular DNA.

To detect the DNA methylation state in the cellular DNA, the oligonucleotide was designed to stop just upstream of a cytosine in a CG dinucleotide. What allowed the discrimination of C as opposed to 5mC at that locus was the pretreatment of the cellular DNA with sodium bisulfite. Once again Marianne Frommer was involved in a breakthrough in the study of DNA methylation, having been influenced by time spent working with Adrian Bird in Edinburgh on methylation-sensitive restriction enzymes. Back in Sydney, Australia, she collaborated with colleagues to identify the conditions that allowed sodium bisulfite to act on single-stranded DNA to convert unmethylated cytosine to uracil, but leaving 5mC intact (Frommer et al. 1992). When the bisulfite-treated DNA is then amplified and sequenced, the uracil is read as a thymine, whereas the 5mC continues to be read as a cytosine. Her collaborator Susan Clark showed that the technique could work on as few as 100 cells (Clark et al. 1994), paving the way for the widespread use of the assay in studies of DNA methylation at single-nucleotide resolution.

The SNP microarray approach was designed to discriminate different SNPs at a specific locus in the genome. What bisulfite treatment was doing was creating alternative nucleotides at a locus: a C if the original cellular DNA was methylated and a T if it was unmethylated. It was therefore a relatively short step to proceed from genotyping microarrays to those reading DNA methylation at a locus in terms of the primer extension component of the assay. Less obvious a problem was that only a small proportion of the ~20% of the nucleotides in the genome that are cytosines are located at CG dinucleotides and are therefore major targets of DNA methylation, so that sodium bisulfite treatment has the effect of changing this large subset of the genome from Cs to Ts (Fig. 6.3). The DNA methylation microarray oligonucleotide cannot therefore be designed to the native sequence of the genome and instead has to anticipate the C→T transitions occurring at non-CG sites.

The first DNA methylation microarray developed by Illumina tested 1536 CGs located near 371 genes, testing one to nine CGs at each gene (Bibikova et al. 2006). Their system expanded to test 27,000 CGs, then more than 450,000, and currently more than 850,000 loci. Even 850,000 probes interrogate only ~3% of the ~28 million CG dinucleotides in the human genome, and therefore this testing is representative of what is described above as a genome-wide survey assay. Unlike the use of restriction enzymes, the ability to design oligonucleotides for a bisulfite microarray approach requires that choices are made in their design. To address this, Illumina engaged experts in the field for their input and designed accordingly.

If sequencing can read the C or T at a site and infer the methylation state of the DNA prior to bisulfite conversion, why not just skip the limited coverage of the microarray and sequence the bisulfite-converted DNA? The first genome sequenced in this way was the small (135-Mb) genome of the plant *Arabidopsis thaliana*, using the then-current sequencing technology that generated 36 nucleotide reads, of which the first 5 nucleotides (nt) were part of the library preparation; therefore, there were only 31 nt per sequence. They generated 3.8 billion nt of sequence in total in order to be able to sample deeply multiple strands of DNA at each cytosine in the genome. As plants have substantial proportions of DNA methylation at cytosines that are not located at CG dinucleotides, they analyzed DNA methylation at CHG and CHH (H=not G) sites also (Cokus et al. 2008). Around the same time, the meDIP approach had been combined with sequencing and used to study the human genome (Down et al. 2008), with the first bisulfite sequencing study of human DNA shortly thereafter (Lister et al. 2009).

Although these studies generated astonishing levels of detail that had never been apparent previously, they also required prohibitively high investments in sequencing. This was in part a reason why it was felt that investment in reference epigenomes would be of value, prompting initiatives such as the Roadmap Epigenomics Project, the International Human Epigenome Consortium, Blueprint, and the continuation of the ENCODE project,

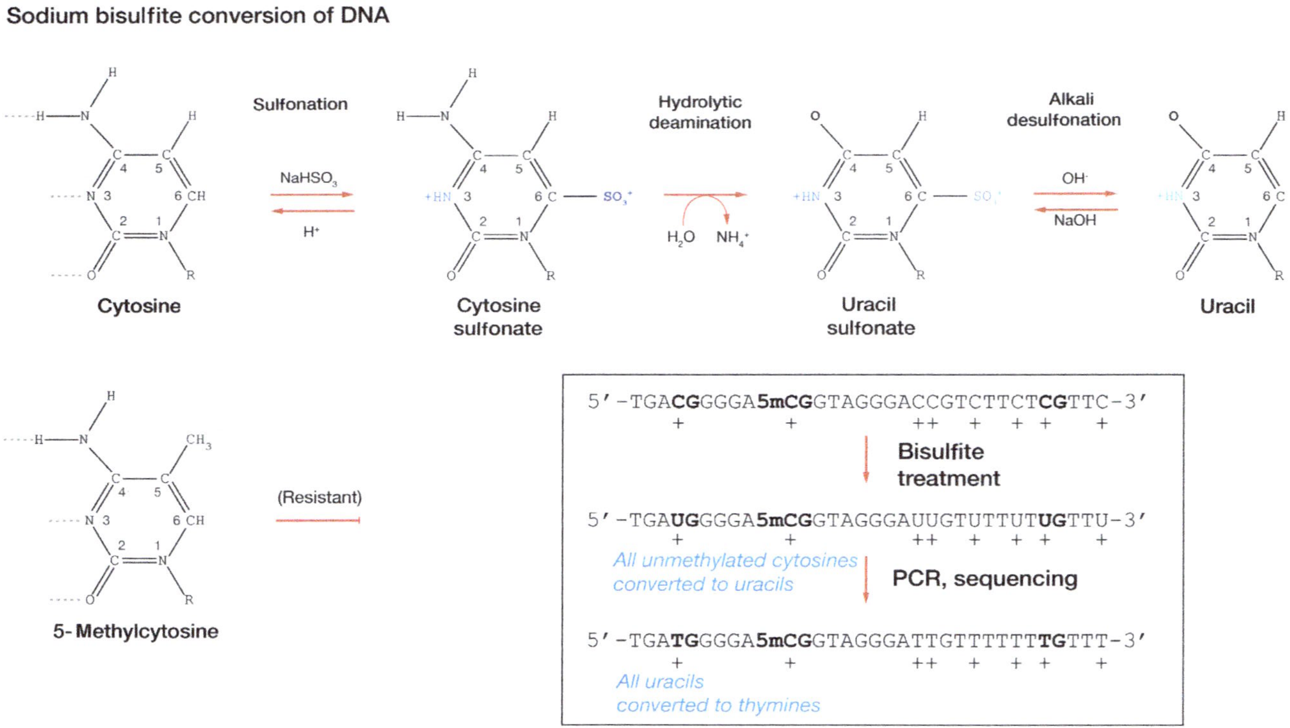

Sodium bisulfite conversion of DNA

Figure 6.3. The resistance of 5-methylcytosine to sodium bisulfite treatment permitted an assay to detect DNA methylation at nucleotide resolution. Unmethylated cytosines are converted to uracils, which are then replaced by thymines when the DNA is amplified using PCR. Any remaining cytosines are those that started as 5mC. (*Inset*) An example of the kind of sequence generated is shown; + signs show where unmethylated cytosines were located. The bisulfite-converted sequence is clearly less complex than the original DNA sequence, which creates alignment challenges after sequencing data are generated.

with the idea that the DNA methylation and other transcriptional regulatory data generated would serve to guide more focused and cost-effective projects within the budgetary scope of individual investigators. A decade later, the costs associated with generating these kinds of data have dropped to the point that reference data can usually be generated by the researcher.

Bisulfite-based microarrays persist as a lower-cost approach suitable for cohort studies, and sequencing-based survey assays such as reduced representation bisulfite sequencing (RRBS) (Meissner et al. 2005) remain valuable for many applications. The bisulfite sequencing era may now have peaked. A nonchemical approach that uses the selective deamination activity of the human enzyme APOBEC3A has advantages as a way of generating bisulfite-like results with less destruction of input DNA (Schutsky et al. 2017). Techniques have been developed for the detection of other modified nucleotides such as 5hmC, 5fC, and 5caC, as well as 6mA, including affinity, enzymatic, and direct detection approaches.

Currently, the most detailed DNA methylation assays are those provided by long-read sequencing. Two examples are shown in Figure 6.4: nanopore-based technology (commercialized by Oxford Nanopore Technologies [ONT]) and the HiFi sequencing assay commercialized by PacBio. The nanopore platform has the advantage of being also able to test for modifications in RNA, whereas the HiFi platform, in generating separate nucleotide and kinetic data, may be better able to discriminate between base modifications and the presence of a DNA sequence variant, currently a problem for nanopore approaches (Stefansson et al. 2024).

One major value of these long-read technologies should be the ability to identify whether a change in DNA methylation is due to the local presence of a DNA sequence variant, as both sequence and variant information are generated simultaneously for each sequenced strand of DNA. This technological advance promises to overcome the issue of DNA sequence variation confounding the interpretation of DNA methylation studies, by revealing functional variants influencing local DNA methylation.

Data analysis: With the focus on sequencing-based assays, there are two major ways of thinking about the data generated. One is the *counting* aspect to the study (i.e., how many times a sequence from one part of the genome is represented in the output compared with other sequences), whereas the other is the *qualitative* information in the sequence reads themselves.

In the restriction enzyme–based approaches, the information gained is all to do with counting—if, for example, a methylation-sensitive restriction

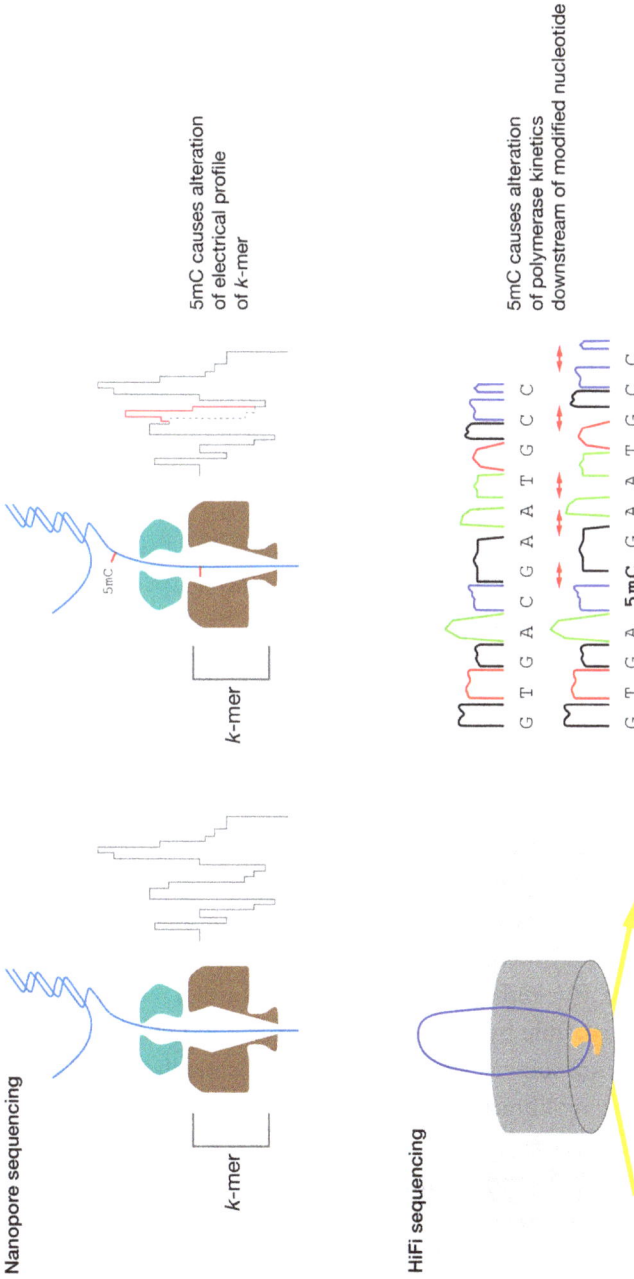

Figure 6.4. New long-read sequencing technologies can identify modified nucleotides. Nanopore sequencing relies on a characteristic pattern of the current across the nanopore when specific sets of nucleotides are occupying the nanopore. The presence of a modified nucleotide alters this pattern. HiFi sequencing separately detects nucleotide sequence and the kinetics of nucleotide incorporation to infer the presence of a modified nucleotide.

enzyme was used to cut the genomic DNA, so that the library for sequencing should be enriched for those sequences, then the alignments of these reads to the genome should start to pile up at unmethylated loci. As a measure of confidence in the results, you would want to see the restriction site for the enzyme at such a site of multiple aligned reads. If the goal is to identify loci where the DNA methylation is different between samples (an intergenomic comparison), then you want to understand how much the representation of reads at a site compares with those at the same site in the other sample, but also whether the number of reads aligned to the genome overall might be influencing the results: If one sample has one million aligned reads and the other has two million, then you would expect to have twice as many reads aligned to a specific site in the second sample, requiring a corresponding adjustment. Such adjustments are referred to by the blanket, nonspecific term "normalization," which can include many possible manipulations of the data and need to be described explicitly in reports of studies so that others can scrutinize and replicate the work.

When an affinity approach is used, once again the major output of interest is quantitative counting of reads representing specific loci that are distinctive for their methylation, such as the methylated DNA enriched in the meDIP assay. Affinity assays tend to require large amounts of starting material, which introduces another qualitative study that can be performed when evaluating libraries generated from randomly sheared DNA—whether any reads have the same start and end positions in the genome. Although this can occur by chance, the proportion of these reads increases dramatically when the amount of DNA available to generate the library is limited. The sequencer is going to generate the same number of reads no matter how many molecules of DNA were available to generate the library. If the library was generated from an excessively limited amount of DNA molecules, you end up sequencing some of them more than once, and they are recognizable by having the same start and end positions in the genome. These are sometimes called PCR duplicates, having been amplified by PCR during library construction, and are eliminated from any counting or other analyses.

With bisulfite sequencing, the information is more qualitative—instead of just counting how many times reads align to a locus, the information of value is within the sequence itself, the proportion of reads that have a C compared with those with a T at the same site. The cytosines that were unmethylated will be read as Ts, and the methylated (or hydroxymethylated) cytosines as Cs, allowing the proportion of DNA sequences in the cell sample that are methylated to be calculated as a proportion of all the reads at the locus. If the sample did not undergo sufficient bisulfite conversion, the overall level of DNA methylation genome-wide would increase, and mammalian samples would start to increase the proportion of 5mC detected at sites other than CG dinucleotides. Spike-in control samples that are premethylated or unmethylated should be used when preparing libraries to provide evidence for such a technical issue.

Finally, as will be discussed below for chromatin immunoprecipitation and gene expression, looking for SNPs within the sequence reads is of value, not only to generate some genotyping data that confirm the sample to be from the correct individual but also at heterozygous loci to determine whether the DNA methylation is different on each chromosome. This is relatively difficult but not impossible in bisulfite-converted DNA, as many SNPs at cytosines will be converted by the bisulfite to thymines. Long-read sequencing looks to be the most promising way of generating high-quality sequencing and DNA methylation data for each strand of DNA sequenced.

Variation of DNA Structure

The number of assays that exist to identify and map noncanonical DNA structures in the genome is limited. An antibody has been developed to bind to RNA:DNA hybrids, allowing them to be identified using an affinity-based assay. A tricky issue inherent to this assay is that you can't make a library for sequencing directly from an RNA:DNA hybrid, but such a duplex molecule is part of the workflow in creating libraries from RNA, after the step to generate a complementary DNA (cDNA), allowing components of this trusted assay to be incorporated in generating these RNA:DNA hybrid libraries.

There has been some success in mapping single-stranded DNA in mammalian genomes (Khil et al. 2012), but this is not a commonly used assay. The interaction and stacking of consecutive guanines within a DNA strand, forming the so-called G-quadruplex, has also been studied using an antibody specific for this structure (Hänsel-Hertsch et al. 2018). Triplex RNA:DNA:DNA structures have been isolated by digesting free RNA and then RNA in RNA:DNA hybrids using RNase H, followed by sequencing of the remaining RNA associated with DNA, presumably enriching for those in triplices (Sentürk Cetin et al. 2019). The field of mapping noncanonical DNA structures remains underdeveloped but is increasingly showing promise.

Data analysis: The assays described above involve counting of read enrichment at specific loci. We have found RNA:DNA hybrids to be enriched for polypurine-rich sequences (Nadel et al. 2015), which confers a thermostability on the structure. Sites of enrichment for G-quadruplexes should have several guanines in *cis* in a short region of the genome, another way that the underlying sequence can help to generate confidence in the results obtained. The type of DNA sequences at which unusual structures are formed can complicate their mapping—RNA:DNA hybrids are found at the highly repetitive ribosomal DNA sequences, which are typically excluded from mapping studies and require specific alignment strategies (Nadel et al. 2015). This represents an example of the need to understand what information we are discarding when performing our genomic data analyses.

Covalent Modifications of RNA

Once again, the approach that is mostly used to map a modification is based on the use of an antibody specific for that structure. RNA immunoprecipitation (RIP) with an antibody to 6-methyladenine (6mA) has been used to define the locations of this modification in RNA molecules (Molinie and Giallourakis 2017), but direct detection during sequencing is now replacing these affinity-based assays (Stephenson et al. 2022).

Data analysis: Once again, the affinity-based assay is based on counting the number of reads aligned to a specific locus and determining whether they are enriched significantly compared with what would be expected by chance. In antibody-based affinity assays, a useful control experiment is to omit the antibody (or to use an antibody with a different binding specificity) as a parallel control. When this DNA is sequenced, there will be some variability from locus to locus in the genome, in part because of differences in the relative ease of amplification of some loci compared with others, and in part because of things like copy number variations in the sample being tested. A duplication of a region, for example, will cause more reads to be generated by that region, which might have misled the researcher into thinking that the antibody was enriching that region as a whole. With single-molecule nanopore sequencing, the results should be much more accurate, representing the modified nucleotide directly within each RNA strand sequenced.

Chromatin Components

Post-Translational Modifications of Histones

The overwhelmingly most common approach to map histone modifications is the use of specific antibodies targeted to the covalent modification of interest for chromatin immunoprecipitation, now referred to as ChIP, followed by sequencing of the DNA enriched at these fragments of chromatin.

The variation in the ChIP assay largely resides in how the chromatin is prepared, as the immunoprecipitation component of the assay is standard. The most common approach traditionally used has involved the cross-linking of chromatin using formaldehyde followed by sonication to shear the chromatin into fragments of a few hundred nucleotides in size. The sheared chromatin is then exposed to the antibody, which binds selectively to those fragments containing the modification of interest. The nonbinding chromatin is washed away, and the DNA is purified from the chromatin proteins for sequencing and mapping. An alternative way of preparing the chromatin involves its digestion with micrococcal nuclease (MNase) to cleave between nucleosomes, following which the antibody is used to immunoprecipitate the sample. There is a concern that fixation leads to a lot of spurious signaling and that the sonication preferentially shears at sites of open chromatin, leading to both false positive and false negative results in a ChIP-seq assay.

As an update to ChIP-seq, Steve Henikoff developed the CUT&RUN assay (Skene and Henikoff 2017). CUT&RUN was based on the yeast chromatin immunocleavage (ChIC) technique developed by Ulrich Laemmli (Schmid et al. 2004), involving the antibody detecting the post-translational modification bringing a protein A–micrococcal nuclease (MNase) to the locus to digest away the local fragment of DNA, allowing it to be enriched for sequencing. CUT&Tag (cleavage under targets and tagmentation) replaces the MNase with the kind of hyperactive Tn5 transposase used for library generation, allowing direct PCR amplification of the local DNA (Kaya-Okur et al. 2019).

Data analysis: The above are counting assays but involve the identification of specific loci of enrichment, or peaks. A control sample lacking the antibody or using a nonspecific antibody can be used for comparison, allowing loci with increased numbers of reads in the immunoprecipitated sample to be identified. For many histone modifications, there will only be a few adjacent nucleosomes that have this change and will generate a very focal enrichment of aligned sequences, a narrow peak. For other modifications associated with regions being actively transcribed (see below) or broader do-mains of chromatin organization like heterochromatin, the contiguously en-riched regions will be much more extensive and will form broad peaks. The type of peak calling analysis required and the amount of sequencing needed differs for narrow and broad peaks.

Open Chromatin

These regions of open chromatin in the genome are defined by being unoc-cupied by nucleosomes. However, to describe them as open might give the impression that they are just naked DNA, when in fact they are characterized by being actively bound by TFs and the site of recruitment of other proteins. Although open is probably a misnomer, the DNA locally is more accessible to enzymes and less stably or consistently bound by proteins, allowing the development of several assays to identify these sites in the genome.

The original approach used to define these loci involved exposing chro-matin to DNase I. If the DNA was not packaged as chromatin, all the DNA in the genome would be equally digestible. Bound up by nucleosomes and other proteins, the DNA is variably accessible, with the open chromatin regions defined as hypersensitive. DNase-seq was a groundbreaking ap-proach to identify open chromatin by enriching DNA from these loci for sequencing.

A second major assay resulted from a mistake when one postdoctoral fellow forgot to tell his colleague to reverse the cross-links in his input chro-matin before phenol–chloroform extracting the material (de Souza 2007). Normally in reversing cross-linking, you digest the protein component of the chromatin, allowing the DNA to be cleaned using phenol–chloroform, which leaves the DNA in the aqueous layer and segregates the proteins and digested peptides into the phenol. By leaving the proteins attached to the DNA, most DNA ended up in the phenol and was lost, but what was se-lectively enriched in the aqueous layer was the DNA not stably bound to proteins (i.e., the open chromatin component of the genome). This mistake allowed the development of the FAIRE (formaldehyde-assisted isolation of regulatory elements) assay, which was a second valuable approach (Giresi et al. 2007) complementing DNase-seq studies.

The most commonly used approach used today is probably ATAC-seq (assay for transposase-accessible chromatin using sequencing), which is also based on an enzyme acting on open chromatin preferentially, but this time using a transposase to insert a sequence into the genome from which a sequencing library can be generated (Buenrostro et al. 2013). The transposase-based approach for making libraries had already been developed for purified DNA; the innovation was exposing the transposase to DNA still packaged as chromatin. ATAC-seq works sufficiently well that it is now being used in single-cell assays, providing great insights into heterogeneity of regulatory landscapes within cell populations (Buenrostro et al. 2015).

Data analysis: Open chromatin regions are small, of the order of hundreds of base pairs, allowing exactly the same kind of analytical approach as for ChIP-seq for narrow peaks. A useful control is the library created from the same cell sample but from purified DNA, so that the relative efficiency of the transposase for different sequences, and any effect of copy number variation, can be excluded from causing spurious results.

Histone Variants

Although histone variants like centromeric histones, macroH2A, and H2A.Z can be mapped using ChIP-seq, being distinct proteins, you also can add a "tag" to the protein and express that in a cell line (or experimental animal). Such tags are designed to be bound by affinity reagents, so that when you harvest the chromatin from these cells you can then bind the histone variant specifically and capture its associated DNA, allowing its sites of enrichment to be identified. When studying a histone variant that differs in amino acid sequence from canonical histones by only a few amino acids (Goldberg et al. 2010), it is much less likely to be distinguishable by an antibody, making tagging approaches much more practical.

A further technique that can be used is DNA adenine methyltransferase identification (DamID). This is a versatile tool for mapping protein–DNA interactions in the genome, using DNA adenine methyltransferase tethered to the DNA-interacting protein of interest, generally expressed as a construct in cultured cells. When the protein interacts with the DNA, it brings the Dam enzyme into proximity to the binding site, allowing it to methylate adenines in the genome. Methyladenine is not present in eukaryotic genomes, making this experimental modification highly specific for the targeting event. When the adenine in the GATC sequence is methylated, it can be selectively cut by the DpnI restriction enzyme. It is therefore possible to digest and PCR-amplify the DpnI-digested DNA from the test sample (with a control sample expressing Dam but without fusing it to a protein, to get an indication of the background nonspecific digestion and enrichment). The enriched sites have been mapped originally using microarrays but more recently using sequencing. Although DamID has been used for mapping of numerous different protein–DNA interactions, it has also been used for histone variant studies, including the H1 histone variant (Braunschweig et al. 2009).

Data analysis: Using tagging approaches or the DamID assay, enrichment can be sought in the same way as for ChIP-seq studies. Some histone variants are very broadly distributed in the genome, including macroH2A (Sato et al. 2019), requiring that the analysis be directed toward the identification of wider peaks.

A further issue that affects the mapping of some histone variants is the nature of the DNA sequence to which they bind. Centromeric DNA in humans binds to highly repetitive α-satellite sequences, which would be normally left out of typical alignment results.

Chromatin Insulation and Chromatin Looping

Chromatin insulators are defined in terms of their functional properties. To map them in the genome, it needs to be demonstrated that these elements block enhancer influence on promoter activity, which has not been performed. Instead, the DNA-binding proteins that have been identified to be components of functional insulators have been mapped genome-wide (in *Drosophila*), allowing candidate insulators to be identified (Nègre et al. 2010).

Chromatin looping is related to insulation of domains from each other and is amenable to genome-wide mapping. In Chapter 4 we were introduced to Job Dekker's chromosome conformation capture (3C) assay, involving restriction enzyme digestion of DNA while still organized as chromatin, allowing the dangling ends to re-ligate to whatever is nearby, thereby revealing interactions occurring distantly within a chromatid in vivo to be revealed.

The 3C technique was published in 2002 and has been the foundation for many subsequent modifications. The original 3C assay used quantitative PCR to compare interaction frequencies, which is not very scalable. By adding a microarray, the 4C (chromosome conformation capture-on-ChIP) assay was developed. Like 3C, these 4C assays were designed to start with a locus of interest, asking where in the flanking regions were interactions occurring more frequently. The (somewhat laboriously named) chromosome conformation capture carbon copy (5C) assay used a different primer design approach to get past the need to have a single locus of interest, instead allowing all interactions within a region to be tested. Taking 5C a step further to look genome-wide was the Hi-C assay, which requires enormous amounts of sequencing but has generated comprehensive maps of chromatin looping. In Figure 4.8, we showed the Micro-C assay that uses MNase to generate the highest resolution yet for these chromatin looping assays. As an intriguing alternative, genome architecture mapping (GAM) was developed by micro-dissecting slices within nuclei, and then sequencing to test which DNA sequences end up in the same nuclear section, allowing inference of proximity (Beagrie et al. 2017).

When performing a sequencing-based or other genomic assay, with all the ways that technical issues can mislead you, it is very valuable to perform a validation study to confirm that the predictions of the assay are accurate. Fluorescence in situ hybridization (FISH) has been used for decades to look at proximity of different sequences in interphase cells. Because the concordance between 5C and FISH is not consistent, Wendy Bickmore's group has sounded a cautionary note that we should be careful about overinterpreting these genomic assays and take into account local chromatin structure and other factors when trying to understand apparent proximity of loci (Williamson et al. 2014).

Data analysis: The critical steps in Hi-C data processing have been nicely summarized by Han et al. (2018), starting with alignment to the genome of the pairs of reads from each end of the ligated DNA, followed by implementation of decisions about which pairs of reads to keep, with a subsequent organization of contact events into genomic "bins." The final step of normalization is designed to allow more accurate data to be generated, with the outcome of these processing steps a map of interactions that can be explored by researchers.

Transcription Factors

Transcription factors (TFs) bind to DNA at loci that can be mapped using ChIP, as described for other DNA-binding proteins above. A difference with TFs is that they bind more transiently than nucleosomes, bouncing on and off the DNA. Steven (Steve) Henikoff has been a source of insights while assembling this section, describing the history of understanding how formaldehyde cross-linking works for DNA binding proteins, a topic he has also discussed in published work (van Steensel and Henikoff 2003). Back in 1985 it was noted that formaldehyde cross-linking works less well in the nucleosome-free regions of the genome where TFs bind (Solomon and Varshavsky 1985). In 1999, the kinetics of formaldehyde were studied by Vaughn Jackson, revealing that histones cross-link to each other within 5 minutes at 4°C, only ending up cross-linking to the DNA after 2 hours. The standard ChIP protocol involves 10 minutes of formaldehyde exposure, too little time for protein–DNA cross-links, indicating that the effect of cross-linking is predominantly to bind histones together, entangling the DNA locally and probably capturing nearby TFs by cross-linking them to the nearest histones through the primary amines on their tails.

The fragment of DNA in the piece of chromatin immunoprecipitated is generally a few hundred base pairs in size, substantially larger than the site of binding of the TF. To help identify the exact binding site, the immunoprecipitated sample is exposed to an exonuclease that chews away the free ends of the DNA, stopping when the DNA is bound tightly to a protein. This ChIP-exo approach has been useful in the more precise mapping of binding sites of transcription factors than conventional ChIP-seq (Rossi et al. 2018). CUT&RUN or CUT&Tag now allow precise mapping of TF binding sites without the need for cross-linking or shearing of chromatin (Skene and Henikoff 2017).

Data analysis: These assays are fundamentally generating data that resembles the narrow peak output of ChIP-seq, with the analytical approaches associated with these assays suitably applied here also. As a way of increasing confidence in the results, an enrichment of the DNA sequence motif representing the binding site of the TF should be found at the peaks observed.

RNA

RNA Expression

There are many ways of thinking about the expression of RNAs in the genome. At its simplest, a bulk population of cells is harvested, and the messenger RNA is enriched and converted into DNA for sequencing with a short-read technology, counting the number of reads from each annotated gene in the genome. This is the basis for comparing gene expression levels between comparable cell types that are thought to have different properties, the cellular reprogramming model from Chapter 5.

However, if RNA expression studies are like jazz, what is described above is the basic melody or what musicians call the "head," to be played the first time around but the foundation for all subsequent solos by inventive members of the group. In this section, the breadth of inventive solos will be described. An excellent review of the topic is available and is recommended for further reading (Stark et al. 2019).

A necessary first step is the conversion of RNA to double-stranded DNA for sequencing. This can use an oligonucleotide of thymines (oligo-dT), which anneals to the polyadenylated end of the RNA and can be used with an RNA-dependent DNA polymerase to create a complementary DNA (cDNA). This is the basis for mRNA-seq, a useful sampling of the genome for quantification of expression of protein-coding genes. As many noncoding RNAs and the core histone genes are not polyadenylated, this oligo-dT approach will fail to generate these transcripts in the library generated. In addition, the polymerase is working from the extreme 3′ end of the RNA molecule, tending to overrepresent this part of the transcript and underrepresent the 5′ end, with less ability to detect alternative splicing events occurring in the upstream exons. To add coverage of the more 5′ regions and to represent RNA molecules lacking polyadenylated 3′ ends, random oligonucleotide primers are added to the reaction when generating the cDNA.

When the cDNA has been generated, and the protocol has progressed to the next step of generating a second strand of DNA to turn the RNA:DNA hybrid into double-stranded DNA, dUTP can be included instead of dTTP. This has no effect on the ability to make the double-stranded DNA but makes the dUTP-containing strand of DNA uniquely susceptible to digestion by uracil-DNA-glycosylase (UDG). By digesting the DNA prior to PCR, only the original cDNA strand will be amplified. The orientation of the DNA in terms of the adapters ligated to the ends prior to PCR allows it to be determined whether the RNA was transcribed from one strand or the other, a directional RNA sequencing assay. This is valuable in quantifying gene expression more accurately, as spurious reads that align to the opposite strand can be identified and removed. In addition, if there are loci where there are transcripts being generated from both strands and overlapping (or different cells in the population tested express overlapping transcripts from each strand), by having directional information you can discriminate these events and have more accurate quantification of the type and level of gene expression at a locus.

The sequencing of RNA using long-read technology allows the entire transcript to be read as a single molecule, giving direct insights into the splicing patterns of a gene without having to infer these patterns from short reads.

Small RNAs need a different approach. Typically, these studies involve ligating an adapter to the 3′ hydroxyl group followed by ligation of a different adapter to the 5′ phosphate, and PCR amplification designed for the short inserts of tens of nucleotides for the small RNA population. One problem is that the size of the product with inserts is not much greater than the population of primer dimers, the molecules where the 5′ and 3′ adapters have ligated directly to each other, wasting some of the sequencing throughput.

Sometimes you do not get the whole cell—just the nucleus—such as when you isolate neuronal nuclei from brain tissue. The representation of RNA molecules succeeds in generating insights into neuronal physiology (as opposed to other cell types), but there remain differences between nuclear and whole-cell RNA profiles, with more intronic representation for nuclear preparations, for example.

This insight about the relative representations of nuclear as compared with whole-cell RNA profiles raises the interesting question about what a level of expression of an RNA molecule signifies. We are measuring a steady state reflecting the balance between production and degradation of the RNA molecule. The factors involved in RNA degradation are numerous and complex (Houseley and Tollervey 2009) and include the influence of the RNA (and template DNA) sequence itself, with some nonsense amino acid variants capable of triggering nonsense-mediated decay of the RNA. The rate of production of RNAs can be measured using the precision nuclear run-on sequencing (PRO-seq) assay, which is based on adding biotin-labeled nucleotide triphosphates (biotin-NTPs) to the living cell for a short pulse of time and then isolating RNA and capturing the newly synthesized RNAs using streptavidin to bind to the biotin-containing RNAs.

Finally, the context of the RNA in the cell may be of interest to the researcher. Mentioned earlier was the identification of RNAs in triplex RNA:DNA:DNA structures. Chromatin-associated RNAs represent another area of interest, potentially linking noncoding RNAs to transcriptional regulation (Li and Fu 2019).

Data analysis: Analysis of RNA expression is generally performed with the goal of identifying differences in gene activity between samples. The process of analysis starts with alignment of the RNA sequences to a reference genome followed by quantification of the number of sequence reads per annotated transcript. There follows a normalization step to account for multiple types of influences that can skew results, generating an output of genes for which there is high confidence in expression differences. How the researcher progresses through these steps involves choices between many alternative types of tools, and there can be the addition of qualitative information such as alternative splicing or allele-specific expression, but the general workflow is relatively easily understood.

INSIGHTS INTO GENOMIC ORGANIZATION: THE "EPIGENOME"

The beauty of having a reference human genome available is that we can map every one of the assays described above to the genome and look for patterns. This patterning is what is referred to as the epigenome and is very much based on the definition of epigenetics being a layer of information above or upon (epi-) the DNA sequence (-genome).

A starting point for studying these patterns is the gene itself; a gene can be represented by exons with intervening introns and will have a start and an end position. If a gene uses different start and end sites and different exons, these alternative transcripts can all be mapped at the locus. At the transcription start site is the promoter of the gene, where CpG islands are found for about half of the genes in the human genome (Glass et al. 2007).

To create the simplest possible framework for thinking about organization of the epigenome, it is convenient to divide patterns into three types. The first is what could be called *short regulatory sequences*, including promoters, enhancers, and other regulatory elements. The second is the pattern found that results from the RNA polymerase running through a region, or *transcription-dependent modifications*. The third is a higher scale of organization that can encompass many genes and could be called *broad domain* organization.

Short Regulatory Sequences

The epigenomic assay toolshed described above reveals short regulatory sequences in certain ways. A promoter is typically characterized by very low levels of DNA methylation, a nucleosome-free region that is detectable by assays mapping open chromatin, and patterns of post-translational modifications of the histones in the nucleosomes flanking the open chromatin region, best characterized by the histone H3 lysine 4 trimethyl (H3K4me3) modification. An enhancer, on the other hand, can be close to the promoter or more distantly located and is characterized by H3K4me1 and acetylation of lysine 27 (H3K27ac). Like promoters, enhancers are transcribed, generating noncoding enhancer RNAs (eRNAs), which can be mapped in transcriptomic assays, providing another means of identifying the location of these elements. Thus, both promoters and enhancers can be mapped using a combination of assays identifying open chromatin, mapping specific histone modifications, regions of low DNA methylation levels, and the proximity of gene or noncoding RNA transcripts.

Less studied are silencers in the genome. It has been known for some time that some TFs have repressive activities. REST (RE1-silencing transcription factor), also known as neuron-restrictive silencer factor (NRSF), is a TF that was found to silence neuronal genes in nonneuronal cell types, binding to the neuron-restrictive silencer element (NRSE). The TF called YY1 got its name (Yin Yang 1) for its ability to activate or to repress gene expression in different situations. These are only two of many TFs found or predicted to have repressor function. As these repressive TFs bind to DNA, they create loci of open chromatin, but they otherwise differ from promoters and enhancers in a number of ways. Their DNA methylation levels appear to be higher than at enhancers (Doni Jayavelu et al. 2020), and they lack specific histone modifications nearby apart from a nonspecific increase in H3K27me3, which is present in much broader regions of the genome and doesn't mark the silencer element as specifically as the H3K4 methylation that reveals promoters and enhancers (Huang et al. 2019). Surprisingly, there have been no studies demonstrating the expression of ncRNAs from silencers or NRSF sites, creating the equivalent of eRNAs at enhancers. Whether this is because they have not been studied in sufficient detail or whether they are distinct from enhancers for not being expressed is unclear at this point.

Finally, we have the insulator element, defined as a sequence with the potential to block the effect of an enhancer upon a promoter. In mammalian cells, the only characteristics known to define insulators are binding by the CTCF protein and the presence of DNase hypersensitivity, indicating an open chromatin conformation locally. Although these are features of insulators, CTCF binding at other sites of open chromatin can occur without conferring insulator activity. The discussion of insulators is now inextricably linked to issues of chromatin topology, as will be discussed below.

Transcription-Dependent Modifications

The short regulatory sequences help to get RNA polymerase onto DNA to perform transcription, running through a region of the genome contiguous with a promoter (gene) or enhancer (eRNA). The act of transcription is enough to create patterns in the genome. When the RNA polymerase transcribes a gene, this is associated with local recruitment of an enzyme called SETD2 in mammals, which does two things: It catalyzes the methylation of histone H3K36 and also recruits the DNMT3B enzyme through its PWWP domain to cause increased DNA methylation in the gene body (Baubec et al. 2015). Mechanistically, this is thought to have a role repressing cryptic promoters located within the bodies of genes (Maunakea et al. 2010). The more proximal part of the actively transcribed gene body is not enriched by H3K36me3 but is instead enriched in H3K79me3 (Huff et al. 2010), which is due to the recruitment of the DOT1L enzyme by the RNA polymerase itself (Kim et al. 2012). Using microarrays, Richard Meehan showed that hydroxymethylation of DNA (5hmC) is positively associated with gene expression levels (Nestor et al. 2012). Provocatively, it has recently been shown in yeast that histone acetylation occurs only after the locus is transcribed (Martin et al. 2021). The picture revealed by epigenomic assays of transcribed regions is therefore distinctive compared with nontranscribed loci, heterogeneous within the body of the gene, and dependent to some extent on the level of transcription. Therefore, although there is no single set of marks that defines a transcribed locus, some like H3K36me3, H3K79me3, and 5hmC are relatively specific, and it should be borne in mind when interpreting epigenomic data that they may be the consequence and not the cause of transcription.

Broad Domains

Finally, we get to patterns that are apparent in regions that extend beyond individual enhancers, silencer, or genes and can even become visible cytologically with light microscopy. In Chapter 4 we learned of mammalian centromeres extending over hundreds of kilobases, heterochromatin and euchromatin extending over tens of megabases or more to generate cytogenetic banding patterns, and insights into topologically associating domains (TADs) that encompass many genes at a time. It is becoming clear that intranuclear organization is much more complex than a simple binary difference between euchromatin and heterochromatin and that the solubility of the nuclear material influences the patterns of 3C signals (Belmont 2014). Epigenomic assays reflect these complexities. Mapping centromeric histone variants reveals their organization in large blocks at centromeres, whereas mapping histone modifications (H3K9me3, H3K27me3) and histone variants (macroH2A) associated with heterochromatin formation reveals large contiguous blocks of signals, but these marks are interspersed within euchromatic domains as well (Becker et al. 2017). Bas von Steensel used the DamID assay to identify which genomic sequences contact the nuclear lamina (the lining of the inner nuclear membrane, defining the genomic regions located at the periphery of the nucleus). He found that mammalian cells have 1000–1500 lamin-associated domains (LADs) of median size 500 kb, usually heterochromatically organized (van Steensel and Belmont 2017) and strikingly lacking in DNA methylation (Berman et al. 2012). Large blocks of hypomethylated DNA had been identified previously and described as partially methylated domains (PMDs) where gene expression was markedly reduced (Lister et al. 2011). It was subsequently shown that PMDs are heterochromatically organized and late-replicating, extending our prior finding that late-replicating heterochromatin had decreased DNA methylation (Suzuki et al. 2011). Broad domains are therefore identifiable using epigenomic assays mapping heterochromatin constituents, DNA methylation, DNA replication timing, and nuclear lamin association, with substantial interrelatedness of many of these features.

As a final generalization to help put epigenomic assay data into a practical framework, genomic organization can be considered to have a default state of nucleosomal organization and DNA methylation. In addition, quantitative mass spectrometry indicates that most histone H3 proteins have either mono- or demethylation of K9 and K27 with a further ~10% having H3K27me3 and ~20% H3K9me3 (Huang et al. 2015), making the methylation of these lysines a relatively default state also. Oddly enough, if you map the DNA methyltransferases using an epigenomic approach, you find that they are located at the unmethylated loci in the genome, at enhancers of gene expression (Rinaldi et al. 2016). From a biochemical point of view, it could be surmised that the enzymes are only trying to establish the default pattern at the few remaining loci lacking DNA methylation. Focusing on how epigenomic assays show departures from what could be considered default patterns is a valuable way of thinking about the regulatory organization of the genome.

HOW EPIGENOMIC PATTERNS CHANGE

Different cell types have different epigenomic patterns. Although there are other ways of distinguishing cell types—including morphology and cell surface markers—the innate molecular properties of a cell type are sufficiently distinctive that they can also define cell types, even when the cells look identical and have the same surface markers (Jaitin et al. 2014).

Naturally, this prompts the question how one epigenomic pattern transitions to another as cells differentiate. In Chapter 4 we went through how TFs can access sites in the genome and activate them, a de novo event at the heart of a cell changing its properties. From the point of view of epigenomic assays, ChIP-seq for specific TFs would reveal their redistribution with differentiation, something nicely demonstrated by Alex Meissner and his colleagues for mouse embryonic stem cell differentiation to three germ layers, testing 38 TFs (Tsankov et al. 2015).

The changes in patterns of TF binding during cell differentiation would be expected to result in altered patterns of chromatin accessibility, and they do (Miraldi et al. 2019). However, it's when we get back to DNA methylation that we see something unexpected that is probably less appreciated than it could be. Emily Hodges was studying DNA methylation using whole-genome bisulfite sequencing (WGBS) in hematopoietic development and was able to identify the loci where DNA methylation was changing as cells differentiated. An advantage of DNA methylation studies is that they allow you to look at each informative CG dinucleotide, whereas chromatin studies generate less precise signals. What she showed was unexpected: Instead of isolated loci going from completely methylated to completely unmethylated or vice versa, many of the changes were at loci that were constitutively completely unmethylated but extended or contracted to make broader or narrower regions of unmethylated DNA during differentiation (Hodges et al. 2011).

This was reminiscent of the observation by Rafa Irizarry and colleagues several years earlier of DNA methylation being (a) constant and low in CpG islands but (b) flanked by less CG-dense loci where DNA methylation changed dynamically in disease and between cell types, "CpG island shores" (Irizarry et al. 2009), which we subsequently showed to have the chromatin modifications typical of enhancers (Wijetunga et al. 2014). What Hodges' work suggested was that the changes in DNA methylation had a directional quality, extending out and back from the hypomethylated core region (which she found also to include non-CpG island loci) in a way that she described evocatively as "tidal." We are led to conclude that many transcriptional regulatory events occur at loci where the chromatin is being held in an open and accessible state, allowing loading of TFs at the edges of those loci when new changes are signaled.

Not all loci behaved in this manner, however. Some loci were indeed found to open and close in the absence of adjacent hypomethylated loci. Dirk Schübeler noted that the loss of DNA methylation at sites of de novo TF binding was accompanied by the acquisition of 5hmC at those loci (Feldmann et al. 2013), indicating the active role for TET enzymes in mediating the loss of 5mC (Sardina et al. 2018). Another finding across all these studies was that loci where chromatin or DNA methylation changes were occurring were those more likely to have changes in gene expression, and it was commonly observed that the histone modifications typical for promoters and enhancers also characterized these loci.

This gets us back to an earlier, influential observation about chromatin changes during differentiation. Bradley (Brad) Bernstein used a tiling microarray approach back in 2006 to study chromatin organization in mouse embryonic stem cells at large developmental loci whose strong evolutionary conservation was not explained by protein-coding sequences. When he mapped the H3K4me3 mark, typically associated with active promoters, he was surprised to see that a subset of these loci also appeared to be the target of the Polycomb-mediated H3K27me3 repressive mark. Why should activating and repressive marks be at the same loci? The trivial explanation was that the embryonic stem cells were heterogeneous—some cells having H3K4me3, others H3K27me3—but he showed that there were nucleosomes that contained both marks, indicating that the activating and repressive marks could co-exist at a locus. Furthermore, when the cells were differentiated in vitro to neural progenitor cells, he showed that genes that activated in this cell type lost the repressive mark when their promoters started off with both, whereas at genes that were silenced in the neural cells only the repressive H3K27me3 mark remained. He described these promoters as "poised" and referred to them as "bivalent chromatin domains" (Bernstein et al. 2006).

Brad Bernstein (2019, pers. comm.) did not initially trust these results:

The moral of this story is trust your data. It's been a recurring theme in my research life as we've developed new methods that provided clearer and more comprehensive views of how genomic DNA is organized in the nucleus. In the case of bivalent domains, we had programmed a set of microarrays that densely tiled large developmental loci. In what was called a ChIP-on-chip experiment, we mapped different histone modification states in different kinds of cells. I had a talented technician [Dana Huebert] working for me at the time. She generated maps for embryonic stem cells, and we saw overlapping active and repressive modifications over scores of master developmental genes. This was a surprising result because these modifications were thought to be mutually exclusive at the time. I concluded that we'd messed up the experiment and told her we needed to repeat everything from scratch. Again, the same result—overlapping "bivalent" chromatin marks. I realized that the bivalent genes were all transcriptionally inactive in the stem cells. But they're "poised" to come on later, depending on the developmental fate of the embryonic stem cells. So, I hypothesized that bivalent chromatin helped keep these genes in check but poised for future activation. I'd say the model has held up pretty well.

Several times since then, new technologies have guided us to a surprising new biology. ChIP-on-chip gave was to ChIP-seq, and some of the first ChIP-seq maps seemed to suggest that GC-rich ("CpG islands") DNA sequences determined where Polycomb repressors settle in the genome. This again was unexpected, because dogma held that these repressors were recruited by transcription factors that should recognize more complex sequences motifs. So, I kept looking for other explanations. But I was finally convinced when we put random pieces of GC-rich DNA into the genome of embryonic stem cells and, lo and behold, they recruited Polycomb! It's exciting to bring powerful emerging technologies to study fundamental questions in biology. Sometimes you'll be tempted to ignore the more unusual outcomes. But my experiences with stem cells, and more recently cancer, have led me to think very carefully before discounting even the most unexpected result.

Epigenomic profiling is therefore capable of defining the endpoint differences during cell differentiation, but in practice it is more likely that an in vivo–derived sample will contain cells at intermediate stages of differentiation, complicating interpretation. The ability to perform ATAC-seq looking at chromatin organization in single cells is now revealing the regulatory dynamics during this differentiation process, a valuable added insight (Buenrostro et al. 2018) and a powerful addition to the range of epigenomic assays we can apply in understanding changes of patterns during differentiation.

INSIGHTS INTO THE PROPERTIES OF CELLS

The unit of organization of molecular genomic states is the cell.

If you study epigenetics and are considering getting a tattoo, the line above might be worth considering placing somewhere prominent. There is a great temptation to reduce the results of epigenomic assays to quantitative variables that assume homogeneity of the cells tested. By understanding how cellular heterogeneity influences the outcomes of epigenomic assays, the integration of different types of data can be performed more effectively.

Epigenomic Assays Test Collections of Cells and Alleles

All the assays described above represent ways of testing samples composed on many cells pooled together. The ChIP, ATAC, meDIP, RIP, RNA expression, and chromatin looping assays are all examples of assays that read out the cumulative properties of all alleles in the bulk cell population. The problem becomes one of the tyranny of the majority—bulk patterns disguise mosaic events; cell subsets with distinct properties are probably universally present in all bulk populations, but these assays are unable to define what is happening per cell.

The chromatin assays are, at best, semiquantitative. Although it is possible to distinguish chromatin properties present in all cells on both alleles from properties present on only one of the pair of alleles or on neither allele (first noted for ChIP-seq in 2010 [McDaniell et al. 2010]), a sample of cells in which 40% of cells having a specific chromatin state at a locus is very difficult to distinguish from another sample in which 60% of the cells have that chromatin state. Chromatin states are typically tested to look for a binary outcome of a peak or no peak: Is the state present at this locus in the genome and not in other loci (an intragenomic comparison)? These chromatin assays are poorly suited to intergenomic comparisons, in which the signal strength can be compared across samples as is typically performed for RNA expression studies. The equivalent binary outcome for RNA-seq would be to test whether a gene is on or off in a cell sample, an intragenomic outcome rarely sought.

Colocalization Does Not Mean Co-Regulation

One problem that arises when interpreting the results of multiple epigenomic assays performed on the same pool of cells is that patterns may colocalize but be occurring in different cell subsets. This was studied by Paul Soloway and colleagues using single-molecule approaches, testing how DNA methylation (5mC) colocalized with the repressive H3K9me3 and H3K27me3 chromatin modifications. They found that although 5mC and H3K9me3 could be found on the same fragments of chromatin, 5mC and H3K27me3 did not co-occur on the same fragments, therefore their tendency to colocalize is due to separate targeting in different cell subtypes (Murphy et al. 2013a).

DNA methylation, when tested using bisulfite or EM sequencing, reads out individual allelic states. An assay may show one cell sample to have 40% of alleles with 5mC at a specific CG dinucleotide, and another cell sample 60% at the same CG, an outcome that is readily distinguishable. An individual cell cannot, however, have 40% DNA methylation; at a diploid locus there can be DNA methylation on neither (0%), both (100%), or one of the two alleles (50%). To get a value of 40% DNA methylation means that a mosaic subset of cells has one or both alleles methylated at that site, cumulatively representing 40% of the chromosomal alleles present. To simplify this idea, it can be assumed that most of the time the two alleles have the same DNA methylation pattern, so a 40% DNA methylation pattern means that 40% of the cells are methylated at this site on both alleles, and 60% are unmethylated in that cell sample.

DNA methylation is therefore a digital (binary) readout of the alleles tested, whereas chromatin and RNA expression studies can be associated with variable numbers of sequences per locus in a given cell. As a useful clue, if DNA methylation shows intermediate values (e.g., between 20% and 80%), this can be taken to indicate the presence of cell subtypes in the population tested, with distinctive patterns of regulation at that locus (Wijetunga et al. 2014).

Does intermediate DNA methylation always indicate cell mosaicism? Probably not, as evidenced by the PMD example described above, and DNA replication effects during the cell cycle appear to influence the re-establishment of DNA methylation also (Charlton et al. 2018). Furthermore, if there is a heterozygous, unrecognized DNA sequence polymorphism at the locus tested, that can generate the artefactual appearance of intermediate DNA methylation. The context in which the observation of partial DNA methylation is made therefore needs careful attention.

The very low proportions of 5hmC, 6mA, and m^1A in mammalian genomes can also be interpreted in terms of allelic events. Taking 6mA as the example, in Chapter 4 we noted that if 6mA is indeed present in the human genome (which is doubtful [O'Brown et al. 2019]), it has been measured to occur at approximately one in every 2000 adenines genome-wide (Xiao et al. 2018). If this 6mA is randomly distributed across all adenines, then you would expect one cell in every 1000 to have a single allele marked by

6mA. If instead, to make up a number, this mark preferentially selects one in every 100 adenines in the genome, then you would have on average a (1/20) 5% level of 6mA at each of these targeted adenines. This would mean 5% of the cells have the 6mA modification. In correlating the presence of 6mA at a locus with a functional property of the cells studied, this kind of logic becomes important because it indicates the proportion of cells in which to expect co-occurring events when performing integrative epigenomic studies to understand the effects of 6mA.

There is, of course, another factor to consider, illustrated by the 5hmC modification, the influence of the dynamic turnover of the modification. The level of 5hmC at a locus tends to be low (<10% [Li et al. 2015]) and seems to represent the activity of TET enzymes as they constantly demethylate new 5mC, generating 5hmC that is subsequently further oxidized and replaced with unmethylated cytosine (Chapter 4). Therefore, although the proportion of alleles currently containing 5hmC might be <10%, the proportion of alleles that recently had 5hmC might be quite substantial and would be reflected by a correspondingly higher proportion of cells in which this modification had recent functional effects.

Deconvolution of Epigenomic Data

If cell subtypes present in the population tested have such consequential effects on interpretation of epigenomic data, it would be valuable to be able to identify these cell subtypes and their proportions. Fortunately, emerging techniques are allowing this, permitting a deconvolution (taken to mean an "unmixing" of signal) of cell subtypes present.

Many deconvolution techniques were developed to allow gene expression profiles from tumors to be tested to estimate the types and proportions of inflammatory cells present, including the CIBERSORT approach (Newman et al. 2015) that tests for 22 different blood cell types. DNA methylation data can be similarly tested using the *estimateCellCounts* function in the *minfi* software package (Aryee et al. 2014) but only for six blood cell subtypes.

Studying Individual Cells

If bulk assays have problems with their interpretability, single-cell epigenomic studies represent the logical approach to understand cell subtype effects.

Assays are now developed for gene expression, open chromatin, ChIP for histone modifications, and DNA methylation, allowing new insights to be generated into heterogeneity within populations. By understanding the gene expression signatures of single cells, reference data for deconvolution of bulk assays are being generated (Baron et al. 2016; Kong et al. 2019a). As will be discussed in the next chapter, once we can account for cell subtype effects in variability of results of epigenomic assays, we can then identify events more likely to represent genuine cellular reprogramming. The recent development of spatial transcriptomics allows us to identify where within a tissue the cell subtypes with specific molecular properties are located, linking these cutting-edge technologies back to our foundations in histopathological studies of diseases.

HOW THE EPIGENOME ALTERS THE GENOME

Finally, we should recognize that the genomic patterns obtained from these epigenomic assays reveal influences on DNA sequence variability. This is possibly a difficult idea when coming at the idea of epigenetics as a layer of information residing on top of the DNA sequence, which may suggest that the DNA sequence is merely read and regulated and not altered by epigenetic processes.

In discussing CpG islands above, we have already described the increased mutability of 5mC, causing 5mCG to become TG and escape recognition by DNA repair processes. This is the best understood way that an epigenetic modification can lead to DNA sequence variation.

However, we now also appreciate that the patterns revealed by other epigenomic assays reveal distinct patterns of mutability of the genome, not just the rate of mutation but also the type. This has been the focus of an excellent 2015 review (Makova and Hardison 2015), in which the chromatin contexts influencing both germline and somatic mutations are described. Interesting patterns emerge from these studies. The H3K36me3 modification, mentioned earlier as being enriched in transcribed loci, actively protects genes from new mutations (Huang et al. 2018). Presumed *cis*-regulatory loci (open chromatin or H3K4me1-enriched) have higher than average germline levels of somatic insertions, deletions, and substitutions of nucleotides (Makova and Hardison 2015). The cell therefore appears to want to keep our protein-coding sequences unmodified but permits loci regulating gene expression to be preferentially altered. In the next chapter, we will discuss how variation of the sequence of the genome influences the epigenome, creating a circle of influence that drives variability of epigenomic and cellular properties between individuals.

RECOMMENDED FURTHER READING

Preissl S, Gaulton KJ, Ren B. 2023. Characterizing *cis*-regulatory elements using single-cell epigenomics. *Nat Rev Genet* **24:** 21–43. doi:10.1038/s41576-022-00509-1

A review of single-cell approaches that allow one or more assays to be performed on individual cells, providing insights into transcriptional regulation in cell subtypes.

Stergachis AB, Debo BM, Haugen E, Churchman LS, Stamatoyannopoulos JA. 2020. Single-molecule regulatory architectures captured by chromatin fiber sequencing. *Science* **368:** 1449–1454. doi:10.1126/science.aaz1646

An example of the new kinds of assays that promise to generate much better insights into the epigenome than previously, Fiber-seq uses N^6-adenine methyltransferase exposed to nuclear DNA to add 6mA modifications at accessible DNA, reading out the results with long-read sequencing.

CHAPTER 7

Identifying Epigenetic Mechanisms of Disease

W e have now arrived at the part of the story where we can apply our insights to practical questions. A common reason why people study epigenetics is because of the presumed role of "epigenetic" changes in human disease. In previous chapters we have been building up to the point that we can address this question, starting with a discussion of what we mean when we say "epigenetic" changes in human disease. Almost all the time this is taken to mean nongenetic influences, acting through cellular reprogramming, involving permanent changes to the transcriptional regulatory apparatus.

In Chapter 8 we will focus on cancer and in Chapter 9 on the possibility of transmission of disease risk through nongenetic mechanisms across generations, leaving the current chapter to cover the remainder of human disease.

ARE ANY DISEASES REALLY KNOWN TO BE CAUSED BY EPIGENETIC MECHANISMS?

The typical review or textbook list of human diseases commonly described to be caused by epigenetic mechanisms is shown in Table 7.1. These diseases are centered on disorders involving genomic imprinting, mutations of epigenetic regulators of transcription acting in *trans*, local mutations acting in *cis*, and environmental effects on the cell, reviewed comprehensively by Huda Zoghbi and Arthur (Art) Beaudet (Zoghbi and Beaudet 2016). In almost all these examples, the primary event can be traced to a genetic mutation, whether a deletion, uniparental disomy, a protein-coding mutation in a gene, or the expansion or contraction of a repetitive sequence in the genome.

Table 7.1. Examples of human diseases attributed to epigenetic mechanisms and their causative genetic mechanisms, where known

Imprinting disease	Genetic mechanism(s)	Epigenetic mechanism
Prader–Willi syndrome	15q11-q13: Large deletion, uniparental disomy, small (imprinting center) deletion	Genomic imprinting
Angelman syndrome	15q11-q13: Large deletion, uniparental disomy, small (imprinting center) deletion, *UBE3A* mutation	Genomic imprinting
Beckwith–Wiedemann syndrome	11p15.5: Chromosomal rearrangements, somatic mosaicism, uniparental disomy, *trans*-acting mutations affecting multiple imprinted loci, point mutations in imprinting centers, *CDKN1C* mutation, and undefined	Genomic imprinting
Silver–Russell syndrome	11p15.5: Chromosomal rearrangements, somatic mosaicism, uniparental disomy (for Chromosome 7 also), *trans*-acting mutations affecting multiple imprinted loci, point mutations in imprinting centers, *CDKN1C* mutation, and undefined	Genomic imprinting
Pseudohypoparathyroidism	*GNAS* mutation, copy number variants at *GNAS* regulatory elements, uniparental disomy (20q), and undefined	Genomic imprinting
Rubinstein–Taybi syndrome	Single-gene mutations of *CREBBP* and *EP300*	*Trans* effect of epigenetic regulator
Kabuki syndrome	Single-gene mutations of *KMT2D* and *KDM6A*	*Trans* effect of epigenetic regulator
Rett syndrome	Single-gene mutation of *MECP2*	*Trans* effect of epigenetic regulator
α-Thalassemia mental retardation syndrome X-linked (ATRX) syndrome	Single-gene mutation of *ATRX*	*Trans* effect of epigenetic regulator
Immunodeficiency, centromeric region instability, and facial anomalies (ICF) syndrome	Single-gene mutation of *DNMT3B*, *ZBTB24*, *CDCA7* or *HELLS*	*Trans* effect of epigenetic regulator
Fragile X syndrome	Increased number of CGG trinucleotide repeats in 5′ UTR of *FMR1* gene	*Cis* effect of repeat expansion
Facioscapulohumeral dystrophy (FSHD)	Reduced number of D4Z4 repeats on 4q35	*Cis* effect of repeat contraction

Data from Zoghbi and Beaudet (2016).

It is therefore fair to say that these diseases are all due to genetic mutations, but they end up being called epigenetic because they reveal existing or induce new transcriptional regulatory changes. Describing them as epigenetic is reflective of the definition of epigenetic properties in terms of transcriptional regulation, the layer of information regulating the DNA sequence. When it comes to genomic imprinting, there is one added thought process involved. As described in Chapter 3, genomic imprinting involves a memory of the chromosomes of their last exposure to gametogenesis, allowing the definition of epigenetic properties as cellular memory (or persistent homeostasis, as described by Nanney) to encompass the unusual properties of imprinted loci. It is therefore completely understandable that these definitional underpinnings would cause these diseases to be called epigenetic. What needs to be remembered, however, is that the primary event in each case is a genetic mutation.

To describe these conditions as caused by epigenetic mechanisms implies that they are primarily due to the alteration of the properties of epigenetic regulation of the cell. Instead, these conditions are primarily due to genetic mutations and rely on epigenetic mechanisms (X-chromosome inactivation, genomic imprinting, etc.) to be working normally for the phenotype to be produced. The possible implication that epigenetic changes occur to cause these human diseases is not supported.

As an aside, we do not know the mechanisms involved in some of the cases of Beckwith–Wiedemann syndrome (BWS). Paternal duplication or uniparental disomy of 11p15.5 is the macro-scale mutational event underlying many cases, but smaller mutations at the *H19* imprinting center or in the coding sequence of *CDKN1C* account for many other cases, leaving some clinically apparent cases lacking an obvious genetic basis. Given that we cannot diagnose DNA mutations in all cases of BWS, it remains theoretically possible that some cases represent primary abnormalities with the normal epigenetic regulation of 11p15.5. Mosaicism for mutations causing BWS is a relatively frequent finding (Romanelli et al. 2011), suggesting that some cases may be difficult to diagnose if the mosaic cells are very sparse in the tissue sampled diagnostically (typically blood).

This leaves environmental influences as the only category in which the perturbation of the intrinsic properties of the cell might mediate disease in a way that does not involve DNA sequence changes. It is widely accepted that environmental influences on cells and tissues alter the epigenome and contribute to disease. This may not always be a reasonable assumption, as will now be discussed.

HUMAN DISEASE CAUSED BY ENVIRONMENTAL EXPOSURES MEDIATED THROUGH TRANSCRIPTIONAL REGULATORY MECHANISMS

The overwhelmingly common assumption about epigenetic mechanisms of human diseases is that environmental influences, associated epidemiologically with human diseases, mediate their effects through transcriptional regulators and the reprogramming of cells. In the special situation of an environmental exposure having occurred in the past, and not currently, but the disease risk remains as a memory of that past exposure, the property attributed to epigenetic regulation of being able to confer memory to cells makes epigenetic mechanisms even more compelling.

Obviously, some exposures damage DNA and lead to disease acutely or temporally remotely through DNA sequence mutations, with cigarette smoking a common exposure in this category. The concept of epigenetic mediation of risk is mostly based on the idea that there can be nonmutational mechanisms that cause disease during or following an exposure.

Another contributor to the interest in environmental perturbations and epigenetic mediation of phenotypes comes from the idea that the epigenome is at its most easily perturbed during development. The dramatically dynamic changes of DNA methylation through mammalian development (Monk et al. 1987) were being noted by Marilyn Monk around the same time that the epidemiologist David Barker was first correlating childhood with adult diseases and suggesting "that poor nutrition in early life increases susceptibility to the effects of an affluent diet" (Barker and Osmond 1986). This idea of a window of susceptibility to environmental exposures during development leading to a risk of adult disease was first described by Barker as a "programming" during fetal life (Barker 1992). Over time, the idea was extended beyond the fetal stage of life to become more broadly referred to as the developmental origins of health and disease (DOHaD) (Wadhwa et al. 2009). This will be discussed in more detail in Chapter 9.

MOLECULAR CANDIDATES FOR MEDIATING ENVIRONMENTAL INFLUENCES

The first and still the most widely studied candidate molecular epigenetic mediator of environmental influences is DNA methylation, as described above. However, as the biochemical insights into genomic regulation started to unfold in recent years (Chapter 4), it became apparent that not only was DNA methylation potentially influenced by the environment, but some other chromatin properties appeared to be environmentally responsive also, so a quick overview is worth highlighting at this stage.

Environmental factors that increase the production of S-adenosyl methionine (SAM)—in particular, folic acid and vitamins B_6 and B_{12} in the diet—have been linked to increased methylation of both DNA and histones throughout the genome (Mentch and Locasale 2016). Ascorbic acid (vitamin C) is a co-factor for the TET enzymes that cause hydroxymethylation of DNA and has been associated with a genome-wide increase in 5-hydroxymethylcytosine (5hmC) and loss of 5-methylcytosine (5mC) (Blaschke et al. 2013). Short-chain fatty acids such as those produced by gut fermentation of dietary fiber have histone deactylase (HDAC) inhibitory properties and have been linked to increased histone crotonylation at gene promoters (Fellows et al. 2018). Ethanol also has effects to induce expression of genes involved in signal transduction and learning and memory, the exposure associated with new loci of H3K9ac and H3K27ac formed at promoters and enhancers (Mews et al. 2019).

Hypoxia has multiple effects: It stabilizes the euchromatic histone-lysine N-methyltransferase 2 (EHMT2, G9a) protein, which leads to increased methylation of H3K9 at gene promoters (Casciello et al. 2017) and inhibition of the KDM5A lysine demethylase with effects at loci with H3K4me3 and H3K36me3 (Batie et al. 2019). Glycolysis to generate glucose through hexosamine biosynthetic pathway creates uridine 5′-diphospho-N-acetyl-glucosamine (UDP-GlcNAc). GlcNAcylation appears to modify mammalian histones (Dupas et al. 2023), whereas GlcNAcylation of EZH2 appears to enhance the protein stability and enzymatic activity and has an effect to increase heterochromatin formation (Lo et al. 2018), and GlcNAcylation of TET enzymes enhances the activity of TET1 and increases the amount of 5-hmC at loci where this already exists (Hrit et al. 2018). Lactic acid generated by anaerobic metabolism appears to cause lactylation of histones at active gene promoters, possibly leading to their activation (Zhang et al. 2019), whereas serotonin can be added to chromatin as serotonylation of histones at gene promoters, also possibly helping to boost expression of those genes (Farrelly et al. 2019). Finally the nicotinamide adenine dinucleotide (NAD) co-factor has an uncertain relationship with the activity of the SIRT1 deacetylase (Imai and Guarente 2016) but has been explored as a link between metabolism and histone acetylation.

These are extremely fascinating links between diet, metabolism, intercellular signaling, and some of the regulators of transcription. These all fall into the category of regulators that lack sequence specificity, so it is especially intriguing how some of these effects like the HDAC inhibition can lead to selective effects at a subset of promoters (Fellows et al. 2018). This puzzle will be explored again below in terms of the potential contributory role of TFs, but the focus will now turn to the study of DNA methylation as the primary means of exploring not just environmental exposures but disease phenotypes in general.

DNA METHYLATION AND THE EPIGENOME-WIDE ASSOCIATION STUDY

The question about how the environment influences cells to cause phenotypes is possibly the most hypothesis-based current rationale for epigenetic studies of human diseases. Sometimes the reason for embarking on the study is that the genetic basis for what appears to be a very heritable disease is not being revealed through ongoing genome-wide association studies (GWASs), and probably some studies are just exploratory, looking to see whether adding epigenetic information improves insights into the disease being studied through more mainstream genetic approaches.

Whatever the reasons, the outcome has been an explosion of what have been called epigenome-wide association studies (EWASs), taking a cue from the GWAS approach that surveys the common variants of the human genome looking for alleles that are nonrandomly associated with the phenotype. In an EWAS, the typical approach is to study DNA methylation in peripheral blood leukocytes, partly because of the insights into the biochemical heritability of this molecular mark through cell division, partly because of its assumed potential to be influenced by the environment, and probably mostly because the genome-wide survey assays for DNA methylation described in Chapter 6 are readily available and relatively cost-efficient.

These surveys of DNA methylation test from tens of thousands to millions of CG dinucleotides in the human genome. DNA methylation is measured on a continuous scale from 0 to 100%, whereas other assays testing molecular transcriptional regulatory mediators are often more binary (e.g., peak vs. no peak of a chromatin state at a given locus). A positive EWAS outcome is therefore defined in terms of a sufficiently large change in the level of DNA methylation consistently in a specific direction (increased or decreased) in enough individuals in the groups compared to allow statistical significance to be attributed to the finding as nonrandom.

Typically, these studies identify more than one site with differences in DNA methylation. The use of these results bifurcates at this point. One application is the development of biomarkers, using the pattern of DNA methylation in a predictive way. The second is the understanding of the pathogenetic mechanisms of disease.

Biomarker studies have represented a very intriguing outcome of analysis of EWAS data. Probably the most robust example is the change in DNA methylation in peripheral blood in smokers, first identified in 2011 (Breitling et al. 2011). An effect is also seen when a mother smokes during pregnancy, and the child's peripheral blood leukocyte DNA is tested, with DNA methylation at the *AHRR* gene altered in both personal and in utero exposures (Joubert et al. 2016). This is a fascinating observation, as it indicates a memory of past exposure reflected by changes in DNA methylation. DNA methylation of blood cells is also a very useful biomarker when studying diseases for which cigarette smoking may be a contributor, as shown by Kelly Bakulski and colleagues in a study of lung cancer (Bakulski et al. 2019). Other robust biomarkers include peripheral blood DNA methylation patterns that reflect the age of the individual (Hannum et al. 2013; Horvath 2013) and blood leukocyte DNA methylation as a biomarker for cancers (we will look at examples in Chapter 8).

When using DNA methylation as a biomarker, the goal is limited to building a robust association between the change in DNA methylation and the exposure, phenotype, or outcome of interest, without needing to understand why the change in DNA methylation occurred in the first place. If, however, the goal is to understand the mechanism of the disease phenotype, the typical approach is to start by trying to understand why specific loci in the genome were chosen for their changes in DNA methylation. This frequently results in a two-step approach in which the first step is to link the sites of DNA methylation with nearby genes and then look for coherence of the properties of those genes, whether using ontological terms describing the individual genes or the known interactions of their protein products.

What is then inferred is the response of the cells studied to an environmental exposure or a disruption of properties inherent to the cells tested as a contribution to the phenotype being tested. With a goal at the outset of understanding exactly these issues, there is a satisfying sense of closure when

DNA methylation data indicate enrichment for a pathway that is plausibly involved in the phenotype.

Although this sounds like a relatively fail-safe approach, we have come to realize that this EWAS design has several unforeseen problems, to the point that we cannot be confident in our interpretation of EWAS results as representing the kinds of mechanisms described above. The reasons behind this uncomfortable realization are described below, but we will then return to a more positive perspective by talking about how embracing these challenges can bring us to a stronger position in terms of understanding disease pathogenesis using epigenomic assays.

PROBLEMS WITH EWAS DESIGN AND INTERPRETATION

The big problem is that a change in DNA methylation can have multiple causes. In Chapter 5 we noted that the implicit cellular model being tested in epigenetic association studies is that of *cellular reprogramming*, a change of transcriptional regulation that alters the properties of a canonical cell type. We also assume that we are looking for changes that are independent of DNA sequence variation between individuals, and that the changes are contributing to the pathogenesis of the disease. As it turns out, these assumptions are not reasonable, making EWAS interpretation extremely difficult. There are also practical issues with our study designs and execution that will be the first issues to discuss.

Surrogate Tissue Sampling

When studying samples from humans, the trade-off is always the same—you can study a lot of people easily if you sample accessible tissues that may not mediate the disease, or you can study many fewer individuals if you sample the tissue likely to mediate the disease. The approach to sample accessible tissues is highly preferable when developing a biomarker, as you want to be able to use accessible tissues when using the biomarker in practice. Typically accessibility means that the sample can be obtained with minimal risk, pain, or injury to the patient over and above routine medical care. Taking a sample of blood meets these criteria, but it should also be noted that if the standard of care routinely requires more invasive tissue sampling (e.g., performing a diagnostic biopsy of a tumor), then the additional testing of this tissue does not pose additional risk to the patient. Biomarkers used for cancer fall into this category.

When the goal is instead to understand the mechanism of the disease by using accessible, surrogate tissue samples, problems arise. The major issue is that there is no clear rationale expressed why a surrogate sample should be informative about a disease occurring in another tissue in the body. Implicitly, we are asked to assume that the changes reflecting cellular reprogramming in the disease-mediating tissue are the same as those occurring in the sampled, accessible tissue, which is most often peripheral blood leukocytes. Maybe there is a very good argument to be made that an exposure to the inaccessible heart, brain, or kidney is likely to have a similar effect on peripheral blood leukocytes. The temptation is to treat DNA methylation in the same way that we treat genotyping—the genotype in peripheral blood is highly likely to represent the genotype in an inaccessible tissue (unless there has been a somatic mutation in one or the other), whereas the DNA methylation pattern in peripheral blood is going to be different to that in an inaccessible tissue; the question becomes whether this pattern changes as part of a disease process in the same way in both tissues. Such explicit hypotheses are not typically part of the rationale for the use of surrogate tissue sampling but really should be.

Cohort Sizes

The related issue is cohort size. It is obviously a lot easier to get samples from a lot of individuals when you do not need to perform a major procedure on each of them. The very large studies of DNA methylation have therefore been focused on peripheral blood leukocytes, whereas studies of less accessible tissue can be very limited in number—for example, numbering only in the tens for studies of fatty liver disease (Ahrens et al. 2013; Murphy et al. 2013b; de Mello et al. 2017; Gerhard et al. 2018; Hotta et al. 2018; Wu et al. 2018; Zhang et al. 2018). With small numbers of individuals tested and up to hundreds of thousands or millions of loci measured for DNA methylation, the statistical challenges to assign significance to findings become immense.

Reverse Causation

Another reason why it is valuable to state a hypothesis explicitly when performing an EWAS is because we reveal what our a priori assumption is about the direction of causality. Implicit in many EWASs is the model of the DNA methylation changes causing the phenotype. However, as Ewan Birney and George Davey Smith have taught me (Birney et al. 2016), our typical cross-sectional study design used in EWASs inherently involves one group already having the disease, which raises the possibility that the disease itself causes the changes in DNA methylation. This "reverse causation" has been shown to occur in studies of obesity-related phenotypes, with changes in DNA methylation of peripheral blood leukocytes in individuals with increased body mass index (Richmond et al. 2016a; Wahl et al. 2017) or abnormal blood lipid profiles (Dekkers et al. 2016). There are interesting analytical techniques that can be used to explore the direction of causality when associating things like medical and molecular phenotypes, including mediation analysis and Mendelian randomization (Richmond et al. 2016b), and alternative study designs (such as longitudinal studies) can be better choices when reverse causation is a concern.

Cell Subtype Proportional Composition

In Chapter 6 we discussed how small changes in DNA methylation in a bulk assay must reflect a change in the proportions of one or more cell subtypes, because an individual cell cannot change DNA methylation by <50%. In Chapter 5 we saw how Andy Houseman and his colleagues causally linked these modest changes of DNA methylation to changes in cell subtype proportions (Houseman et al. 2012). Having had these discussions, we probably do not need to go any more deeply into the issue whether cell subtype proportion alterations cause DNA methylation changes—this is now well-accepted.

How would we know whether the change in DNA methylation is due to cell subtype proportion changes? There are ways of testing cell subtype composition directly, and there are ways of inferring cell subtypes present from the data generated on bulk samples.

The most widely used test of cell subtype composition on our planet is the differential blood count, a routine clinical laboratory test. This test measures proportions of lymphocytes, granulocytes (neutrophils, eosinophils, and basophils), and monocytes. Although better than nothing, this is a very high-level summary of major classes of cell subtypes and is insensitive to some of the more subtle changes that occur in disease and other physiological states. For example, if you were to compare lymphocytes between neonates and older people, there would be a shift to an increased proportion of memory T lymphocytes in the older individuals (Lappalainen and Greally 2017). The differential blood count will not identify such changes, but they could cause changes in DNA methylation.

A detailed study of the cells in blood or other tissues can be performed by using antibodies that recognize different cell subtypes and performing flow cytometry. Quantitative histology is a further possibility that could be enhanced by new artificial intelligence approaches (Serag et al. 2019), but it remains challenging to generate accurate cell subtype proportion estimates, especially in solid tissues.

The techniques that allow single-cell studies (mentioned in Chapter 5) appear to generate a degree of resolution beyond what antibodies and flow cytometry can provide (Jaitin et al. 2014). The cell subtypes are defined by clustering of patterns of gene expression and can be linked to the kinds of antibody profiling used in flow cytometry through techniques like CITE-seq (Stoeckius et al. 2017). Dissociation of solid tissues to generate representations of all cell subtypes present is challenging, so it must be said that there is no perfect, direct means of identifying and measuring the proportions of all cell subtypes present in a solid tissue at present.

An alternative approach is inferential, using data from an assay of the bulk tissue made up of its many subtypes and using an analytical method to deduce the presence and proportions of subtypes present. We can focus on two here for simplicity, recognizing that this field is rapidly evolving. An approach developed to use gene expression data is the cell-type identification by estimating relative subsets of RNA transcripts (CIBERSORT) algorithm, used commonly to measure the representation of white blood cells within tumors. CIBERSORT uses data from individual cell subtypes believed to be present within the sample tested. Each cell subtype has patterns of gene expression that distinguish it from other cell subtypes, which is used by CIBERSORT to make estimates of the proportions of these cells in the bulk tissue tested. The reference individual cell subtype expression data can be generated by purifying individual cell subtypes using antibody-based enrichment (Newman et al. 2015) or can be based on the use of single-cell transcription data (Kong et al. 2019a; Newman et al. 2019).

Deconvolution can also be performed using DNA methylation data. This was pioneered by Houseman in 2012 for blood leukocyte subtypes, specifically granulocytes, monocytes, and four subtypes of lymphocytes: B cells, natural killer cells, and CD4 and CD8 T cells (Houseman et al. 2012). Whereas CIBERSORT originally used gene expression profiles from 22 different blood cell subtypes (Newman et al. 2015), reference DNA methylation data from blood have remained focused on these six relatively broad cell subtypes. We found the accuracy of prediction of CD4/CD8 ratios to be better for DNA methylation than gene expression–based deconvolution (Kong et al. 2019a), indicating that pursuing DNA methylation for cell subtype deconvolution may be an avenue worth exploring for the field.

There are also analytical approaches that have been used with the goal of removing cell subtype proportion effects on the data generated. One representative approach uses surrogate variable analysis (SVA), which in essence looks for major sources of systematic variability, assumes these to be technical or due to cell subtype proportion changes in origin, and removes this source of variability from the data, with the hope that what remains is more likely to represent cellular reprogramming (Leek et al. 2012). This works, but only because cell subtype effects tend to be a major influence on DNA methylation variation. The less of an influence cell subtype is on variability, the more that this kind of approach will start to remove the biological variability sought as the experimental outcome (Kong et al. 2019a).

Genetic Effects on DNA Methylation

Now we get to the very awkward problem. The comfortable traditional idea of epigenetic regulation being the information sitting above or upon the inert, dull DNA and telling it what to do is very attractive. The problem is that the sequence of the DNA exerts strong effects on how it is organized. This has been discussed to some extent in Chapter 6 with the description of CpG islands, but in the context of disease studies we are less concerned with that kind of variation within the human genome and more interested in DNA sequence polymorphism between human genomes.

The figure commonly quoted is that 0.1% of the genome differs between individuals, which is consistent with the ~4 million single-nucleotide polymorphisms (SNPs) and small insertions or deletions (indels) found when comparing a diploid human genome to the reference genome (Supernat et al. 2018). As it turns out, these sequences can influence gene expression and many of the molecular regulatory mechanisms described in Chapter 4.

The first observations in this area of research were in the early 2000s and founded on studies of yeast and *Drosophila*, reviewed at the time by Vivian Cheung and Richard Spielman (Cheung and Spielman 2002) who went on to do some of the groundbreaking studies mapping loci influencing gene expression in *cis* and in *trans* in the human genome (Morley et al. 2004). These loci began to be referred to as gene expression quantitative trait loci (eQTLs) (Schadt et al. 2003).

If DNA sequence variants could be influencing gene expression, it was a reasonable bet that they could also be influencing the molecular regulators of gene expression. In 2007, a study performed looking at the variability of DNA methylation between individuals at the imprinted *H19/IGF2* locus revealed that much of this variability was heritable and due to SNPs acting in *cis* (within the chromosomal region) (Heijmans et al. 2007). A clever insight from Benjamin (Ben) Tycko was that allele-specific methylation (ASM) outside imprinted domains was associated with local SNP patterns (Kerkel et al. 2008), strongly indicating an effect of the local haplotype on DNA methylation. A formal study by Jordana Bell and colleagues of 77 lymphoblastoid cell lines integrating gene expression and genotyping data allowed the DNA methylation counterpart of the eQTL approach above, identifying *cis* methylation quantitative trait loci (meQTLs) (Bell et al. 2011). Around the same time Rick Myers' group reported that bisulfite sequencing data revealed certain loci to have strikingly distinct patterns of DNA methylation on homologous parental alleles distinguished by individual SNPs, outside of known imprinted domains (Gertz et al. 2011).

Now we have lots of types of functional variants being reported—for example, those influencing microRNA expression levels (Huan et al. 2015), variants associated with post-translational modifications of histones (Pelikan et al. 2018), or loci of open chromatin (Degner et al. 2012). How these sequence differences are read so that they lead to functional effects has been assumed to involve changing the ability of the sequence to be bound by TFs (Karczewski et al. 2011; Deplancke et al. 2016; Behera et al. 2018; Madsen et al. 2018), although it was also proposed that the effects may be mediated by changing the structural properties of enhancer RNAs (Ren et al. 2017). These studies to identify functional variants have been valuable in refining the results of GWASs, which link phenotypes with haplotypes of tens to hundreds of kilobases in size (Edwards et al. 2013). Of the many (up to 1000) variants in the haplotype, only a small number end up having functional properties in the tissue type, presumably mediating the phenotype, allowing attention to be focused on these loci when dissecting the genetic mechanism of the phenotype.

In Chapter 5 it was mentioned that the proportion of variability of DNA methylation between individuals has been estimated to be between

18% (Bell et al. 2012) and 66% (Cheung et al. 2017). A DNA methylation change as the outcome of an EWAS is therefore highly likely to involve a component of genetic influence. As the large majority of EWASs do not perform genotyping on the individuals studied, how much of an influence DNA sequence variation plays is unknowable and carries a high likelihood of the results being overinterpreted.

We have many reasons why EWASs to date are uninterpretable. Out of this realization comes the opportunity to rethink our traditional EWAS approach and come up with a new approach that has much more potential to yield insights into the mechanisms of phenotypes but represents a challenge to how many people would define epigenetics. In walking through that logic, the first step is to embrace rather than discard influences that we currently regard as confounding.

HARVESTING THE CONFOUNDERS

Using Epigenomic Assays to Report Cellular Genomic Models

This one is relatively easy, because the field has already started to move toward the viewpoint that cell subtype changes that occur in individuals with a phenotype represent an important insight into the phenotype. It is also worth reminding ourselves that cell subtype proportion changes are only going to influence DNA methylation results when comparing two groups if the cell subtype variation is biased between the groups—the cell subtype proportions are different in those with the phenotype compared with those who are unaffected. Of course, this is a potential insight into the pathogenesis of a disease phenotype; this is exactly why histopathology and flow cytometry has been used in understanding diseases.

Remembering what was discussed in Chapter 5 for a minute, we note that the typical idea about how epigenetic events occur in diseases is through cellular reprogramming. The alternative idea was also proposed that cell fate decision changes (polycreodism) are in fact a very Waddingtonian way of thinking about disease and merit a description as epigenetic. Although this is not at all a mainstream idea, there is no shame in seeing how this concept fits the disease or other phenotypes that you might be studying.

There are examples of how people have sought and reported the cell subtype differences after having performed DNA methylation studies. Often overlooked in the 2012 Houseman study was the group's systematic analysis of the blood leukocyte DNA methylation changes associated with different diseases. They found that in people with head and neck squamous cell carcinoma, they had a lower proportion of $CD4^+$ T lymphocytes but more natural killer (NK) cells and granulocytes; women with ovarian cancer had more granulocytes but fewer B and $CD4^+$ T lymphocytes and NK cells. Trisomy 21 (Down syndrome) was associated with fewer B lymphocytes, and obesity in African–Americans showed an increase of granulocytes and a decrease in NK cells in the blood (Houseman et al. 2012). The same approach applied to DNA methylation data from umbilical cord blood leukocytes derived from neonates exposed in utero to arsenic revealed evidence for an increased proportion of $CD8^+$ T lymphocytes (Koestler et al. 2013). Peripheral blood from schizophrenia patients was found to have DNA methylation patterns that revealed them to have relatively fewer $CD8^+$ T cells and more monocytes than control individuals (Montano et al. 2016). A known poor prognostic indicator in cancer patients is a high neutrophil:lymphocyte ratio in peripheral blood, which was shown to be calculable from DNA methylation data (Koestler et al. 2017). Some investigators are beginning to report cell subtype variation as one of the outcomes of EWASs, and they are not just using this information as a measurable confounding variable to be eliminated from further consideration.

Using Epigenomic Assays to Report DNA Sequence Effects

If we can get away from the idea that the study of epigenetics must always be insulated from genetic, DNA sequence-based mechanisms of phenotypes, we can align the field of epigenetics with the field that is generating excellent insights into human diseases by identifying noncoding functional variants in the genome. As is the case with cell subtype composition differences, a functional variant can only cause DNA methylation to be different between compared groups if the allele increasing DNA methylation is overrepresented in one group, with the other allele correspondingly increased in the other group. What is identified is a genetic association with the phenotype and, what's more, a functional variant. In a study of small numbers of individuals, as is frequently the situation for EWASs, these kinds of associations can be due to chance, but to eliminate them as merely confounding is potentially wasting valuable insights into the genetics of the condition. Once again, this is an opportunity to harvest information from what would commonly be thought of as a confounding influence. To think of the influence of genetic sequence on DNA methylation as confounding only occurs in the context of a need to prove a nongenetic cellular reprogramming hypothesis, which is very confining.

Both cell subtype composition and DNA sequence variation appear to have large influences on DNA methylation and, by extension, probably all other epigenomic and transcriptomic assays. How do the two coexist and have such large effects? Is it possible that they are interdependent in some way, with DNA sequence variation influencing cell subtypes? That could be a way for a sequence polymorphism to create the appearance of an meQTL but by causing cell subtype proportions to vary, something we introduced in Chapter 5. This has not been studied in the context of an EWAS, but certainly there have been loci identified that influence cell subtype proportions in peripheral blood (Ferreira et al. 2009; Okada et al. 2011; Roederer et al. 2015). When such loci are located close to a gene expressed distinctively in an influenced cell subtype, it could end up getting mistaken as a *cis*-acting eQTL, but in general they will be associated with expression of genes that are located elsewhere in the genome and would be considered *trans*-acting eQTLs instead. What goes for eQTLs should also be the case for meQTLs: If a locus influences the proportion of a cell subtype in a tissue and is close to a CG that is distinctively methylated in that tissue, it will look like it's causing DNA methylation to change. In studies of eQTLs and meQTLs that don't consider cell subtype proportions, some are possibly acting through unrecognized effects on cell subtype proportions (Fig. 6.2). In a very nice demonstration of the value of taking into account cell subtype proportions in eQTL studies, the group of Kerrin Small used a deconvolution approach to estimate the proportions of four subtypes of cells in adipose tissues (adipocytes, macrophages, CD4$^+$ T lymphocytes, and microvascular endothelial cells), increasing the number of eQTLs identified, and defining the cell subtype in which they were active (Glastonbury et al. 2019).

Of interest to those studying epigenetics from the perspective of understanding how the environment influences cells, response eQTLs are now being studied. These were first identified in dendritic cells infected in vitro with *Mycobacterium tuberculosis*. The authors used genotype information and gene expression studies before and after infection, finding that the set of eQTLs was overall similar but that a subset changed following infection, described by the group as response eQTLs (Barreiro et al. 2012). In subsequent studies response eQTLs have been identified following exposures of human monocytes (Fairfax et al. 2014; Kim et al. 2014; Quach et al.

2016), dendritic cells (Lee et al. 2014), and monocyte-derived macrophages (Nédélec et al. 2016). There is now a parallel example of studying chromatin accessibility (by ATAC-seq) and observing how these patterns change with infection in vitro (Scott-Browne et al. 2016), revealing the loci mediating these host cell responses, where polymorphism could mediate interindividual differences in susceptibility to the infection. After cataloging the enhancer landscape in naïve T lymphocytes differentiating to T helper 1 and 2 (Th1, Th2) cell types, enrichment was found in disease-associated variants implicated in autoimmune and inflammatory disorders (Hawkins et al. 2013).

Although these studies are taking approaches that focus on understanding DNA sequence effects and consequently do not fit neatly into prevalent ideas about epigenetic mechanisms of disease being independent of DNA sequence effects, they nonetheless reveal the value of letting gene expression or its regulatory mechanisms reveal insights into the effects of DNA sequence variation in disease. If epigenetics studies were to embrace rather than discard the information from meQTLs, eQTLs, chromatin accessibility, or other functional variants identified during a study looking for cellular reprogramming, the results would be even more informative in terms of gaining insights into why the individual developed the disease.

TFs AND CELLULAR EPIGENETIC CHANGES, MEMORY, AND PHENOTYPES

When the enhancer landscape was surveyed in the differentiating T lymphocytes, and disease-associated variants were observed to be preferentially located within them (Hawkins et al. 2013), the other information generated was an insight into the DNA sequence motifs enriched at these loci. What is varying between people with diseases compared with those lacking them is not the property of the TF to bind different sequences in different people— that appears to be invariant. Instead, the DNA to which the TF binds varies between people, revealing these loci as functional variants and leading to changes in chromatin accessibility, DNA methylation, and gene expression.

In Chapter 4 the historical shunning of TFs as members of the epigenetics club was brought up. Fortunately, this view is now shifting to the recognition that the interaction between TFs and DNA is more complex than one of chromatin states and DNA methylation telling TFs where to go, and that TFs have a role in setting up the positioning of sequence-nonspecific regulatory mechanisms as a cell differentiates or responds to a perturbation.

Taking this a step further, we can start to look at epigenetic studies of phenotypes with more of a TF-centric perspective and gain some insights that would not have been considered when using a mindset that downplays their role as secondary.

First, both cellular reprogramming and cell fate changes are exactly what TFs do for a living, and these represent two of the models being studied in epigenetic studies of disease. A study in which my group was involved led by Marija Kundakovic tested the chromatin organization and gene expression in the ventral hippocampus of mice during the high- and low-estradiol times of the mouse estrus cycle. Marija confirmed findings that these physiological hormonal fluctuations are associated with behavioral changes, and that both gene expression and loci of open chromatin changed systematically with the cycle. When we looked for motif enrichment in the sites changing chromatin accessibility, the Egr1 TF emerged as the likely mediator, acting downstream from estradiol binding to a cell surface receptor and activating a cell signaling pathway that ultimately resulted in Egr1 being activated to create a transcriptional response in the nucleus (Jaric et al. 2019). This result illustrates a few things about cellular reprogramming. One is that this excellent model of cellular reprogramming is plausibly mediated by a TF, but also that TFs themselves can act downstream of other influences—in this case, cell signaling pathways (the separate nuclear receptor group of TFs [Sever and Glass 2013] are directly bound in the nucleus by ligands, bypassing cell signaling). Developmental biologists who study cell differentiation (epigenesis) are used to the idea of morphogen gradients, in which molecules secreted by cells in the developing embryo influence the differentiation of cells nearby, acting through cell signaling pathways and TF activation. In Chapter 2 we talked about Waddington spending time with Hans Spemann, who developed the idea of the "organizer" in the developing embryo as a group of cells that could act to influence the differentiation of adjacent cells. As it turns out, this is due to morphogens being released by the organizer cells (Wartlick et al. 2009), an idea that captured the interest of Alan Turing (1912–1954), prompting him to propose theoretical models for patterning in nature (Turing 1952). The idea that exogenous stimuli activate cell signaling and act through TFs to reprogram cells is therefore deeply rooted in biology and serves as a model for epigenetic studies of cellular reprogramming. An influence that alters cell fates to change the composition of a tissue during development (our polycreodism model) is likewise dependent on TFs mediating the differentiation process. Furthermore, as mentioned above, the use of epigenomic assays to reveal functional DNA sequence variants appears to be dependent on the DNA sequence variation altering how TFs bind, something we ourselves studied with a CRISPR targeting approach, finding direct evidence for this model (Johnston et al. 2019).

With this new TF-centric perspective, we can take a new look at epigenetic studies of disease and start to realize that there are other mechanistic models we should be considering. The same enzymes that acetylate/ deacetylate (Park et al. 2015) or methylate/demethylate (Carr et al. 2015) histones also act on TFs, so effects attributed to histone modifications may be mediated, at least in part, by changing the properties of TFs. Folic acid in the diet may not be acting through effects on methylation of DNA or chromatin proteins, as the folic acid receptor has been found to act as a TF (Boshnjaku et al. 2012), leading to downstream effects including STAT3 activation (Hansen et al. 2015). Furthermore, we tend to describe the pathway response to a cellular perturbation in terms of the new properties of transcription or transcriptional regulators in the genome, but if these new properties are, in fact, dependent on loci being selected by TFs, which themselves act downstream of cell signaling pathways, the primary pathway response could be said to be that acting to regulate the TFs involved and the secondary pathways those activated by the TFs. Once again, we find ourselves limited by too restrictive a view of what epigenomic assays should be telling us, by ignoring valuable information that doesn't fit into current concepts—in this case, the role of TFs. If we want to understand phenotypes and their molecular mechanisms, TFs and cell signaling need to be part of our thought processes.

DESIGNING AND EXECUTING AN EPIGENETIC ASSOCIATION STUDY

We have now looked into our souls and reflected on where we could have done better in epigenetic studies of human diseases. Now we can emerge with some guidelines that will help improve our approaches in future studies.

Biomarker or Mechanism?

The first step is to decide whether our goal is to develop a biomarker, or whether we want to understand mechanism. With a biomarker, you don't care why the needle is moving, you just need to know that when the needle moves (reflecting changes in DNA methylation, gene expression, or another epigenomic assay), there is diagnostic or predictive utility that is reliable. Obviously, it is desirable to know why an assay has an association with a phenotype when asking it to perform as a biomarker; if a DNA methylation change is diagnostic of a condition but is informative because of meQTLs that are present to a greater extent in Northern European white individuals, the test may perform poorly in people with other ancestries. We don't yet understand the mechanism that allows DNA methylation biomarkers of age to perform robustly (Hannum et al. 2013; Horvath 2013) nor the mechanism of DNA methylation changes in peripheral blood of those exposed to cigarette smoke. The smoking-associated changes in DNA methylation at the *GPR15* locus were initially thought to be due to a change in proportion of a specific T-lymphocyte subtype in blood (Bauer et al. 2015) but were later realized to be associated with the induction of expression of the gene across many T-lymphocyte subtypes (Bauer et al. 2019).

Study Design

If performing a study seeking to understand the molecular mechanism of a phenotype, then we have to tackle study design as a first step. Cross-sectional designs are convenient but are susceptible to reverse causation, as described earlier. That by itself is not necessarily a reason not to do the study—phenotypes can have complications: Obesity has effects throughout the body and not just in the fat-containing adipocytes, so when DNA methylation changes in peripheral blood as a consequence of obesity-related phenotypes (Dekkers et al. 2016; Richmond et al. 2016a; Wahl et al. 2017), we may be getting an insight into the mechanism of complications of the phenotype through such reverse causation effects. Otherwise the study will need to be designed as longitudinal, testing people before and after the onset of a phenotype, or during and following cessation of an exposure, or using analytical techniques like mediation analysis or Mendelian randomization (Richmond et al. 2016b).

Tissue Sampling

The cells mediating the disease really need to be tested when trying to understand the mechanism of a disease. Peripheral blood leukocytes are very good for biomarker studies, probably for the reasons described above concerning the changes in cell subtype proportions associated with diseases elsewhere in the body.

How do we get the cell type mediating a disease? It is a daunting prospect, requiring careful coordination with those performing biopsies or postmortem studies, probably across multiple institutions, with harmonization of sample collection and storage practices to allow comparisons to be as free as possible from technical variability. If this sounds forbidding, it is probably no different to the emotions felt by the early developers of GWASs, who came to realize that a study of tens or hundreds of individuals was not going to be informative, and that many thousands of study participants would need to be recruited. If the goal is to use epigenomic assays to find loci responding to extrinsic perturbations, including the polymorphic functional variants that cause responses to differ between individuals, then we may be able to exploit the advances being made to differentiate human pluripotent cells in vitro to different mature cell types and organoids. Rodent and other animal models should be a robust system for some time, a good example being the description of studies by Terrence (Terry) Furey and colleagues of eQTLs and chromatin QTLs (cQTLs) in livers, lungs, and kidneys of collaborative cross mice (Keele et al. 2020).

Knowing the Cell Subtype Composition of the Tissue Studied

The immediate future for this issue looks very promising with the development of powerful single-cell genomic approaches and the potential that these approaches can be used for deconvolution of cell subtype proportions from bulk assays of the same tissue type. Orthogonal approaches such as flow cytometry, quantitative histological imaging, or spatial transcriptomic approaches would only strengthen the certainty about the association of cell subtype changes with the phenotype being tested.

Revealing Interactions with DNA Sequence Variability

Unless working with isogenic mice, the genotype of the individual studied needs to be known when performing epigenetic studies. Typically this involves surveying the genome with a microarray or low-coverage whole-genome sequencing approach, identifying haplotypes and imputing variants not tested or detected that should be present in the haplotypes detected (Gilly et al. 2019). When detailed haplotype information is available, typically from Northern European populations, imputation can predict variants occurring in just 1/1000 individuals (McCarthy et al. 2016). This may not be enough—there is evidence that 90% of the heritability of gene expression in a study of lymphoblastoid cell lines is encoded in sequence variants that are present at a minor allele frequency of <0.01% in the gnomAD database (Hernandez et al. 2019). Low-resolution genotyping and imputation are therefore likely to miss most of the variants that influence transcriptional regulation, making a strong case for the need to perform deep whole-genome sequencing in all the individuals studied.

Choosing the Right Assay

DNA methylation is a great approach precisely because of the way that it varies continuously from 0 to 100%, thus revealing cell subtype effects. The big problem is that it is almost always used as a survey approach and may not be interrogating loci where events are occurring in your cell type of interest. Even whole-genome bisulfite sequencing is limited by the sparseness of CG representation in most of the genome, so that a given *cis*-regulatory site may be represented by only one or two sites capable of being methylated. The problem with chromatin assays is that they are substantially less quantitative, as discussed in Chapter 6. They are, however, more uniformly informative throughout the genome. A combination of a comprehensive DNA methylation assay combined with ATAC-seq to define loci of open chromatin would thus generate complementary information. Adding RNA-seq to help prioritize where regulatory changes are having effects on cell function is a further valuable complementary assay, whereas studies of chromatin looping can help to link *cis*-regulatory loci with specific target genes. The whole-genome sequencing data should be scrutinized to look for mosaicism of somatic mutations, which could influence the epigenomic and transcriptional data and cellular physiology.

Analytical Approaches

The analytical approaches to turn the data from these assays were described in Chapter 6. How they are integrated to identify the mechanism of disease has less to do with algorithmic approaches and more to do with being open to the ideas of "harvesting confounders," exploring the evidence for TF mediation of cellular events and possibilities like somatic mosaicism. There is no point in being prescriptive about specific approaches for integrative analyses, as many will be reasonable to perform.

Verification and Replication

Genome-wide assays are inherently noisy and imprecise. When loci are predicted by these assays to have differences in gene expression, DNA methylation, or chromatin states, it is necessary to verify these predictions using quantitative, orthogonal single-locus assays. Replication means that the predictions from a first cohort are borne out by testing of a second, independent cohort. Both steps are needed to allow confidence in epigenetic association studies.

CELLULAR GENOMIC STUDIES OF HUMAN DISEASE

The message from this chapter is that the field of epigenetic studies of human disease has painted itself into a corner through excessive purism. By requiring a model of cellular reprogramming, we have ignored the potential role of cell subtype proportion changes. By requiring mechanisms independent of DNA sequence variation to mediate effects, we have failed to grasp the opportunity to reveal genetic contributions to phenotypes, while our colleagues in the world of functional variant studies have made excellent progress. By downplaying the role of TFs, we are blinded to the contribution of cell signaling in cellular responses to extrinsic perturbations. In Chapter 5 we tried on for size the new term "cellular genomics" to describe an alternative way of doing these studies for those uncomfortable still calling this new, more inclusive approach epigenetics. Whatever it is called, it represents a way of merging epigenetics with the work performed by our GWAS and functional variant research colleagues.

Our goal is to understand phenotypes. Cellular reprogramming that is maintained after the removal of the provocation as a memory of the exposure is likely to mediate at least some phenotypes. Accounting for rather than excluding cell subtype and genetic influences will allow more mechanistic contributions of the phenotype to be detected, while leaving higher-confidence evidence supporting the cellular reprogramming hypothesis. These studies are going to have to be complex and difficult in terms of design, tissue sampling, cohort sizes, genomic assays, analyses, and interpretation, but the wealth of insights that will result promises to be immense.

RECOMMENDED FURTHER READING

Choudhuri A, Trompouki E, Abraham BJ, Colli LM, Kock KH, Mallard W, Yang M-L, Vinjamur DS, Ghamari A, Sporrij A, et al. 2020. Common variants in signaling transcription-factor-binding sites drive phenotypic variability in red blood cell traits. *Nat Genet* **52:** 1333–1345. doi:10.1038/s41588-020-00738-2

Evidence that the genetic variants associated with red blood cell traits are located in enhancers that contain motifs for transcription factors that respond to extracellular signals. The dysregulation of transcriptional regulators that would be seen in these cells would likely be due to the DNA sequence polymorphism mediating the phenotypes.

Smith GD, Ebrahim S. 2024. Mendelian randomisation at 20 years: how can it avoid hubris, while achieving more? *Lancet Diabetes Endocrinol* **12:** 14–17. doi:10.1016/S2213-8587(23)00348-0

Recognizing how Mendelian randomization was being misused in studies that include EWASs, George Davey Smith sets new guidelines for its use.

Weidemüller P, Kholmatov M, Petsalaki E, Zaugg JB. 2021. Transcription factors: bridge between cell signaling and gene regulation. *Proteomics* **21:** e000034. doi:10.1002/pmic.202000034

A review that argues for greater attention to be paid to the links between cell signaling, transcription factors, and gene expression.

CHAPTER 8

Cancer Epigenetics

CANCER EPIGENETICS IN 1927

The first use of the word "epigenetic" in the context of cancer was probably in 1927.[1] In a review of a book snappily entitled *Das Krebsproblem, Rückblicke und Ausblicke, Grund- und Scheinprobleme der Krebsforschung, -Behandlung und -Verhütung: Klinish-biologische Darlegungen von Dr. Alfred Greil* (*The Cancer Problem, Retrospectives and Outlooks, Basic and Illusory Problems of Cancer Research, Treatment, and Prevention: Clinical-Biological Statements by Dr. Alfred Greil*), the reviewer, Charles Kofoid, described how Dr. Greil proposed his own hypothesis about the biology of cancer and compared it with 30 other hypotheses (Kofoid 1927). As Kofoid wrote:

> He founds his hypothesis on neither exogenous nor endogenous determinants, visible or invisible cancer germs, but rather on fundamental processes of epigenetic evolution and variation. Cancer is thus not of parasitic origin, but rather a phase of the epigenetic evolution of cells.

We have to assume that the use of adjective epigenetic in 1927 refers to the embryological concept of epigenesis (Chapter 2) and that epigenetic evolution is therefore a way of describing how cells change their differentiation status. Histopathology was, at the time, the primary means of investigating the differences in neoplastic compared with normal cells. The terms used by pathologists to describe cancerous or precursor cells histologically were based on concepts of differentiation, with metaplasia, for example, describing a change in mature cell types within a tissue, and anaplasia describing a de-differentiation of cells. It is therefore not surprising to see cancer described using a term then in use by embryologists to describe cell differentiation, when Waddington was still a 22-year-old with a geology degree starting graduate school at the University of Cambridge.

---

[1] Thanks to Matthew Hall (2020, pers. comm.) from the National Center for Advancing Translational Sciences at the NIH for letting me know about this.

EPIGENETICS, CANCER, AND A EUGENICIST

Julian Huxley (1897–1975) left behind a complex record. Brother of the writer Aldous, he was a gifted and prolific communicator whose interests in biology and philosophy were broad and encompassed an enthusiasm for eugenics, which taints his legacy. He is credited with coining words like "clade" and "cline" and the use of "ethnic group" to replace the word "race."

He wrote the review in *Nature* of Waddington's *Principles of Embryology* (Waddington 1956a), strongly suggesting in the review that Waddington bring out a new edition of the book and call it *Principles of Epigenetics* instead, cheekily entitling the review "Epigenetics" (Huxley 1956). A year later Huxley was linking cancer biology with viruses and with epigenetics (Huxley 1957). What did he mean by this? As he wrote:

> I am adopting Waddington's (1956) useful term epigenetics to denote the analytic study of individual development (ontogeny) with its central problem of differentiation.

> The method by which tissues and organs differentiate in the course of normal development is at the moment the main blank space in biology's map. Waddington (1956) gives a valuable account of the present state of our knowledge and of our ignorance. Certain concepts, such as induction, evocation, competence, individuation and morphogenetic field, have introduced a preliminary degree of order into the subject, but we know little of the precise steps taken by epigenetic processes, of the biochemical factors involved, and above all, of what determines the replicable specificity of differentiated tissues.

He also made it explicit that he was talking about processes not involving genetic changes:

> The character of a tissue…is in most cases irreversibly determined, though the determination may be non-genetic (epigenetic).

He followed this up a year later with a book on the biology of cancer, prominently featuring discussions of epigenetics (Huxley 1958). He was especially focused on the abnormal differentiation of tissues (metaplasia) and de-differentiation of cells in a tissue. His perception of epigenetic processes in cancer was one based on cell differentiation and fate and saw cancer as a model for other types of diseases.

Whereas Aldous Huxley achieved immortality by appearing on the album cover of *Sgt. Pepper's Lonely Hearts Club Band* by The Beatles, Julian

Huxley's eugenic views are now recognized to be abhorrent. His legacy is multifaceted, however, including founding and leading the United Nations Educational, Scientific and Cultural Organization (UNESCO) and stimulating the formation of what became the World Wildlife Fund, a complexity well-described by Weindling in his thoughtful appraisal in 2012 (Weindling 2012). Huxley's influence in biology waned over time; just as Waddington's epigenetic landscape idea dwindled in its influence over subsequent decades, so also did Huxley's idea of framing the differentiation abnormalities in cancer as epigenetic.

DNA METHYLATION AND THE MODERN REBIRTH OF CANCER EPIGENETICS

Melanie Ehrlich described to me her postdoctoral fellowship work with Julius Marmur at the Albert Einstein School of Medicine as being on "a weird, naturally occurring DNA, biochemically speaking." She studied the DNA of bacteriophage SP-15, which has "bizarre, naturally programmed DNA base modifications" (Marmur et al. 1972; Ehrlich 2015) giving it a strikingly low melting temperature, reflecting its nonphosphodiester backbone phosphate (Ehrlich and Ehrlich 1981; Kropinski et al. 2018). As a counterpoint, she also studied bacteriophage XP-12 DNA, which was found to have the highest melting temperature for any naturally occurring DNA and a complete replacement of cytosine by 5-methylcytosine (Ehrlich et al. 1975). She also describes the emerging realization at the end of her graduate work (at the State University of New York at Stony Brook) that methylation of a thymine at position 5 is critically important for binding of the *lac* repressor (Lin and Riggs 1971), suggesting a regulatory role for these kinds of DNA modifications.

Ehrlich (2020, pers. comm.) also recounts the generosity of colleagues in Taiwan as a critical contribution to her early discoveries:

Among the major early experiences in my career, one stands out. When I was a new Assistant Professor at Tulane Medical School, I found an article by a Taiwanese group about a strange phage, XP12, which infects *Xanthomonas oryzae*, isolated from the water in a paddy field in Taiwan, that has 5-methylcytosine replacing all of the genomic cytosine (Kuo et al. 1968). I contacted the senior author of that paper, Dr. Tsong-Teh Kuo, who generously sent me the phage and its host. Our culture of the host became phage-resistant within a year, leaving us unable to progress with our work. I re-contacted Dr. Kuo, who was kind enough to send me another sample of the host, an act of generosity that allowed me, as a junior faculty member, to generate the results for my early independent publications.

Ehrlich also describes as influential (Ehrlich 2005) articles from Boris Vanyushin and colleagues, working in Moscow State University in what was then the Soviet Union. Vanyushin was showing some of the first evidence for differences in DNA methylation levels between species and between tissues within a species (Vanyushin et al. 1973). Ehrlich went on to publish a detailed study unequivocally demonstrating significant differences in DNA methylation among human tissues (Ehrlich et al. 1982). Also percolating in the mixture of intriguing ideas about DNA methylation in the 1970s was the work of John Pugh and Robin Holliday (Holliday and Pugh 1975) and Art Riggs (Riggs 1975) (see Chapter 3). For example, Ehrlich cites their discussion of "hypotheses about maintenance versus *de novo* methylation and the involvement of this methylation in differentiation and X chromosome inactivation" as one of the forces shaping the field during this period (Ehrlich 2005).

In the late 1970s, Andy Feinberg, a postdoctoral fellow at University of California San Diego (UCSD), was replicating work from 1959 by John Tyler Bonner from Princeton University. Bonner was an early leader in the study of the slime mould *Dictyostelium discoideum*, a fascinating amoeba that lives as a unicellular organism while it has a source of food (bacteria) available, but when the food supply is exhausted the individual cells aggregate and form a multicellular organism organized into posterior spore cells and anterior stalk cells. Removal of part of the multicellular structure leads to its complete regeneration, but what Bonner showed was that an engraftment of cells from, for example, the anterior unit into the posterior position caused the engrafted cells to migrate preferentially to the part of the structure from which they were originally derived. This implied a nongenetic cellular memory, which was confirmed by the UCSD group using more modern approaches (Feinberg et al. 1979).

Andy Feinberg (2019, pers. comm.) found himself at a career crossroads:

After I finished the postdoc, I didn't know what I wanted to do because I was originally "programmed" to do neuroscience and maybe pair that with neurology clinically, but I thought both were too opaque to progress in those days. I spent a year working for the U.S. Government in the National Health Service Corps in the inner city of Baltimore, knowing I'd go back to science but wanting to do some good for people directly for a little while. The place I worked had a number of public health types and I met my wife that year too; she was a doctoral student at the School of Public Health. I decided to do an MPH focusing on quantitative areas, plus tuition was free and they paid me a stipend. During the year by sheer chance, I happened to wander by a room in the hospital where Donald Coffey was giving a lecture (with food, importantly) on cancer, so I listened. He talked about plasticity of cancer cells in their ability to switch states, so I went up to him afterward and said, "Dr. Coffey, that's epigenetics. I worked on that in slime molds." He wanted to introduce me to Bert Vogelstein who was also working on a related species, *Physarum*, but was interested in cancer research. So, I gave a little seminar on Dicty [*Dictyostelium*], and he invited me to do a second postdoc on cancer, where I first got to test my epigenetic hypothesis by looking at DNA methylation. That's how it started, serendipity at many levels.

Robin Holliday's review speculatively linking DNA methylation and cancer (Chapter 3) had been published several years previously (Holliday 1979) around the same time that decreased DNA methylation was being revealed in chemically induced (Lapeyre and Becker 1979) or spontaneous (Walker and Becker 1981) liver tumors in rats. McGhee and Ginder were demonstrating decreased DNA methylation at the chicken β-globin gene specifically in the erythroid cells in which the gene is expressed (McGhee and Ginder 1979). By 1981, Vanyushin had published his finding that DNA methylation was depleted in cancer cells (Romanov and Vanyushin 1981), with the excitement of the era captured nicely in a comprehensive review from Melanie Ehrlich that year (Ehrlich and Wang 1981).

In 1983, the two influential publications that have been taken as the starting point of the modern era of cancer epigenetics were published. Feinberg and Vogelstein studied primary samples from patients who had not yet received chemotherapy or radiotherapy, focusing on colon and lung carcinomas, comparing them with nonmalignant adjacent tissue, and targeted three specific loci (the genes for human growth hormone, γ-globin, and α-globin) rather than global DNA methylation content. Three of the four colon cancer samples and the one lung cancer sample showed decreased DNA methylation at the loci tested (Feinberg and Vogelstein 1983a). Within a few months Ehrlich and her group had published a study of 103 tumors and 43 normal tissues and measured global DNA methylation, showing decreased levels especially in the metastatic samples studied (Gama-Sosa et al. 1983). These two studies emphasize the value of multiple, orthogonal tests of the same question, reinforcing each other's conclusions and creating a foundation for the entry of others into the field.

Although these genome-wide DNA methylation changes in cancer were intriguing, they did not by themselves provide a link to mechanism of neoplasia. Demethylation of growth hormone and globin genes did not shed light on how the cancer cell could have acquired its uncontrolled growth. Feinberg and Vogelstein performed another study, this time testing seven colon carcinomas and one small cell carcinoma of the lung, using *c-Ki-Ras* (*KRAS*) and *c-Ha-Ras* (*HRAS*) probes to study the DNA methylation patterns at these known oncogenes. As in their prior study, they observed hypomethylation in tumor compared with normal tissue, in six of the eight tumors at the *HRAS* locus and two tumors at the *KRAS* locus (Feinberg and Vogelstein 1983b). They noted recent reviews (Razin and Riggs 1980; Felsenfeld and McGhee 1982) that were linking local DNA methylation with gene inactivation and speculated that the loss of DNA methylation at specific critical loci could cause their activation and help to transform the cell.

THE TWO CAREERS OF RUTH SAGER

In Chapter 3 we read how John Pugh was inspired in 1973 to cycle into Central London to hear Ruth Sager (1918–1997) speak about her work on the restriction–modification system in *Chlamydomonas*. Sager studied this alga while on faculty at Hunter College in New York, but in 1975 she was recruited to Harvard Medical School. By 1978 she was fusing tumorigenic and nontumorigenic cells from different species, going on to deduce that not all mutations in cancer were dominant (gain of function of oncogenes), but that others were recessive (loss of function of tumor-suppressor genes) (Sager 1986). She focused on the Maspin (*SERPINB5*) gene which is down-regulated in invasive breast cancer (Sager et al. 1996) and which represented one of the first human genes to be shown to have promoter DNA methylation changes associated with its tissue expression patterns (Futscher et al. 2002) as well as in malignant transformation (Domann et al. 2000). Although possibly not recognized today for her historical role in the field of cancer epigenetics, Feinberg describes her as one of the most influential voices of the era.

A colleague who helped provide Feinberg and Vogelstein with samples for their studies was Steve Baylin, whose career had been focused on figuring out why tumors from nonendocrine tissues could develop the property of secreting hormones (Baylin and Mendelsohn 1980). With the hypothesis that he would see demethylation of the calcitonin (*CALCA*) promoter in the medullary thyroid carcinoma and lung cancer cell samples in which calcitonin expression was found, he tested the DNA methylation at the 5′ end of the gene using the same restriction enzyme and Southern blot approach used by Feinberg and Vogelstein. In the medullary thyroid carcinoma cells, derived from calcitonin-producing cells of the thyroid, he did indeed see such a loss of DNA methylation, but the majority of the lung cancer samples revealed the unexpected finding of increased DNA methylation (Baylin et al. 1986).

The idea that global hypomethylation in cancer could be accompanied by individual loci gaining DNA methylation provoked the idea that gene silencing through DNA methylation could be the equivalent of a loss-of-function mutation. For such a loss-of-function event to predispose to neoplasia, the target locus should be a tumor-suppressor gene. In 1989, the group of Bernhard Horsthemke studied DNA from 21 freshly obtained retinoblastoma samples, testing DNA methylation at a CpG island at the 5′ end of the retinoblastoma (*RB1*) gene, finding one sample to have increased DNA methylation in a pattern consistent with one of the two alleles at the locus being affected (Greger et al. 1989). The end of the manuscript is admirably balanced:

> We suggest that hypermethylation of tumor suppressor genes is an infrequent but potential event in the development of human neoplasia, and that loss of hypermethylation may be involved in the spontaneous regression of some tumors.

RB1 encodes the retinoblastoma-associated protein, which has a number of functions, including interactions that help to maintain heterochromatin (Isaac et al. 2006), but it is mostly appreciated as the first tumor-suppressor gene to be discovered (Berry et al. 2019). With Horsthemke's observation of what appeared to be monoallelic hypermethylation of the *RB1* promoter in a retinoblastoma sample, we had evidence for a new paradigm in cancer— the change in regulation of genes acting in parallel to the mutation of genes and contributing to the neoplastic transformation of cells.

When people saw the result of the DNA methylation studies of the *RB1* promoter, why did they assume that the increased DNA methylation in one case was the oncogenic event, when 20 other cancer samples did not show this pattern? It should be remembered from Chapter 6 that Adrian Bird had already defined CpG islands in terms of their property of lacking DNA methylation. In an entertaining 2009 interview (Gitschier 2009), Bird reveals how 1973 was a pivotal year in his training, the same year that John Pugh had his epiphany listening to Ruth Sager, the undergraduate Hunt Willard was being introduced to X inactivation, and Art Riggs was discovering restriction–modification during his sabbatical. In 1973, Bird was doing postdoctoral training in Zurich in the laboratory of Max Birnstiel, isolating a gene the only way possible at that time, by using density gradient centrifugation to isolate frog ribosomal DNA. On sabbatical in the same laboratory were Hamilton (Ham) Smith and Edwin (Ed) Southern. Smith decided to make the *Haemophilus parainfluenzae* II (HpaII) restriction enzyme, which Bird then used to digest the ribosomal DNA, testing the result with the new technique of agarose gel electrophoresis combined with Ed Southern's recently developed approach to transfer the DNA within the gel to a membrane, since then known eponymously as the Southern blot. It was already known that the extrachromosomal, amplified ribosomal DNA was different from the chromosomal ribosomal DNA in terms of 5-methylcytosine content, which was present only in the chromosomal DNA (Dawid et al. 1970). Consistent with this prior finding, Bird found that only the extrachromosomal DNA could be digested with HpaII, revealing the enzyme's dependence on DNA methylation (Bird and Southern 1978). By digesting vertebrate genomic DNA with HpaII, a population of small fragments could be identified (Cooper et al. 1983), which Bird described as "a horrible blob" (Gitschier 2009), but which he then started to describe more kindly as HpaII tiny fragments (HTFs). These loci had to have two properties: multiple clustered HpaII sites (5′-CCGG-3′), and a lack of methylation of DNA at these sequences. Bird initially described these as "HTF islands" (Gitschier 2009), but a year later Marianne Frommer, who had been on sabbatical in his laboratory, described them as "CpG islands" instead in a paper with Margaret Gardiner-Garden that defined CpG islands in terms of base composition properties (Gardiner-Garden and Frommer 1987). Although Bird and Frommer did not discuss the name change, Bird decided to keep things simple (2022, pers. comm.):

> Later on I decided—on a whim really—that her name was probably better and we switched for all our future papers. With hindsight I am sorry we didn't stick with HTF, but it's not a big issue.

In this context, it should now be apparent that the expectation at the time was that the default state of the *RB1* locus should be one of hypomethylation, which was also supported by Horsthemke's control study of peripheral blood leukocyte DNA in the retinoblastoma patients, which also showed the expected pattern of hypomethylation (Greger et al. 1989).

The emerging picture was one of global loss of DNA methylation, but with individual CpG islands within the genome paradoxically gaining DNA methylation. Steve Baylin took the approach of testing multiple loci in parallel in the same cancer cell lines and found concordantly increased DNA methylation at the promoters of *CALCA* and *HRAS* and a CG-rich probe also mapping to the p arm of Chromosome 11 (de Bustros et al. 1988). This prompted the idea that the hypermethylation of CpG islands might be happening genome-wide in cancer. It was a technological breakthrough a decade later led by Baylin and Jean-Pierre Issa that allowed a more general survey of the genome, methylated CpG island amplification (MCA), revealing numerous loci at which DNA methylation was being acquired (Toyota et al. 1999b). Not all tumors showed this pattern, however, prompting the description of this subset as having the "CpG island methylator phenotype" (CIMP) (Toyota et al. 1999b).

The causality of DNA hypermethylation in gene silencing was established in part using new DNA methyltransferase inhibitors such as 5-aza-2′-deoxycytidine, showing that demethylation as a result of this treatment allowed reactivation of the gene with the hypermethylated CpG island promoter. How these DNA methyltransferase inhibitors were discovered is another story worth telling to illustrate how discoveries were made in the early days of the field of cancer epigenetics.

THE DISCOVERY OF PHARMACOLOGICAL INHIBITORS OF DNA METHYLATION

Possibly the graduate student story to end all graduate student stories is that of Peter Jones. Born in Cape Town, South Africa, but raised in Rhodesia (now Zimbabwe), he was the only one of his 30 cousins to go to college (Viegas 2016), attending the University of Rhodesia. His thesis work involved exposing cells in vitro to chemotherapeutic agents, showing that these agents could by themselves cause cellular transformation. The geographical isolation of his University required that he make his own fetal calf serum, and he could only read journals delivered by surface mail 6 months after publication (Jones 2011).

Things looked to have reached a nadir in March 1972 when his advisor Dr. Joseph Taderera, a supporter of Robert Mugabe, was arrested in front of Jones and the whole laboratory for smuggling AK47 assault rifles into the country. Jones feared that his future potentially involved being sent into the bush to fight Taderera and his AK47s (Jones 2011).

Jones was inspired by Taderera as a scientist: "Joe had trained with Robert Auerbach, a pioneer in research on blood vessel formation and cell differentiation, and Howard Temin, a Nobel Prize–winning geneticist and virologist, in Madison, Wisconsin, and he taught me tissue culture and how to dream big," Jones says. "The fact that we grew mammalian cells and conducted carcinogenesis experiments in the middle of Africa, where the next closest university was 500 miles away, was a testament to his resourcefulness and tenacity" (Viegas 2016).

Jones, Taderera, and head of the department Arthur Hawtrey had just had a paper accepted to the *European Journal of Cancer* (Jones et al. 1972). One of the reviewers decided to write to this young student in Rhodesia (Jones 2011):

> The letter, which is the only one I have ever received from an anonymous reviewer, was from Bill Benedict in the US, complimenting me on our work describing how DNA-synthesis inhibitors used to treat cancer could cause oncogenic transformation themselves. By breaking the wall of reviewer anonymity, he gave me the chance of a lifetime.

Benedict offered Jones a postdoctoral fellowship in his laboratory, where a serendipitous event occurred (Jones 2011):

> My wife and I arrived in Los Angeles in 1973 with a five-month-old baby and three green cards to join Bill at Children's Hospital, which was located on the mythical Sunset Boulevard. I continued work on chemotherapeutic drugs and this is when the scientific turning point happened without warning. We treated a mouse embryo cell line with the drugs and then kept the cells as monolayers for six weeks to see if they turned into foci of cancer cells. Maintaining cells for that extended period required twice weekly medium changes, which meant frequent contamination by moulds, yeasts and bacteria. One Monday morning, while we were changing media, a large mould seemed to be growing in a dish exposed to 5-azacytidine (5azaC), a new drug from Czechoslovakia. When I examined the presumed mould, I was amazed to see a huge syncytium of multinucleated cells visible to the naked eye. It was almost an Alexander Fleming type moment, because the 'contamination' represented a total switch of phenotype into muscle. I was elated because we seemed to have the first drug capable of completely reprogramming cells—maybe the 'philosopher's stone' of developmental biology.

Nice story, but what has this to do with epigenetics? Jones again (Jones 2011):

We knew that 5azaC had to get into DNA to cause reprogramming, but what was the molecular mechanism? I accepted a faculty position at USC back in LA in 1977 and began work with my second student Shirley Taylor to find out. We had not the vaguest idea how it might work until I was being interviewed by Bob Stellwagen for what I thought was an unnecessary hurdle required to get a joint appointment in the Department of Biochemistry. He listened to my description of 5azaC and then coolly asked, "Have you thought of DNA methylation?" I, in my ignorance, answered, "What's that?" In four weeks we had the answer, and as a result of that question posed during a painful interview process, we showed that 5azaC was a potent inhibitor of methylation and linked DNA methylation and differentiation for the first time.

Jones recounts how Hal Weintraub, at the Fred Hutchinson Cancer Center in Seattle, called and asked Jones to collaborate on the cloning of a putative master regulator of muscle cell differentiation. Jones, in what he acknowledges was one of the biggest mistakes of his career, declined because he had one of his own students already working on such a project. Weintraub called a year later and said, "We've got it," referring to the cloning of the first determination gene to be characterized, *MyoD1*. Jones (2020, pers. comm.) describes Weintraub as a true pioneer in epigenetics research, having developed the DNAse I assay for chromatin accessibility among many other achievements before his untimely death from brain cancer.

All of this serves to illustrate how the field of cancer epigenetics came into being in the 1980s, setting a course still being followed today. The definition of "cancer epigenetics" by Baylin and Jones (2011) helps to reveal the mindset that drove and continues to drive this field:

> ...myriad abnormalities that are based on somatically heritable alterations that are not due to primary DNA sequence changes.

This idea of epigenetic properties being both heritable (through cell division) and not due to DNA sequence changes represented a mainstream view of what epigenetics means in terms of cell properties and remains current in the field of cancer epigenetics.

Cancer epigenetics is especially exciting because it represents a research focus in epigenetics in which observations about the dysregulation prompt specific therapeutic interventions. However, just as we have had to rethink some of our assumptions about epigenetics and human diseases in general (Chapter 7), it is a useful exercise to frame the discussions in this chapter in terms of some of the more updated insights in the field of epigenetics in general, with the goal of allowing a more productive focus for cancer epigenetics research.

WHY DOES DNA METHYLATION REDISTRIBUTE IN THE CANCER EPIGENOME?

We have known about global loss of DNA methylation in cancer since the early 1980s and the CpG island methylator phenotype since at least 1999. That has given us decades in which to puzzle out the mechanism of each, with the possibility that the two phenomena share a common underlying mechanism. The results are surprising.

Global DNA Hypomethylation

We are still unsure why global loss of DNA methylation occurs in cancers. It is worth restating that not all cancers undergo global loss of DNA methylation. It is also worth noting that the DNA methylation levels in some noncancerous cells (like placental tissue), are decreased and look strikingly like the hypomethylation of cancer, prompting Wendy Robinson (Robinson and Price 2015) to speculate that there may a mechanistic connection:

> Cancer cells may exploit pathways normal in placental development for invasion of surrounding tissue and avoidance/modulation of the host/maternal immune system.

Some cancers have hypomethylation of DNA because of major *trans* effects of mutations in genes that regulate DNA methylation. These include mutations of DNA methyltransferases (Lee et al. 2020), which makes sense as their loss should cause the loss of maintenance or de novo DNA methylation. Oddly enough, although TET enzymes normally mediate the removal of DNA methylation, and their mutation might consequently be expected to lead to increased DNA methylation, their mutation in cancer has been found by Anjana Rao to be associated with a perplexing global loss of DNA methylation (López-Moyado et al. 2019).

Mutations don't tell the whole story, though. Cancers with a global loss of DNA methylation do not universally have mutations in genes for DNMT or TET enzymes, which has led to speculation about alternative mechanisms for the hypomethylation, including mutations of other genes whose protein products partner with DNA methyltransferases, or influences extrinsic to the cell like diet or infection (Ehrlich 2009).

One observation of note is that the loss of DNA methylation is not global but preferentially affects specific regions of the genome. Work from researchers at Johns Hopkins University has been consistent in showing that cancer hypomethylation preferentially occurs in blocks of DNA of hundreds of kilobases in magnitude (Hansen et al. 2011; Timp et al. 2014), with a similar pattern found in B lymphocytes following Epstein–Barr virus infection and immortalization (Hansen et al. 2014). They describe how some of these large blocks are also noted to be lamin-associated domains (LADs) or the targets of H3K9 methylation, suggesting a link in regulation. Another potential property of these domains is late replication, which defines regions that have been found to undergo partial demethylation in cancer and with age (Zhou et al. 2018), suggesting that mitotic stability of DNA methylation during cell division is decreased in these regions and could account for at least some of the demethylation in transformed cells. Another potential contributor is the Dnmt3a DNA methyltransferase. Rudolf Jaenisch and Frank Lyko used a mouse model of *Kras* mutation to induce lung tumors and showed the selective loss of DNA methylation in LADs, but there was progression to a global loss of DNA methylation throughout the genome when the mice were also made Dnmt3a-deficient (Raddatz et al. 2012). Their conclusion was that the active (non-LAD) fraction of the genome was protected by the action of Dnmt3a and speculated that the regions targeted for loss of DNA methylation in cancer (Raddatz et al. 2012)

> …represent genomic regions that have lower functional requirements for faithful methylation maintenance and that maintenance methylation in these domains is generally inefficient.

It is generally assumed that hypomethylation of the cancer epigenome in some way contributes to oncogenesis. Two mechanisms that are frequently invoked are mobilization of transposable elements and increased chromosomal breakage. There is evidence for somatic retrotransposition events in cancer (Lee et al. 2012), but the number of events appears to be small (usually 0–15 per cancer sample), and it is not yet clear whether this number significantly exceeds what might be expected in noncancerous somatic cells. Furthermore, there is now emerging evidence that activation of transposable elements may have the beneficial effect of increasing the immunogenicity of the cancer cell (Kong et al. 2019b; de Cubas et al. 2020), which may serve as some counterbalance to the pro-neoplastic mutagenic effects of retroelement activation. Increased chromosome breakage has long been associated with hypomethylation of DNA, prompted in part by the chromosome instability that occurs with *DNMT3B* mutations in the human ICF (immunodeficiency, centromeric region instability, facial anomalies) syndrome (Hansen et al. 1999) and correlations in colorectal carcinoma (Rodriguez et al. 2006), although associations based on the use of DNMT1 inhibitors like 5-aza-2′-deoxycytidine (5-Aza-dC) need to be interpreted with caution, as the 5-Aza-dC integrates into DNA and forms adducts with DNMT1, which leads to chromosome instability (Maslov et al. 2012).

It is noteworthy that the only clear mechanisms for loss of DNA methylation in cancer are mutational, whether of TET or DNMT3A enzymes, and not due to some sort of autonomous decision by transcriptional regulators to reprogram the cell. The mechanistic insights into the related event of hypermethylation of CpG islands warrant their own separate discussion.

CpG Island Hypermethylation

Two questions arise when wondering about the mechanism of CpG island hypermethylation in cancer: Why do the CpG islands acquire DNA methylation, and how is this acquisition selectively targeted to these loci? Some effects appear to be due to *cis*-acting DNA sequence variants within the CpG island that increase the chance of local DNA methylation, as has been found for the *MLH1* promoter (Hitchins et al. 2011). To get multiple CpG islands to acquire DNA methylation probably requires a *trans*-acting influence, as it is difficult to imagine how these loci could all consistently acquire somatic mutations as a coordinated group. A great example of a *trans* effect is from the mutations of the *IDH1* or *IDH2* genes, which lead to accumulation of D-2-hydroxyglutarate, which in turn inhibits TET enzymes and appears to cause the CIMP found in gliomas and acute myeloid leukemias (Weisenberger 2014).

The use of an RNA interference screen led Michael Green and his group to identify the MAFG protein as a possible mediator of CIMP (Fang et al. 2014). The MAFG protein binds with a heterodimeric partner BACH1 to the CpG island, subsequently recruiting CHD8 and the DNMT3B DNA methyltransferase, which leads to local deposition of DNA methylation (Fig. 8.1). Peter Laird and colleagues had shown as early as 2006 that there was a strong correlation between the V600E missense mutation of *BRAF* and CIMP in colorectal carcinoma (Weisenberger et al. 2006). Green's

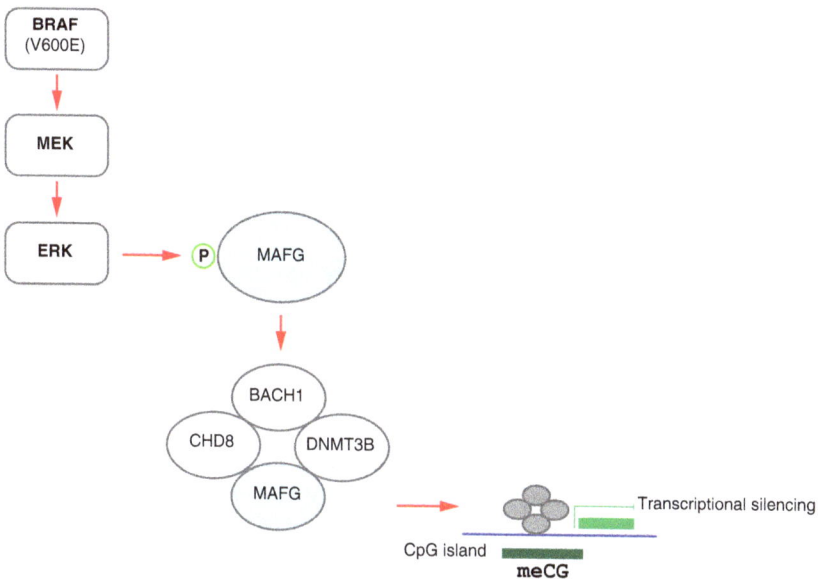

Figure 8.1. A mechanism for the CpG island methylator phenotype (CIMP). The primary event is mutation of *BRAF* (the oncogenic V600E missense variant), which activates MEK and ERK signaling, leading to the phosphorylation of the MAFG transcription factor, a modification that prevents its ubiquitinylation and degradation. The increased availability of MAFG allows it to bind to target CG-rich sequences at CpG islands, recruiting BACH1 and CHD8 but also the DNA methyltransferase DNMT3B, which methylates the DNA of the CpG island.

group found that the BRAF V600E mutation, by its effect on MEK signaling, led to increased phosphorylation and stability of MAFG, increasing the amount of MAFG protein present in the cell and its availability to bind to CpG island promoters (Fang et al. 2014). The specific sequence to which the MAFG transcription factor binds is the 5′-TGC TGA C TCA GCA-3′ palindrome, the Maf-recognition element (MARE) (Yamamoto et al. 2006), which is surprisingly devoid of CG dinucleotides and only ~54% (G+C)—not what might be expected to direct an event to the most CG/(G+C)-rich sequences in the genome.

Although there are probably other, undiscovered mechanisms of CIMP, the *IDH* and *BRAF* examples show how mutations of two types of genes working through very different mechanisms can have the convergent outcome of increasing DNA methylation at CpG islands. The challenge now is not only to find other mutations that could be leading to CIMP but to determine whether there exist any nonmutational causes for CIMP. At present, the most parsimonious explanation for acquisition of DNA methylation in cancer at an individual site is through a *cis*-acting mutation, and in a CIMP manifestation at multiple sites through *trans*-acting mutations in genes like *IDH* and *BRAF*.

Dysregulation of Other Epigenetic Regulators in Cancer

Recognizing that DNA methylation is only one of numerous molecular genomic regulatory mechanisms described as epigenetic, and the genomic distribution of one is often correlated with the patterns of others (Chapter 6), it is unsurprising that studies of other epigenetic regulators have revealed cancer-specific changes. Chromatin studies have demonstrated extensive alterations of histone modifications in different cancers (Sawan and Herceg 2010), with correspondingly major changes in the chromatin accessibility landscape, mostly targeting enhancers (not promoters) in the cancer cell (Corces et al. 2018). Transcription is likewise altered, with evidence for the emergence of novel promoters (Demircioğlu et al. 2019). In Chapter 6 we were introduced to the idea that DNA methylation exists to suppress the expression of transposable elements (Yoder et al. 1997). It should not be surprising that in cancer, which can be associated with global loss of DNA methylation, the expression of transposable elements can increase, although this can also be associated with local rather than global loss of DNA methylation at the transposable elements (Kong et al. 2019b).

MUTATION AS THE PRIMARY EVENT CAUSING DISTINCTIVE CANCER EPIGENOMES

How much of the epigenomic dysregulation in cancer should we attribute to mutations, rather than "somatically heritable alterations that are not due to primary DNA sequence changes" (Baylin and Jones 2011)? That is not yet an easy question to answer, although it is probably reasonable to be cautious and assume that most global changes and many local changes are due to somatic mutations in the cancer cell. If mutation effects can be excluded, then what remains are changes that are higher confidence for fitting the autonomous, nongenetic definition of Baylin and Jones. There is a lot of evidence to support mutational mechanisms for both global and local changes in epigenetic regulation of transcription that has accrued with the increased application of sequencing in different tumor types.

Mutations of Transcriptional Regulators

By 2011, it was clear that "the cancer epigenome" was not only distinctive in ways that extended beyond DNA methylation changes, but many changes were attributable to mutations in genes regulating DNA methylation and chromatin states (Baylin and Jones 2011). Numerous different transcriptional regulators have been found to be mutated in cancer, not just the genes encoding enzymes regulating DNA methylation, but also enzymes modifying histones post-translationally, the proteins that bind to and "read" these post-translational modifications, the enzyme complexes that remodel the positioning of nucleosomes in the genome, and even histone proteins themselves (reviewed by Shen and Laird 2013; Baylin and Jones 2016).

Cancer genomes have many mutations, so to find that a specific group of genes has mutations is difficult to assess in terms of biological significance. An analysis that tested whether mutations in genes encoding chromatin regulatory factors overrepresents driver mutations found supportive evidence in multiple tumor types (Gonzalez-Perez et al. 2013). It has also been noted that disruption of a specific chromatin modification like H3K27me3 can be the result of mutations in multiple different genes (Shen and Laird 2013), indicating that individually rare mutations can have convergent outcomes in terms of transcriptional regulation and epigenomic alterations. This increases the challenge in linking mutations with epigenetic changes unless very large cohorts are studied, revealing recurrent mutations in the same diverse genes in small proportions of patients.

An interesting mutational mechanism causing targeted epigenomic changes in cancer arises from the activities of fusion genes. Cancer mutations can include chromosomal rearrangements that place the beginning of one gene upstream from part of the protein-coding sequence of a second gene, which can lead to a new, fusion protein being expressed in a cancer cell. Most fusion genes in cancer juxtapose coding sequences without creating a frameshift and strikingly feature the involvement of transcription factors (Jang et al. 2020). When one of the two components of the new protein has the ability to bind to specific DNA sequences or chromatin modifications, the other protein component finds itself targeted to unaccustomed regions of the genome. Examples are reviewed by You and Jones (2012), including the *BRD4* gene encoding a bromodomain protein that binds to ("reads") acetylated lysines and the *MLL* gene that acts as a lysine methyltransferase to create the H3K4me3 modification. By acting in new regions of the genome, fusion genes can create local transcriptional regulatory changes that contribute to the development of cancer.

Mutations Causing Metabolic Effects with Secondary Epigenetic Effects

Another way of creating genome-wide changes in chromatin states or DNA methylation is to undergo mutation of a gene critically important for cellular metabolism, with secondary effects on epigenetic regulators.

Probably the most striking example in this category is the mutation of genes encoding the isocitrate dehydrogenases 1 and 2 (*IDH1*, *IDH2*; the focus of an outstanding 2012 review by Yang et al. [2012]). Somatic mutations in these genes had a distinctive pattern—rarely germline, always heterozygous, and almost always a missense mutation in the enzyme's active site, consistent with a dominant gain of function. These mutations occur predominantly in gliomas, acute myelocytic leukemia, chondrosarcomas, and cholangiosarcomas and change the function of the enzymes, which can no longer convert isocitrate into α-ketoglutarate, but instead convert α-ketoglutarate to D-2-hydroxyglutarate (Dang et al. 2009). The latter is structurally similar to α-ketoglutarate but blocks its ability to act as a co-factor for certain dioxygenases, which include enzymes mediating histone demethylation and the TET enzymes that generate 5-hydroxymethylcytosine from 5-methylcytosine (Xu et al. 2011). The effects of these *IDH* mutations differ by tumor type, with more acquisition of DNA methylation in gliomas than other tumors (Unruh et al. 2019). Histone methylation is increased in cells with *IDH* mutations, as would be expected in a situation of inhibition of demethylases (Xu et al. 2011). Another target for the D-2-hydroxyglutarate "oncometabolite" are the prolyl hydroxylases (PHDs), which regulate hypoxia-inducible factors 1α and 2α (Xu et al. 2011), transcription factors important in mediating the hypoxia response. Mutations of the *L2HGDH* gene encoding the L2 hydroglutarate dehydroxygenase enzyme occur in kidney cancers and lead to the accumulation of L-2-hydroxyglutarate, which acts like D-2-hydroxyglutarate to block α-ketoglutarate-dependent dioxygenases (Shim et al. 2014). It is a reasonable guess that other mutations and oncometabolites remain to be discovered that could have comparably promiscuous effects on DNA methylation, chromatin modifications, and transcription factors. For example, a currently intriguing observation has been the association between the metabolism of folic acid and the bromodomain proteins that bind to acetylated histones (Sdelci et al. 2019).

Mutations of DNA Causing Local Changes of Epigenetic Regulation

In Chapter 7 we noted the interplay between DNA sequence variation at transcription factor binding sites and local changes in DNA methylation, chromatin accessibility, nearby gene expression, and other molecular genomic phenotypes. That discussion was focused on variation between individuals—not on variation of DNA sequences in different cell types within the body as occurs with somatic mutations in cancer. In theory, therefore, somatic mutations may not be limited to protein-coding sequences of epigenetic transcriptional regulatory genes but could be acting at regulatory sequences where transcription factors bind. Is there any evidence for this in cancer?

Of course there is. (Otherwise, this would be a short and somewhat wasteful section.) An excellent paradigm for this area of research emerged in two independent reports in 2013, each describing mutations in melanoma samples of the promoter of the telomerase reverse transcriptase (*TERT*) gene (Horn et al. 2013; Huang et al. 2013). These *TERT* mutations lead to the de novo formation of a new binding site for the GA-binding protein (GABP) transcription factor, strengthening the activity of the promoter, and contributing to cellular immortalization by increasing telomerase activity. In 2014, a similar picture emerged for T-cell acute lymphoblastic leukemia (T-ALL), in which a somatic mutation ~10 kb upstream from the *TAL1* gene created a new binding site for the MYB transcription factor and led to the formation of a powerful new enhancer strikingly increasing the expression level of the TAL1 oncogene (Mansour et al. 2014).

To assess whether this was a more general phenomenon in cancer, Mike Snyder's group studied data from The Cancer Genome Atlas, testing 436 samples, and found somatic mutations to be enriched in regulatory elements and at transcription factor binding motifs, with functional effects on enhancer activity (Melton et al. 2015). It remains to be seen how many of these sequence changes in regulatory elements are passenger mutations, incidental to and causing the development of the tumor. New approaches are being developed to address this question that promise to provide more clarity about the pathogenetic role of mutations at regulatory elements (Liu et al. 2020).

If Its Epigenetic Changes Are Due to Mutations, Is Cancer Not an Epigenetic Disease?

All these molecular genomic changes are occurring in cells that are distinctive for having accumulated somatic mutations. As the field of cancer epigenetics was establishing itself in the 1980s and 1990s, we had limited means of testing tumors for mutations and certainly did not appreciate the extent to which DNA sequence variability influences transcriptional regulators like DNA methylation and regulators of chromatin states.

When we first observed the overall decrease in DNA methylation in tumor cells, and with a selected subset of loci paradoxically acquiring DNA methylation, our natural response was to assume that this was "epigenetic"— something happening independently of and acting upon DNA sequence. Baylin and Jones summarized the way we have thought of the roles of mutation and epigenetic modifications in cancer as complementary—if mutation caused loss of function of a tumor-suppressor gene on one allele, epigenetic silencing could provide the second "hit" of the allele on the other chromosome (Baylin and Jones 2016). The idea that epigenetic changes in cancer were themselves the result of mutations was not emphasized to the same degree.

The idea of mutations causing "epigenetic" outcomes like widespread DNA methylation changes sits uncomfortably with the more general assumption that epigenetic events are independent of, and act above/upon, DNA sequence. The epigenome of the cancer cell is strikingly altered compared with nontransformed cells and in ways that are not limited to changes in DNA methylation. A major insight was that many alterations in the cancer epigenome can be tracked back to mutations in regulatory genes acting in *trans* to affect the whole genome.

Is an epigenetic change in cancer no longer interesting if it is caused by a mutation? Absolutely not. We will review how finding these changes is very useful for biomarker development and for defining therapeutic strategies. It is, however, valuable to distinguish between molecular genomic/epigenomic changes that are secondary to mutations of transcriptional regulators and those that are not, as the latter are potentially more likely to represent changes induced by some sort of environmental exposure, which in cancer biology would include viral infections.

EPIGENETIC DYSREGULATION IN CANCER NOT ATTRIBUTABLE TO MUTATIONS

What evidence is there for cancer cells having "myriad abnormalities that are based on somatically heritable alterations that are not due to primary DNA sequence changes" (Baylin and Jones 2011), if this is our definition of cancer epigenetics?

First, we need to remind ourselves that the evidence for heritability (through cell division) of many of the molecular genomic processes described as epigenetic is frequently weak (Chapter 4). If we set aside the need to have proven heritability for each of the "myriad abnormalities," the essence of the definition for epigenetic dysregulation in cancer cells is that it can be dissociated from "primary DNA sequence changes." It's probably best to ignore the difficult heritability property when thinking of cancer epigenetics and instead focus on exploring properties of cancer cells that are not obviously DNA sequence–dependent.

This turns out to be surprisingly difficult, but a few intriguing examples are presented below. In the first example, mutations are present in the cancer cell, but malignant growth is not. The further examples are those in which the untransformed cells in a tissue become more likely as a group to become neoplastic, sometimes in response to a clear stimulus (dietary, viral), but sometimes for no clear reason. In each case there are transcriptional regulatory mechanisms involved, but any DNA mutations present do not appear to be deterministic of neoplasia, making these especially intriguing paradigms for cancer epigenetic studies.

Cancer Dormancy

When a cancer cell migrates within the body and ends up embedded in a new tissue, this is the first step in the process of metastasis, which is ultimately how many cancers end up causing death. A fascinating observation is that the initial deposition of the cancer cell in the new tissue environment can be associated with the cell switching to a quiescent state, referred to as "dormancy." This is a huge problem when trying to kill cancer cells based on their property of increased cell division; these dormant cells are more likely to survive such cytotoxic therapy and be responsible for recurrence of the cancer following chemotherapy.

The history of this field has been nicely recounted by Christoph Klein (Klein 2011), who recounts how Australian pathologist Rupert Willis originally defined "dormant tumour cells" in 1934 (Willis 1934). Early ideas about the mechanism of apparent dormancy included the possibility that dissemination occurred before overt overgrowth, so the disseminated cells fundamentally lacked the ability to proliferate and were not, in fact, switching to a dormant from an active state. Dormancy due to a lack of an adequate vascular supply or due to immune surveillance were also suggested, as well as the development of autophagic properties of the disseminated cells (Klein 2011) or the differentiation of the cancer cells in their new host tissue (Aguirre-Ghiso 2007).

The major focus was, however, on the influence of the new microenvironment being provided by the host tissue for the disseminated cancer cell and the induction of cell signaling resulting from this new surrounding for the cancer cell. Julio Aguirre-Ghiso had been studying the role of mitogen-activated protein kinases (MAPKs), whose role is to transduce signals from cell surface receptors that sense the environment extrinsic to the cell. When the p38α and p38β MAPK activity was high compared with signaling by ERK MAPK molecules, this increased activity was associated with the dormancy state (Sosa et al. 2011). This cell signaling in turn drives a transcription factor (TF) response, up-regulating p53 (TP53) and BHLHB3 and down-regulating c-Jun (JUN) and FOXM1 (Adam et al. 2009). Aguirre-Ghiso's group later added NR2F1 as a further TF mediating dormancy in head and neck squamous cell carcinoma cells (Sosa et al. 2015). In this latter study, patterns of DNA methylation and post-translational modifications of histones were noted to be distinctive at the *NR2F1* locus, and that in cells with high levels of expression of *NR2F1* there were higher global levels of the repressive chromatin marks H3K27me3 and H3K9me3 (Sosa et al. 2015).

Although the field of cancer dormancy is still evolving quickly, the lessons at this point suggest that some of the changes in transcriptional regulation that determine the phenotype of the cancer cell are (a) mediated in part by extracellular influences, (b) act through cell signaling, and (c) interact with transcription factor, DNA methylation, and histone state changes in the cancer cell.

Diet and Colorectal Cancer in Mice

The strange thing about using mice as a model for colorectal cancer is that genetic mutations don't really behave as expected, whereas diet has very striking effects in helping these mice to replicate the human phenotype.

Colorectal cancer well represents the original paradigm for multistep, genetically mediated tumorigenesis. The original Fearon and Vogelstein model (Vogelstein et al. 1988) was of the following sequence of events:

5q mutation or loss (where the *APC* gene is located) → DNA hypomethylation → 12p mutation (where the *K-RAS* gene is located) → 18q loss (location of *DCC* gene) → 17p loss (*TP53* gene) → other alterations.

More updated models now start with the primary event including *APC* mutations and associated WNT-signaling abnormalities, progressing to add *KRAS* and *BRAF* mutations, and then undergoing *CDC4*, *SMAD4*, and *DCC* mutations, with mutations of *TP53*, *TGFBR2*, *BAX*, and *IGGF2R* as late events (Nguyen and Duong 2018). These are the likely major changes over time within the tumors, but the resulting carcinomas appear to cluster into four groups when studied in terms of molecular phenotypes (Guinney et al. 2015), indicating that there is more branching of molecular events than a simple linear progression. This should be easy to visualize for anyone familiar with Waddington's epigenetic landscape.

This all points to *APC* somatic mutation as a first step in colorectal carcinoma. Pathogenic variants of *APC* can be germline, inherited in a family, leading to the cancer predisposition syndrome familial adenomatous polyposis (FAP). A mouse strain called *Apc^Min* was developed by chemical mutagenesis that has a c.2603T>A DNA sequence mutation in the *Apc* gene causing a p.Leu868* amino acid change (Moser et al. 1990). Although truncation of *APC* is the mutation type in the vast majority of families with FAP (Fearnhead et al. 2001), the *Apc^Min* mice do not manifest colorectal carcinoma. Instead, they develop many tiny, benign adenomas, located in the small intestine and not the colon.

We have to assume this was a disconcerting finding for the mouse geneticists. You recreate exactly the kind of mutation found in human patients in a mouse gene conserved 92% at the amino acid level (Zeineldin and Neufeld 2013) and all you get is microadenomas in the wrong part of the digestive tract. Fortunately, the researchers involved didn't give up. Other loci in the genome were shown to modify the number of microadenomas (McCart et al. 2008), a valuable revelation of genetic interactions that could be involved in human cancer.

In striking contrast is the effect of manipulation of the diets of mice lacking any cancer mutation. These were the regular C57Bl/6 mouse strain fed the diet representative of everything wrong with the typical diet of the Western world. As early as 1990, it had been shown that feeding animals a "Western" diet (high fat and phosphate, low calcium and vitamin D) caused hyperproliferation and hyperplasia in the colonic epithelium of young mice and rats (Newmark et al. 1990). They then modified the diet further to reduce fiber and "methyl-donor" nutrients (folic acid, methionine, and choline) and allowed the mice to age to 18 months. Although no mice on a regular diet developed intestinal tumors, 42% on this new Western diet developed tumors, including carcinoma, and all were located in the colon (Newmark et al. 2001). Surprisingly, this is a much better representation of human colorectal carcinoma than the *Apc^{Min}* genetic model.

What was it about this diet that was inducing these tumors? Adding back vitamin D and calcium to the mouse diet reduced but did not eliminate the increased tumorigenesis (Newmark et al. 2001). It appears that this manipulated diet causes inflammation, demonstrated through gene expression and histological studies of the colon (Erdelyi et al. 2009). An observation worth noting was that the mice on the new Western diet were heavier than those on the regular diet, attributable to the increased fat content of the diet (Newmark et al. 2001), with obesity a known risk factor for colorectal carcinoma (Bardou et al. 2013). Alterations in the enhancer landscape, inferred through changes in the pattern of H3K27ac in ChIP-seq experiments, have been found in the colonic epithelium of obese mice (Li et al. 2014). An elegant study of mice fed a high-fat diet revealed the expansion of intestinal stem cells (ISCs) in the crypts of the epithelium, mediated through induction of peroxisome proliferator-activated receptor delta (PPARδ) transcription factor signaling (Beyaz et al. 2016). Combining the insights from these studies, it becomes likely that the epigenomic (chromatin state) changes associated with obesity reflect a combination of new patterns of transcription factor activity combined with altered cell subtype composition within the colonic epithelium. The model proposed for increased susceptibility to tumorigenesis was conceptually simple—by expanding the ISC pool, you create more cells in which mutations can occur that lead to tumor formation (Beyaz et al. 2016).

Viral Effects

The COVID-19 global pandemic has probably turned everyone into an amateur virologist. We have had it vividly illustrated how disease-causing viruses are using humans merely as a mechanism for their self-propagation while they are undergoing genetic mutation and evolving.

For a virus to be involved in causing cancer, it needs to cause a chronic infection. The virus somehow needs to infect host cells without killing them while avoiding being recognized and removed by the immune system, limiting the degree of harmful inflammation and allowing the host to live despite the presence of the virus in their body, a difficult balancing act (Virgin et al. 2009).

There are several viruses capable of chronic infection and associated with cancer in humans (McLaughlin-Drubin and Munger 2008). Electron microscopy in 1964 revealed virus particles in lymphoblasts cultured from Burkitt's lymphoma (Epstein et al. 1964), subsequently revealed to be the Epstein–Barr virus (EBV), with hepatitis B virus revealed in a similar way in 1970 (Dane et al. 1970). The history of tumor virology has been excellently reviewed by White and colleagues (White et al. 2014). The link between chronic viral infection in cancer and epigenetics really started in the early 2000s, with studies of DNA methylation using bisulfite sequencing of tumor-suppressor genes in EBV-positive gastric carcinomas (Kang et al. 2002).

It should not be surprising that widespread dysregulation is found in cells chronically infected with oncogenic viruses, which target both p53 and pRB (Krump and You 2018). Shelley Berger has reviewed the complex interrelationship between p53 and epigenetic (transcriptional regulatory) events in the cell (Levine and Berger 2017), whereas pRB has a large number of interacting proteins involved in chromatin organization and DNA methylation (Chinnam and Goodrich 2011), indicating how dysregulation of each can have downstream consequences on molecular genomic properties of the cell. Furthermore, by targeting cell signaling pathways (Krump and You 2018), we should expect downstream effects on transcription factor activity in the nucleus.

Complicating the virus story is the other property of oncogenic viruses—that they can reduce the DNA damage response in the host cell, allowing the accumulation of mutations that could end up being a major cause of neoplastic transformation. Viral infections cannot therefore be regarded as completely nongenetic in their neoplastic effects, but they represent a fascinating model for understanding a perturbation of cellular physiology involving transcriptional regulatory mechanisms.

Field Effects

The final example is the idea of a field effect, in which all or a region of a tissue develops multiple independent tumors, suggesting that something predisposed cells throughout the "field" to cancerous transformation. The idea was first suggested in 1953 by Slaughter and colleagues (Slaughter et al. 1953), who noted the appearance of multifocal growth of squamous cell carcinomas of the oral cavity, and speculated that

...it would appear that epidermoid carcinoma of the oral stratified squamous epithelium originates by a process of "field cancerization," in which an area of epithelium has been reconditioned by an as-yet-unknown carcinogenic agent.

Although today we would naturally point to tobacco product exposures in the causation of oral epithelial cancers, the epidemiological evidence linking the exposure with lung cancer was only emerging in the 1950s (Wynder 1988). Currently there is a substantial literature describing DNA methylation, transcriptional, and histone changes associated with tobacco and alcohol exposures in oral carcinomas (Ghantous et al. 2018). The human papilloma virus (HPV) is another potential influence in these head and neck carcinomas, and the mutagenic effects of carcinogenic adducts involving compounds present in tobacco is a DNA-based influence on neoplasia of these epithelial cells.

So, what are the examples of field effects in human cancer? Numerous organs and tissues have been described to manifest field effects with associated changes in DNA methylation that are present in histologically normal-appearing adjacent tissues (Giovannucci and Ogino 2005). The key observation is one of multiple independent tumors forming in the susceptible anatomical field. Most examples of cancer field effects are tumors of epithelial cells (Giovannucci and Ogino 2005), for which it is difficult to imagine a spreading of a primary tumor to a discontinuous location within the same epithelium, prompting the idea of multiple tumors being formed independently. Although this kind of multifocal pattern can occur in people who carry a germline variant predisposing them to cancer, a field effect that is not due to a germline predisposition is instead assumed to reflect a perturbation in many or all cells within the field, not accompanied by histological changes, but somehow predisposed to becoming neoplastic.

An example of a field effect that ties together many of the threads discussed above is hepatocellular carcinoma. This can occur in livers that have been chronically exposed to injury from, for example, alcohol, or as part of metabolic-associated fatty liver disease, or from chronic viral infection of hepatocytes. The chronic hepatitis is accompanied by increased proliferation and turnover of hepatocytes, accumulation of morphological changes including fibrosis, and invasion by inflammatory cells. Multiple carcinomas can develop, discriminating between local spreading (intrahepatic metastasis) and genuinely independent multifocal tumors by sequencing (Furuta et al. 2017). Sampling the noncancerous tissue in these livers reveals transcriptional (Hoshida et al. 2009) and DNA methylation (Wijetunga et al. 2017) changes relative to normal tissue, which should be unsurprising when considering the cell subtype changes that must be contributing to these patterns. Of note is that the molecular profiling of this noncancerous tissue helps to predict outcome in these individuals (Hoshida et al. 2009), a potentially valuable use of molecular genomic information in patient care.

EPIGENETIC AND GENETIC INTERACTIONS IN GLIOMAS

A great story from Brad Bernstein's group was published in 2015 (Flavahan et al. 2016), integrating many of the separate concepts discussed above. They were studying gliomas, a tumor of the glial cells of the brain and spinal cord. These tumors have a relatively high frequency of *IDH* mutations, leading to the accumulation of the D-2-hydroxyglutarate oncometabolite, decreased TET enzyme activity, and increased DNA methylation, as described earlier in this chapter. They showed that the binding sites for the CTCF protein gained DNA methylation, leading to the loss of CTCF binding at these locations in these cells. Using an innovative analytical approach, they showed that this loss of CTCF binding was associated with the breakdown of insulation of expression levels between nearby genes. One gene in particular emerged from this analysis, the known glioma oncogene *PDGFRA*. They studied the chromatin topology around this gene and were able to show that a CTCF binding site that insulated *PDGFRA* from adjacent active genes became methylated, lost CTCF binding, and permitted *PDGFRA* to become up-regulated in expression.

This example shows how a primary DNA mutational event (in an *IDH* gene) led to the production of an oncometabolite, inhibition of an enzyme, acquisition of DNA methylation, altered binding of a DNA-binding protein, loss of chromatin insulation, and up-regulation of an oncogene. As the authors acknowledge, this is unlikely to be the only mechanism for *IDH* mutation effects in gliomas, but it highlights how genetic, metabolic, and transcriptional regulatory effects can interact in neoplasia.

An intriguing question that arises is whether a mutation with the capability to drive neoplasia acts differently in a cell in which transcriptional regulation is perturbed compared with a cell with normal regulation? Is it possible that oncogenic mutations occur in everyone, but have more of an opportunity to drive uncontrolled growth in cells that have altered cell signaling, transcription factor activities, and dysregulation of chromatin states and DNA methylation? Or would the ability to preserve DNA repair in cells with transcriptional dysregulation restore the background risk of cancer? These are the kinds of questions that are prompted by thinking of disease in terms of combined effects of DNA sequence and "epigenetic" transcriptional regulatory effects.

The same train of thought underlies the idea proposed by Isidro Sánchez-García and colleagues, that of "epigenetic priming" in cancer (Vicente-Dueñas et al. 2018). They describe how the resetting of transcriptional regulation may be due to an oncogenic mutation or due to an exposure, with secondary events leading to overt tumorigenesis. A further provocative observation was from a study of a mouse colon organoid system with a $Braf^{N600E}$ mutation (Tao et al. 2019). They found that prolonged culture of the organoids caused them to acquire DNA methylation changes similar to those found in colons from older mice and humans, which was associated with increased tumorigenic conversion. We do not understand why DNA methylation changes with age, and it is highly unlikely to be due to pervasive acquisition of similar DNA sequence changes across most cells in a tissue, so this would appear to be another nongenetic means of "priming" cells to become more sensitive to the effects of mutations.

THE EPIGENOME INFLUENCES THE MUTABILITY OF THE GENOME IN CANCER

DNA methylation is mutagenic. This mechanism was described in Chapter 6, involving spontaneous deamination of a cytosine to a uracil, which is easily recognized and repaired by the DNA mutation repair system, but 5-methylcytosine deaminating to a thymine and failing to be correctly repaired at least some of the time. The MBD4 glycosylase selectively binding to these T/G mismatches and excising the thymine (Hendrich et al. 1999; Otani et al. 2013) helps but does not completely fix the problem; this is why methylated CG dinucleotides are hotspots of mutation in genes like *TP53* (Baugh et al. 2018).

Overexpression of the gene for the centromeric histone variant CENPA occurs in many cancers and is associated with ectopic deposition of the protein in the genome, which can then lead to chromosome breakage (Sharma et al. 2019). Late-replicating regions in the genome are more polymorphic between individuals and are also more likely to acquire point mutations (Koren et al. 2012). This reflects a broader pattern of mutation rates and types varying within genomes of many species (Hodgkinson and Eyre-Walker 2011), requiring that when studying mutation patterns in cancer these regional differences be taken into account to avoid overcalling mutations (Imielinski et al. 2017). The chromatin context of the loci appears to be a major factor in protection against acquisition of new variants, with oxidative damage increased at loci of open chromatin and decreased at heterochromatic domains, the bases of chromatin loops, exons, and gene promoters (Poetsch et al. 2018).

Saving the most head-scratching example for last, APOBEC enzymes only act on single-stranded DNA and can create C to T mutations in isolation or in large domains, which, if unrecognized, would be read out in bisulfite assays for DNA methylation as unmethylated. The interactions between the epigenome and mutation are clearly complex and have significant potential for technical problems when performing and interpreting genome-wide studies of DNA sequence and transcriptional regulation.

EPIGENETIC CHANGES IN CANCER USED AS BIOMARKERS

Biomarkers can be used for diagnostic, prognostic, or therapeutic purposes. Diagnostic biomarkers need to work on readily accessible samples (like blood, oral rinses, nasopharyngeal swabs, vaginal fluid, sputum, urine, or feces), because their utility depends on their being applied to large numbers of people presymptomatically. A paradigm for a diagnostic epigenetic biomarker is the presence in plasma of methylated DNA from a promoter of the *SEPT9* gene, permitting the sensitive detection of all stages of colorectal carcinoma (Grützmann et al. 2008). This assay involves bisulfite conversion of cell-free DNA (cfDNA) in plasma followed by real-time PCR quantification of the *SEPT9* locus. The use of other assays testing methylation of cfDNA in serum, plasma, or other body fluids for a wide variety of human cancers has been reviewed comprehensively (Thomas and Marcato 2018). An approach that avoids harsh bisulfite treatment by instead using antibodies to 5-methylcytosine in cfDNA samples from plasma (cfMeDIP-seq) was shown to have the capability to detect circulating tumor DNA (ctDNA) from a number of different cancers (Shen et al. 2018). The use of epigenetic markers extends beyond DNA methylation and includes identifying the fragmentation patterns caused by tissue- and cancer cell–specific nucleosome and transcription factor positioning in cfDNA (Snyder et al. 2016).

Prognostic biomarkers overlap to some extent with the diagnostic biomarkers, again highlighting the example of DNA methylation of *SEPT9* defining worse prognosis in colorectal carcinoma (Shen et al. 2019). A valuable prognostic biomarker in acute myeloid leukemia (AML) is the functional loss of the CCAAT/enhancer binding protein alpha (CEBPA) transcription factor—whether through inactivating mutations or through methylation of its promoter—representing distinct events (Szankasi et al. 2011) that are both associated with better outcomes in these patients (Lin et al. 2011). Therapeutically, it had been hoped that inactivation with associated DNA methylation of the gene encoding O6-methylguanine-DNA methyltransferase (MGMT) would make gliomas more sensitive to alkylating agents and could be used predictively (Esteller et al. 2000), but this test has performed in trials less well than had been hoped (Butler et al. 2020), possibly because of major inconsistencies in how the DNA methylation test is performed between different institutions (Malmström et al. 2020).

Of note, when discussing the use of DNA methylation or other "epigenetic" assays as biomarkers, we care less about why the DNA methylation pattern is changed in the sample and more about the pattern being predictive of the desired outcome. For example, the acquisition of DNA methylation at the *CEBPA* promoter could theoretically be due to the mutation and loss of binding of a transcription factor like RUNX1 or could be due to local somatic mutations in the promoter DNA, but if both distinct mechanisms point to the same prognosis, further mechanistic insights are not needed.

It should be remembered from Chapter 5 that when Kelsey and colleagues were identifying the effect of blood cell subtype proportion variability on DNA methylation, their intent was to study how the different kinds of leukocytes in peripheral blood mirror the presence of difference types of tumors in the body (Houseman et al. 2012). Although the discussion in that chapter was focused on the value of using molecular genomic/epigenomic assays to gain insights into what have traditionally been thought of as confounding influences, and how understanding cell subtype composition is a valuable insight, for the purpose of biomarker discovery it is less important to know the source of the variability. The promising performance of whole-blood DNA methylation profiling in detecting breast cancer (Guan et al. 2019) is likely to have a contribution from cell subtype proportion variation (not that it really matters for use as a diagnostic biomarker).

The studies above are relatively broad in scope. Where the field appears to be headed may be indicated by a study from Frank Lyko's group. He focused on adenomas, the benign precursors of colorectal carcinoma, and found their DNA methylation profiles to cluster into three groups, each of which represented cells at different stages of differentiation in intestinal crypts. This finding revealed how the adenomas maintained enough of the

DNA pattern of their precursor cell type to remain distinguishable. When the group then looked at malignant colorectal carcinomas, the same three groups of DNA methylation patterns could be identified. From a biomarker perspective, what was valuable was the finding that the patients with the more stem-like subgroup of carcinomas had a significantly worse outcome (Bormann et al. 2018). This more focused and mechanistic study ended up revealing a valuable prognostic biomarker with insights into why these tumors are distinctive.

CANCER THERAPIES DIRECTED AT TRANSCRIPTIONAL REGULATORY MECHANISMS

Cancer is the main focus for epigenetic therapy development for historical and practical reasons. In recounting Peter Jones' training earlier, it should be remembered that he was studying how drugs being developed to treat cancer were causing oncogenic transformation, including 5-azacytidine (5-azaC), which had been developed in the former Czechoslovakia as an anticancer drug (Cihák 1974). This prompted the U.S. Food and Drug Administration (FDA) to approve a clinical trial to test 5-azaC as an Investigational New Drug in 1971 (Ganesan et al. 2019), long before its DNA methylation effects were known. The results of this early experience were encouraging for patients with AML but not those with solid tumors (Von Hoff et al. 1976).

Peter Jones (2020, pers. comm.) was disappointed with the results of the initial clinical trials:

> I think the problem was that they were using them as any other chemotherapy agent and they were not that good at killing cells. At the time we made our discoveries, their effects on mammalian DNA methylation were unknown and our results were completely unexpected. I had never heard of DNA methylation at the time!
>
> They only began to get traction when the doses were lowered, and physicians realized that they had to wait to see responses unlike standard chemo.
>
> Most oncologists had given up and it was the unsung hero Lewis Silverman at Mount Sinai who kept going.

The cascade of observations that led to the development of histone deacetylase inhibitor (HDACi) compounds started with the finding that dimethylsulfoxide (DMSO) induced differentiation of mouse erythroleukemia (MEL) cells (Friend et al. 1971). Charlotte Friend (1921–1987) had observed in electron microscopy images what appeared to be virus-like particles in a mouse carcinoma, prompting her to work on the then-heretical idea that cancer could be caused by viruses and transmissible. She succeeded in isolating the Friend leukemia virus as the causative agent of a mouse leukemia and derived the MEL cell line for in vitro studies. To make the MEL cells more readily transfected, she treated them with DMSO and noted that the cells turned red, indicating that they had started to produce globin, a mark of differentiation of this erythroid lineage of cells.

In collaboration with colleagues at the Sloan Kettering Institute for Cancer Research and Columbia University in New York, she started to screen other compounds for the same effect. The group identified hexamethylene bisacetamide (HMBA) as a more potent agent that caused the MEL cells to arrest their growth and switch to differentiation (Marks and Breslow 2007). They then identified an even more potent compound suberic bishydroxamic acid (SBHA), which they further refined to suberoylanilide hydroxamic acid (SAHA), the most potent agent of all tested. The group recognized the structural similarity between SAHA and trichostatin A (TSA), giving the first clue into the mechanism of action of these differentiation-inducing compounds, as TSA was found in 1990 to have HDACi properties (Yoshida et al. 1990). Now SAHA (marketed as vorinostat) is approved as a treatment for cutaneous T-cell lymphoma. As with any HDACi, SAHA has effects to influence acetylation on many nonhistone proteins (Marks and Breslow 2007), making it difficult to conclude that its therapeutic benefits are limited to chromatin targets.

Although there have been further DNA methyltransferase and HDAC inhibitor drugs developed in the interim, other regulators of chromatin states have come into focus for development. One category of drugs to be developed is the lysine histone methyltransferase (KMT) inhibitors. The first to be discovered by Axel Imhof's group in 2005 was chaetocin, emerging from a screen of almost 3000 compounds as the most inhibitory of *Drosophila* methyltransferase SU(VAR)3-9 (Greiner et al. 2005). The focus since then has been on inhibitors of the G9a and GLP enzymes that add methyl groups at the H3K9 position as part of the formation of heterochromatin (Kaniskan et al. 2015).

It required a screen of 175,000 compounds (Knutson et al. 2012) to identify the first inhibitor of EZH2, a Polycomb protein that adds methyl groups to H3K27, another repressive mark. Further studies have now revealed inhibitors of methylation of H3K4, H3K36, H4K20, and H3K79, arginine methyltransferases that act on histones (PRMT1, PRMT3, and CSRM1/PRMT4) (Kaniskan et al. 2015).

When designing inhibitors to lysine demethylases (KDMs), it was helpful that these enzymes are similar to monoamine oxidases (MAOs), which had been targeted for inhibition since the 1950s, recognizing that they degrade norepinephrine, serotonin, and dopamine in the brain. By inhibiting the enzymes and making more of these neurotransmitters available, mood could be improved and anxiety reduced. Repurposing MAO inhibitors like tranylcypromine allowed their KDM activities to be used therapeutically in conditions like AML, restoring sensitivity of refractory tumors to differentiation mediated by retinoic acid signaling (Schenk et al. 2012). More potent KDM inhibitors are now being developed.

Bromodomains, protein structures that bind to acetylated histones, were described in Chapter 4. The first BET (bromo and extra-terminal) inhibitor was discovered fortuitously during a screen looking for mediators of up-regulation of expression of the *APOA1* gene, using the *APOA1* promoter driving a luciferase reporter (Chung et al. 2011). The lead compound was found to be a BET inhibitor, the first example of an inhibitor of what would be often called a histone "reader." Although the clinical trials of these agents in cancer have been less successful than hoped (Mita and Mita 2020), BET inhibitors represent the most studied group of agents for epigenetic therapies beyond cancer, including a clinical trial to boost *APOA1* expression and high-density lipoprotein levels (Ganesan et al. 2019).

CANCER AS A PARADIGM OF HUMAN EPIGENETIC DISEASE

Cancer was the first human disease to be studied extensively from an epigenetics perspective and has progressed all the way to therapeutic interventions in clinical practice. Although it has been tempting to hope that lessons learned in cancer epigenetics can extend directly to other human diseases (Feinberg 2013), the burden of somatic mutations in cancer cells makes them extremely unlike other cells in the body, in terms of both interpreting why molecular genomic changes have occurred and their responses to therapeutic interventions.

What the field of cancer epigenetics has delivered to date that is of unquestionable value is prognostic biomarkers based on assays like DNA methylation studies of cfDNA and novel and effective therapeutic approaches targeting transcriptional regulatory mechanisms in a number of malignancies. This field is fast-moving, intensively studied, and extraordinarily complex and promises to deliver many more advances in clinical care delivery.

RECOMMENDED FURTHER READING

Flavahan WA, Drier Y, Liau BB, Gillespie SM, Venteicher AS, Stemmer-Rachamimov AO, Suvà ML, Bernstein BE. 2016. Insulator dysfunction and oncogene activation in *IDH* mutant gliomas. *Nature* **529:** 110–114. doi:10.1038/nature16490

The story of working out a mechanism for a cancer that starts with a mutation affecting cell metabolism, leading to a global acquisition of DNA methylation, leading to altered insulator protein binding and chromatin topology, leading to activation of an oncogene.

O'Neill K, Pleasance E, Fan J, Akbari V, Chang G, Dixon K, Csizmok V, MacLennan S, Porter V, Galbraith A, et al. 2024. Long-read sequencing of an advanced cancer cohort resolves rearrangements, unravels haplotypes, and reveals methylation landscapes. *Cell Genom* **4:** 100674. doi:10.1016/j.xgen.2024.100674

An indication of the insights and complexity that we will reveal with long-read DNA sequencing of cancer genomes for both genetic variation and changes in DNA methylation.

Yang H, Ye D, Guan K-L, Xiong Y. 2012. *IDH1* and *IDH2* mutations in tumorigenesis: mechanistic insights and clinical perspectives. *Clin Cancer Res* **18:** 5562–5572. doi:10.1158/1078-0432.CCR-12-1773

An exceptional review of the oncogenic effects of mutations affecting cell metabolism.

CHAPTER 9

Inheriting Memories

Acompelling idea in epigenetics is that molecular genomic regulatory mechanisms can mediate what can broadly be called "memories" of past events, manifesting as phenotypes of varying types. The definition of epigenetic properties that focuses on the requirement that molecular properties of the genome can be passed from parent to daughter cells provides a mechanistic framework for nongenetic memories, although, as described in Chapter 4, the evidence for such heritability of molecular states is frequently weak. These molecular marks on the genome are also believed to be responsive to the environment, as discussed in Chapter 7. When you combine these definitions and associated beliefs, you can construct a model in which an environmental exposure irreversibly alters molecular genomic regulators, which are then passed through mitotic and meiotic cells (Russo et al. 1996) throughout the life course and from one generation to the next. This model generates a mechanism for propagation of phenotypic traits that is not dependent on changes in the DNA sequence.

This chapter is dedicated to unpicking the strengths and weaknesses of this model. An influential review in 2014 from Edith Heard and Rob Martienssen juxtaposed the strong evidence for inheritance of transcriptional regulatory states at specific loci in plants with the weakness of evidence for comparable events in mammals (Heard and Martienssen 2014). Plant and nematode systems remain much more compelling in terms of the evidence for nongenetic heritability of molecular genomic states and associated phenotypes, but there has also been substantial energy put into looking for comparable mammalian events, providing the major focus of discussion below.

THE INHERITANCE OF ACQUIRED CHARACTERISTICS IS A LONG-HELD BELIEF

The idea that experience shapes destiny has been attractive since at least the Abrahamic writings of Jeremiah in ~600 BCE. In Chapter 31, verse 29, he writes:

In those days they do not say any more: Fathers have eaten unripe fruit, And the sons' teeth are blunted. (Young's Literal Translation)

In Chapter 2, Aristotle was noted to have written a few centuries later about parents acquiring scars and their children being born with marks on their skin in the same place, although acknowledging that the child's markings could be "confused and not clearly articulated."

It should also be recalled that Jean-Baptiste Lamarck felt that his idea of heritability of acquired characteristics was so obvious that it didn't warrant formal declaration (Burkhardt 2013). Charles Darwin, for all that he is viewed as having disproven Lamarck's proposal and might therefore have been staunchly in opposition to everything Lamarckian, was in fact quite overt in supporting ideas about heritability of acquired characteristics. In his *Origin of Species* (Darwin 1859):

I think there can be little doubt that use in our domestic animals strengthens and enlarges certain parts, and disuse diminishes them; and that such modifications are inherited.

In his later book *The Variation of Animals and Plants under Domestication* (Darwin 1868), Darwin proposed a hypothesis about development and the inheritance of variation that he called "pangenesis." This idea focused on the idea that minute particles were released by different cell types in the body and sent to the gametes, mixing with the gemmules of the other parent to create a blend of parental characteristics in the offspring. Today, an exciting area of research focuses on the modern equivalent of the gemmule—the intriguing exosome. Francis Galton (1822–1911) was enthusiastic about his cousin Charles Darwin's idea and decided to try to test it experimentally, transfusing blood between rabbits with different fur colors and testing to see whether that influenced the fur colors of offspring. To give him credit, he did publish his negative results (Galton 1871).

In Chapter 2 we also learned about Waddington's Lamarckian leanings, and in Chapter 7 we read about Waddington's experiments to expose *Drosophila* larvae to increased temperature (Waddington 1953) or to ether vapor (Waddington 1956b) over a number of generations. These experiments must have been exciting at the time, as he found morphological phenotypes that became heritable in the absence of further exposures. Waddington attributed this acquired heritability to his prior environmental exposures, referring to the development of the resulting fixed and heritable phenotype as "genetic assimilation" (Waddington 1959). The cautionary tale that arises from this study is the replication of the experiment 58 years later by Laura Fanti, Lucia Piacentini, Sergio Pimpinelli, and colleagues from the Sapienza University of Rome. They successfully reproduced the phenotype in heat-stressed *Drosophila* embryos but showed that the effect was, in fact, genetic—that is, due to heat-induced activation of transposable elements causing DNA sequence mutations (Fanti et al. 2017).

Some studies purportedly showing inheritance of acquired characteristics did not take as many decades to refute. Guyer and Smith reported in 1920 how they were able to induce lens defects by injecting serum known to be "sensitized" to lens tissue into pregnant rabbits, inducing the phenotype in offspring, and showing that the phenotype could be maintained for at least six generations thereafter without further serum injections. Peter and Jean Medawar stated that these experiments were not reproducible (Medawar and Medawar 1983), but as is an ongoing risk for negative studies, these validation experiments do not appear to have been published anywhere.

And then there was Trofim Lysenko (1898–1976), who made Lamarckism national policy in the Soviet Union. His most destructive idea didn't directly involve Lamarckism but involved another non-evidence-based concept—that plants from the same "class" would never compete but would aid each other's survival, which did more to fit with the government's communist doctrine than with any biological data. Although pseudoscientific nonsense, it nevertheless prompted the government of the Soviet Union to direct that wheat be planted more densely than previously, an idea adopted enthusiastically by the Mao Zedong communist government in China, leading to famines in both countries in which tens of millions of people died. (The Chinese famine will recur later in this chapter when talking about early-life events leading to later life diseases.)

Lysenko's influence on science in the Soviet Union was profound. A 1936 meeting of the Lenin Academy of Agricultural Sciences was reported in *Nature* (Genetics and plant breeding in the U.S.S.R. 1937) with a tone of incredulity, recounting how political considerations were dismissing Western science, describing "Lysenko's forceful statements that the external conditions under which a plant develops have a profound effect not only on the somatic, but also on the sex cells. Therefore, responses of the constitution of an individual plant to its environmental conditions are inheritable, contrary to the views of 'bourgeois' geneticists." The writers concluded that "Lysenko's theory is nothing but Lamarckism in its simplest form" and marveled at how Lysenko avoided what would be a treasonous refutation of Darwinism by being able to embrace both Darwinism and Lamarckism with some convoluted logic. Disturbingly, there appears to be a modern pro-Lysenko movement in Russia that is coupled with nostalgia for Stalinism, in part justifying this resurrection of reputations by claiming that modern plant epigenetics reveals Lysenko to have been correct all along (Kolchinsky et al. 2017). Modern plant epigenetics research reveals nothing of the kind, and we should just move on from Lysenko and Stalin, neither of whom adds value to contemporary epigenetics research.

The attractive idea of heritability of acquired characteristics has been around for millennia, recurring in different guises in different ages, and has consistently failed to generate scientific support until recently in plants and worms. Of all the areas of epigenetics research today, this is the most provocative and contentious, generating the most public attention and the most scientific skepticism in equal measures. This heightened concern is a good example of what Pierre-Simon Laplace (1749–1827) was conveying in his Sixth Principle of his General Principles of the Calculus of Probabilities, that "we ought generally to conclude that the more extraordinary the event, the greater the need of its being supported by strong proofs" (Laplace 2009). To modern and United States–based audiences, this idea would be recognized as having been popularized by astronomer Carl Sagan in his television series *Cosmos.* He simplified the idea to say that "extraordinary claims require extraordinary evidence," the ECREE aphorism. It would be revolutionary to find evidence for nongenetic heritability of acquired characteristics, but the bar is currently set high for those making claims supporting such findings. Nevertheless, these extraordinary claims are being made. Is there extraordinary evidence to support them?

MEMORIES WITH DIFFERENT DURATIONS MAY INVOLVE DIFFERENT MECHANISMS

We tend to distinguish two types of memories of past events when considering nongenetic mechanisms. One reflects a direct exposure of the individual or the gametes that gave rise to the individual, whereas the other reflects an exposure to an ancestor without a direct exposure to the individual.

The first category is referred to as *intergenerational inheritance.* This occurs when an acquired phenotype or exposure of a parent is passed on to offspring. When the exposure occurs to a pregnant mother, simultaneously exposing the developing offspring, this overlaps significantly with the developmental origins of health and disease (DOHaD), in which an exposure or event during development or early in life is retained as a memory influencing phenotypes later in the life course. DOHaD models can also include exposures early in life after pregnancy and weaning and are worth retaining for discussion in this chapter because there is an overlap in the mechanism of maintenance of memory of a past event between intergenerational and DOHaD models.

The second type of memory is when the exposure to an ancestor influences the phenotype of an individual from such a distant generation that a direct exposure is inconceivable. This is described as *transgenerational inheritance*. The different models are represented in Figure 9.1. This is where we stray into Lamarckism when the acquired phenotype is considered to be nongenetically inherited.

In their excellent review, Baugh and Day add the further category of acclimation, the "adaptive changes in a plastic phenotype within a generation which allow individuals with the same genotype to have different phenotypes as a response to their environment" (Baugh and Day 2020). This property of plasticity of phenotypes perhaps appears intuitive but becomes a problem in the logic of the idea of transmission of memories through the generations. This will be seen later in this chapter when discussing transgenerational memories, in which the assumption is made that later generations not only inherit phenotypes but lose the plasticity to reverse this process.

TRANSGENERATIONAL MEMORY

Starting with transgenerational memory has the advantage of being able to point to excellent examples of such memories being propagated by molecular regulators of transcription, based on studies of plants and nematodes. Of course, we already have a powerful mechanism for transgenerational memory of traits—DNA sequence variants. What sets the examples below apart is their clear independence from effects mediated by DNA sequence variation, the Lamarckian model.

Plants

The evidence for transgenerational memory being mediated by molecular regulators of the genome in plants is especially strong. Barbara McClintock

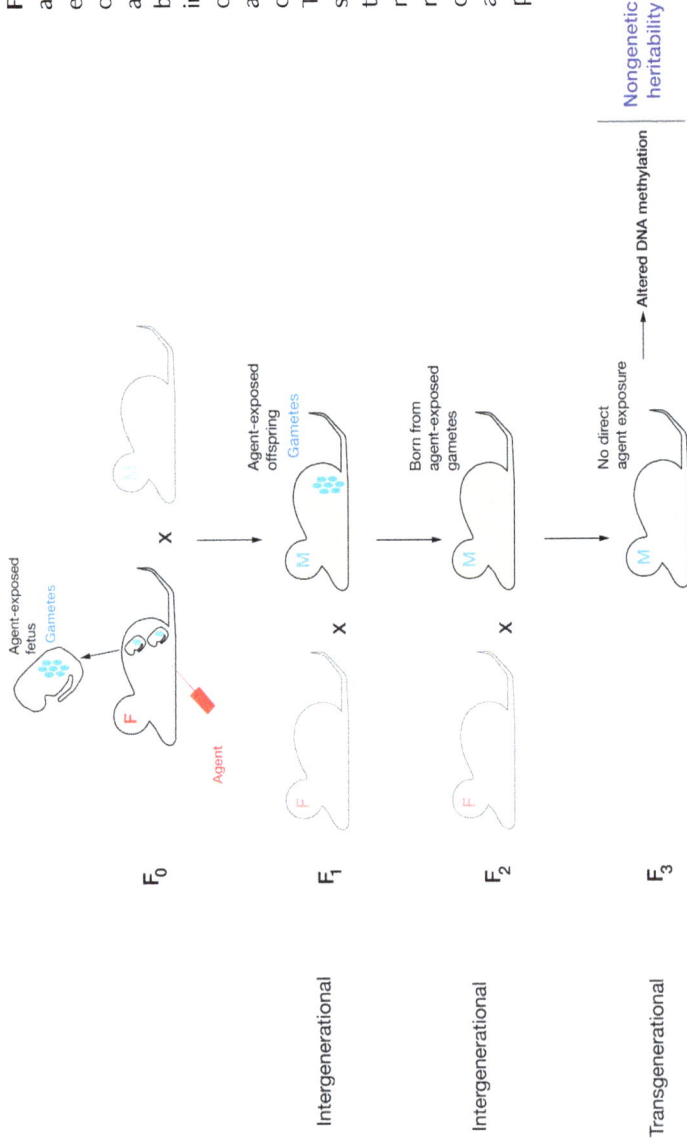

Figure 9.1. Defining intergenerational and transgenerational effects of exposure to an agent. Shown is the complex situation of an exposure to a pregnant dam. The F_1 mice have been exposed directly to the agent in utero, but the F_2 offspring are born of gametes directly exposed to the agent, so any phenotype they have could be attributable to exposure. The F_3 animals, on the other hand, should have no direct exposure to the agent; therefore, any phenotype maintained in these animals may be reflected by transcriptional regulatory changes like altered DNA methylation, and the mechanism of heritability is proposed to be nongenetic.

(1902–1992) was interested in non-Mendelian inheritance of pigmentary patterns in maize, finding that some of these patterns were attributable to the mobilization of transposable elements. In 1983 McClintock received the Nobel Prize in Physiology or Medicine for this work, becoming the first woman to receive the Prize without sharing it with a male co-recipient.

Although transposable element mobilization is a type of genetic mutation that leads to long-lasting phenotypic changes, another model for maize pigmentation was being studied in the 1950s. Alexander Brink (1897–1984) identified a non-Mendelian pattern of inheritance of maize pigmentation (Brink 1956) that became described for this and other loci as paramutation. What happens in paramutation is akin to a genetic hit-and-run phenomenon—homologous, nonidentical alleles interacting with each other when they co-occur in the same individual, with one allele instructing the other to become relatively silenced. In progeny, the instructing (paramutagenic) allele may then be separated from the silenced homolog, but the silenced (paramutated) allele remembers its instructions and remains silenced in that and subsequent generations (Arteaga-Vazquez and Chandler 2010).

If the DNA causing the phenotype is not present, then the phenotype is mediated by nongenetic mechanisms. What's happening at the molecular level to mediate this memory? The molecular processes involved in plant paramutation are heterogeneous, but a few common themes are of interest. The silencing mechanisms for gene expression described in Chapter 4, including DNA methylation (Cubas et al. 1999), are unsurprisingly involved in repression, but what is more interesting is how a signal gets from an allele on one chromosome to its homolog on the other. Work from Vicki Chandler allows a model for some of the better-understood loci at which plant paramutation occurs, involving tandemly repetitive DNA that generates small interfering RNA (siRNA) molecules. Although siRNAs have the capability to cause degradation of mRNAs, they are also able to induce the RNA interference (RNAi) pathway to target heterochromatin formation at loci complementary to the siRNA sequences (Martienssen and Moazed 2015).

This is a nice model for how a signal can be generated by the paramutagenic homolog and sent as siRNA molecules to act on an allele with the same sequence, but how is this signal maintained in the next generation when the paramutagenic allele is no longer there? You could invoke the usual "Epigenetic modifications are heritable and stable" property with the caveats described in Chapter 4, but in paramutation there is a further way of maintaining molecular memory. Plants have enzymes that have the capability to copy RNA to create new RNA molecules, the RNA-dependent RNA polymerases (RdRP). The siRNA generated by the repetitive DNA at the paramutagenic allele not only induces the RNAi pathway, it also gets replicated by RdRP activity in all generations to maintain a pool of siRNAs to continue the RNAi signaling (Arteaga-Vazquez and Chandler 2010) (Fig. 9.2).

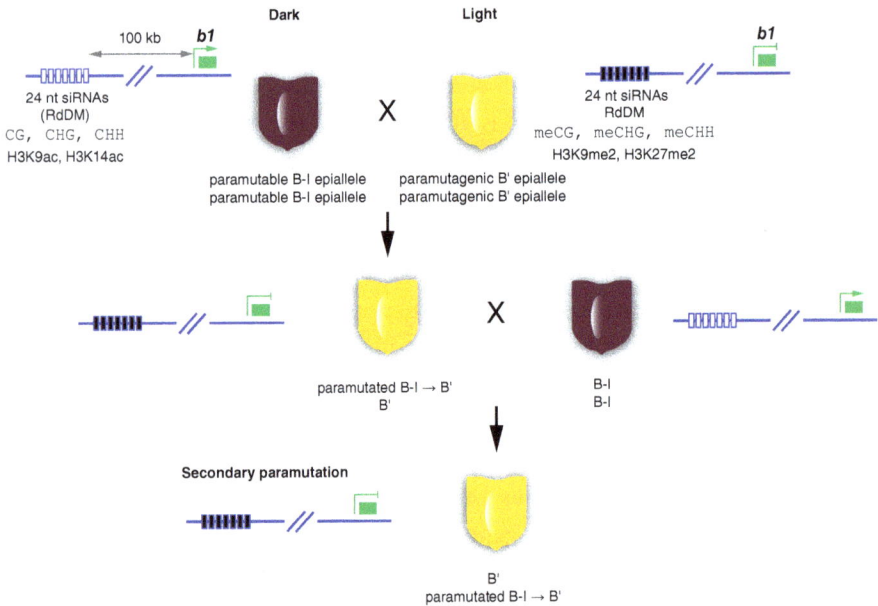

Figure 9.2. An example of the mechanisms of paramutation, focused on the *booster1* (*b1*) locus of maize. As background, the *b1* allele is recessive, so one functional copy should be enough to confer the dark pigment. The *b1* gene in the dark and light plants is the same, the difference is not due to a sequence change in the plant with the light color. Approximately 100 kb upstream of *b1* is a set of seven repeats, each 853 bp, making the locus paramutable, whereas *b1* genes with a single one of the repeats at the upstream locus are not susceptible to paramutation.

The repeats are transcribed, generating a double-stranded RNA that is then chopped into 24-nt siRNAs. These in turn attract RNA-directed DNA methylation (RdDM) to the repeats, and chromatin changes are also induced. The B-I epiallele is associated with activity of the *b1* gene and the dark phenotype, but the B' epiallele has silenced the repeats with DNA methylation (at CG, CHG, and CHH contexts) and repressive chromatin marks, associated with silencing of the *b1* gene.

When one allele is the B-I epiallele and the other is the B' epiallele, the silenced state of the B' repeats propagates to the B-I repeats, the *b1* gene silences, neither allele is active, and the light color results.

If a formerly B-I epiallele becomes a B' epiallele and then gets crossed with another B-I epiallele, secondary paramutation can then occur.

Paramutation is not the only mechanism of heritable, nongenetic influences on gene expression. So-called epialleles have been found in plants where patterns of DNA methylation are found to be heritable without evidence for DNA sequence changes in the surrounding genome (Jacobsen and Meyerowitz 1997). These epialleles are thought to be selected for targeting initially by the presence of siRNAs with complementary sequences, followed by what has been called RNA-directed DNA methylation (RdDM). The maintenance of DNA methylation patterns through meiosis and between generations may be further facilitated by the preservation of existing DNA methylation patterns through early development; unlike in animals, in whom there are global waves of loss of DNA methylation during gametogenesis and in early embryogenesis (Monk 1990).

The other issue that makes it difficult to extend these plant models to mammals is the absence of RdRP enzymes encoded by vertebrate genomes. RdRPs appear to have been lost at multiple times independently during animal evolution (Pinzón et al. 2019). Fortunately, they were retained in the Rhabditida order of the Nematoda phylum, allowing us to observe the striking animal transgenerational phenotypes in the *Caenorhabditis elegans* model organism.

Only 302 Neurons but an Amazing Memory: *C. elegans*

The otherwise unassuming-looking *C. elegans* is spectacular when it comes to transgenerational memories of exposures past. Although Sydney Brenner (1927–2019) is credited with popularizing *C. elegans* as a model organism in the 1970s, these nematodes had been the focus of interest since the turn of the twentieth century (Nigon and Félix 2017). The discovery of RNAi (see Chapter 4) prompted studies in *C. elegans* that led Craig Mello to the realization that RNAi in one generation was often accompanied by its maintenance in subsequent generations (Grishok et al. 2000). In 2006, Ronald Plasterk used RNAi to target 171 genes in *C. elegans* and found 13 to maintain silencing through several generations. He also developed a green fluorescent protein gene reporter system that remained stably silenced for more than 80 generations following initial RNAi exposure (Vastenhouw et al. 2006). The group also used a screen to identify mediators of this effect, recovering genes involved with chromatin states (in particular, histone acetylation), causing them to conclude that the effect of the RNAi to cause multigenerational effects was not through post-transcriptional effects on RNA but instead through silencing of transcription.

It would be logical to predict a model of transmission of RNAi from one generation to the next by dsRNA propagated through the germline. Oliver Hobert (2018, pers. comm.) recounts a second-hand anecdote from the Mello laboratory:

> Historically people injected DNA (for transgenesis) or dsRNA into the gonad. They thought this is how it had to be done. As for dsRNA, you inject into gonad and then you will get silencing effects that last for several generations. At some point Craig noted that *everyone* in his lab, including people in his undergrad class, could get RNAi to work. He thought "those students can't all be *that* good in injection" and he closely watched them. And indeed most of them didn't actually hit the gonad, but hit the head, the butt, the gut, whatever—and it didn't matter, you would get silencing effects in the next generation!
>
> It turned out that dsRNA propagates automatically to the next generation and does *not* need to be induced into the germline—it will travel there via transmembrane dsRNA transporters.

The transmembrane dsRNA transporters described above were found in 2003 (Feinberg and Hunter 2003) and their role in multigenerational inheritance through gametic effects were characterized in 2016 (Marré et al. 2016). Following their inheritance in the next generation, RdRPs then replenish the dsRNA pool so that it can then pass on to a further generation. This phenomenon does not occur for all genes, and even in an isogenic population of worms there can be major differences in the degree to which RNAi is propagated. Oded Rechavi's group has studied these differences and has found evidence (1) for lineages having equal chances of adopting either on or off states for a target gene of RNAi from exogenous dsRNA, (2) that this is related to how strongly each lineage responds to RNAi in general (not just from exogeneous dsRNA), and (3) that this may be related to variable expression of heat shock factor 1 (HSF-1) expression (Houri-Zeevi et al. 2020). He did not, however, describe the physiological relevance of the findings.

Insects

We should recall from Chapter 2 that Waddington had an interest in the effects of the environment on phenotypes. His experiments to expose developing *Drosophila* for several generations to increased temperature (Waddington 1953) or to ether vapor (Waddington 1956b) led to morphological phenotypes that became heritable, prompting him to develop the idea of "genetic assimilation" (Waddington 1959). Waddington's thinking about genetic assimilation was probably best illustrated in an earlier 1942 paper (Waddington 1942), in which he cited work describing callosities of the adult ostrich that were already present in the embryonic stage. As he put it:

> In this case we have an adaptive character (the callosities) of a kind which it is known can be provoked by an environmental stimulus during a single lifetime (since skin very generally becomes calloused by continued friction) but which is in this case certainly inherited. The standard hypotheses which come in question are the two considered by Robson and Richards: the Lamarckian explanation in terms of the inheritance of the effects of use, which they cannot bring themselves to support at all strongly, and the 'selectionist' explanation, which, in the form in which they understand it, leaves entirely out of account the fact that callosities may be produced by an environmental stimulus and postulates the occurrence of a gene with the required developmental effect. A third possible type of explanation is to suppose that in earlier members of the evolutionary chain, the callosities were formed as responses to external friction, but that during the course of evolution the environmental stimulus has been superseded by an internal genetic factor. It is an explanation of this kind which will be advanced here.

In advancing this idea, he cited an unexpected source, Charles Darwin (Darwin 1871), writing in what might be thought of as a relatively Lamarckian way:

> In infants, long before birth, the skin on the soles of the feet is thicker than on any other part of the body; and it can hardly be doubted that this is due to the inherited effects of pressure during a long series of generations.

Waddington believed that he had found evidence for a nongenetic mechanism of inheritance of an acquired characteristic. However, as described earlier in this chapter, researchers from the Sapienza University of Rome showed Waddington's heat stress phenotype to be due to genetic changes, caused by heat-induced activation of transposable elements (Fanti et al. 2017).

There have been numerous other studies of multigenerational effects in *Drosophila*, including the effects of diet, exposures, and age (Xia and Belle 2018). To take an example of these kinds of studies, a 2019 publication looked in detail at a phenomenon known to occur in *Drosophila melanogaster*, the preferential (>90%) laying of eggs where they can use ethanol as a food when the female is exposed to endoparasitoid wasps within the last 10 days (Bozler et al. 2019). The F_1 offspring derived from this preferential egg-laying behavior were not exposed to the wasps, but F_1 females still showed a significant preference (>70%) for laying eggs on ethanol, declining each generation to >40% in F_5 flies, after which significant differences were no longer apparent.

This is an example of one of the more careful and thoughtful *Drosophila* studies of nongenetic transgenerational inheritance. Males were exposed without effect on the next generation, inhibition of memory (in this case meaning the more common use of the word, that mediated by the brain) did not affect the inheritance of the ethanol preference, and the effect of neuropeptide F (NPF) was studied in detail. Reduction of NPF on its own was enough to induce the ethanol preference for egg laying and was associated with apoptosis of the female germline. Inheritance of the maternal copy of the gene encoding NPF was also found to influence the phenotype (Bozler et al. 2019).

As with most transgenerational studies, it becomes difficult to perform definitive experiments that address all the potential sources of variation. It would have been very interesting to see how much of an effect egg laying on ethanol-rich food had by itself, without wasp exposure, on the properties of the offspring. As the *Drosophila* were not explicitly described as isogenic, is it possible that the major apoptotic effects exerted a nonrandom selective effect on genotype? How variable was the amount of the volatile alcohol ethanol in the food? The authors found minimal transcriptional differences between the heads of the F_0, F_1, and F_2 flies, but this by itself does not exclude transcriptional or regulatory changes or cell subtype compositional changes in the brain if the proportion of cells involved is small. This is an example of an intriguing preliminary observation that should be followed by definitive studies, following the ECREE principle.

Rodents

Moving to rodent models, possibly the best-known example of a transgenerational influence by an environmental exposure is the effect of the endocrine-disrupting chemical vinclozolin on rodent spermatogenesis. This was first described by Michael Skinner and colleagues, who had noted that injecting pregnant rats with vinclozolin early in pregnancy (e8–e15) caused the F_1 offspring to have increased apoptosis of spermatogenic cells accompanied by decreased sperm number and motility (Cupp et al. 2003). In follow-up studies, his group mated F_1 males and females from exposed F_0 dams and showed that the sperm phenotypes were maintained, in the absence of further exposure, by repeating this breeding approach to the F_4 generation (Anway et al. 2005). Mouse studies that followed showed the same outcomes as the rat studies, even to the point of showing that the effects were limited to outbred but not inbred animals (Guerrero-Bosagna et al. 2012). The meDIP assay using tiling microarrays was used to identify differential DNA methylation in mature sperm, interpreting the modest changes in fluorescence intensity as evidence for altered DNA methylation in these cells. These studies generated a lot of excitement, given the conclusion that the effects of endocrine-disrupting chemicals were to act upon DNA methylation to create a stable molecular phenotype that resulted in a tissue phenotype involving spermatogenesis.

Applying the same standard of rigor as for the *Drosophila* studies above, a few questions arise. The number of rodents in each group was very limited, generally three to five. As mentioned in Chapter 6, the meDIP assay is influenced by the local density of CG dinucleotides, and the motifs enriched at the putative differentially methylated regions in the mouse studies were notable for being adenine homopolymers lacking the CG dinucleotides that should have been present in a study of DNA methylation differences (Guerrero-Bosagna et al. 2012). These results raise the suspicion that the results may have artefactual influences. As is typical for studies focused on DNA methylation, any effects of the exposure to cause DNA mutations were not considered, despite evidence that vinclozolin causes chromosome breakage in vitro (Hrelia et al. 1996) and is genotoxic in insects (Aquilino et al. 2018). Of particular concern is the failure of three separate groups to replicate these rat transgenerational effects on spermatogenic cells (Gray and Furr 2008; Schneider et al. 2008; Inawaka et al. 2009).

Assuming that future studies were to resolve the replication issue, and the original rat and mouse observations were upheld, the question why these effects are only seen in outbred and not inbred lines becomes intriguing. With very high rates of apoptosis (~90%) in spermatogenic cells, is it possible that vinclozolin is strongly selecting for natural variation in DNA sequences, possibly in combination with mutagenic effects of the chemical that generate new mutations, upon which (what would have to be extremely rapid) selection is acting to propagate an oligospermia phenotype? There is clearly much more to be explored in understanding the striking transgenerational vinclozolin phenotype in rodents.

The next logical step in this chapter would be to move to examples of transgenerational, nongenetic paradigms in humans, but these do not currently exist. Instead, human studies have almost exclusively focused on intergenerational examples and will be described next.

INTERGENERATIONAL MEMORY

Whereas multigenerational nongenetic heritability is all about traits being passed on without direct exposure to the original perturbation, intergenerational heritability can involve the germ cells, embryo, or fetus being co-exposed with the parent, with overlap into the DOHaD idea when an event in early life influences traits later in life. The idea that the memory of a past exposure is retained is what brings DOHaD under the epigenetics umbrella. We have a great example of a memory of our parents leading to changes in expression of genes, DNA methylation, chromatin states, and even phenotypic problems when revealed by genetic events—genomic imprinting. This was described in Chapter 3, but this phenomenon is physiological and does not obviously vary substantially or in response to environmental influences between individuals. Instead, the field of research on intergenerational memory is largely focused on early influences that change the future health of an individual.

We have another definitional problem when it comes to what we describe as epigenetic influences on intergenerational memory. An exposure to a pregnant mother that results in malformations of the offspring could easily be considered an example of a nongenetic memory in the offspring. However, our convention is to describe this as teratogenicity and not an epigenetic phenomenon. Adding to the confusion is the special situation of endocrine-disrupting chemicals (EDCs), which in the wild alter sexual differentiation, certainly a long-term memory of an exposure early in life. Unlike agents typically classified as teratogens, EDCs are studied as examples of agents with epigenetic properties (leaving a memory of their exposure on the developing individual). Retinoic acid embryopathy lives solidly in the definition of a teratogenic effect but is likely to mediate its effects through binding to the retinoic acid receptor with subsequent transcriptional regulatory effects, which certainly fits the common definition of an epigenetic effect. Exposure of pregnant women to the synthetic estrogen diethylstilbestrol (DES) is associated with malformations of the genital systems of female and male offspring, but also an increased rate of vaginal or cervical clear cell adenocarcinoma (CCA), typically a very rare cancer (Reed and Fenton 2013). These types of malignancies are usually associated with characteristic somatic copy number variants and mutations of genes governing oncogenic pathways (Marks et al. 2020). As DES was withdrawn from therapeutic use decades ago, the only study looking for mutations in samples from individuals whose mothers received DES was performed prior to the era of massively parallel sequencing; it failed to reveal any specific mutations in selected exons of three candidate genes (Boyd et al. 1996), although microsatellite instability associated with the exposure was observed. If DES-associated tumors were to be found to have mutations similar to those from similar tumors lacking DES exposure, then that would challenge our model of this example of epigenetic influences being DNA sequence–independent. Instead, we might usefully speculate whether DES-induced malformation of the developing genital tract creates a microenvironment conducive to propagation of cells with spontaneous somatic mutations. It was shown decades ago that injection of teratocarcinoma cells into blastocysts results in mice born with the teratocarcinoma cells present but differentiated within tissues and nonmalignant in their growth characteristics (Brinster 1974). What is clear is that the microenvironment of the neoplastic cell is very influential on its growth and differentiation; by reversing the paradigm, we can speculate that an altered microenvironment caused by the effects of a teratogen like DES may permit the uncontrolled growth of a cell with growth-promoting somatic mutations. This links back to the discussion of "field effects" in Chapter 8, but this time potentially related to events during development, creating a memory of past exposures that predisposes to future malignancy.

Animal Models of DOHaD

Worth noting are the historical foundations established by René Dubos (1901–1982). In his 1965 book (Dubos 1965), Dubos was prescient about a number of topics of current interest, including the body's microbiome and issues of personalized medicine, but his discussion of how events during early development influenced phenotypes later in life was illuminating, as it reveals a lot of the likely influences for those who would go on to develop the DOHaD field. Dubos reminds us that Sigmund Freud was very influential about how "early experiences condition man's whole living experience": The use of thalidomide in the 1950s was a grim reminder of effects of teratogens in pregnancy, and the experiments using electrical shock to induce anxiety in pregnant rats, leading to behavioral effects in offspring even when cross-fostered (Thompson 1957), were established paradigms at the time. The ideas of early exposures leading to later diseases therefore had foundations established earlier in the mid-twentieth century, and it has been suggested (Prescott and Logan 2016) that "Dubos and his colleagues paved mechanistic roads to the contemporary developmental origins of health and disease (DOHaD) concept," with Dubos explicitly addressing how early-life exposures affected later health as early as 1969 (Dubos 1969). Dubos' ideas extended to the issue of environmental exposures and how they shape the developing immune system, which is yet another legitimate inclusion under the DOHaD umbrella.

Probably the most highly influential paradigm for DOHaD was described by Michael Meaney, who used an existing model of rat maternal nurturing that was known to influence the responses to stress in later life of the pups (Weaver et al. 2004). The pups of the mothers who were less nurturing became more readily stressed later in life, an interesting observation that the Meaney group sought to explain through studies of the regulation of the glucocorticoid receptor (*Nr3c1*) gene in hippocampal tissue from the offspring. They studied DNA methylation at 17 CGs at one of several overlapping transcription start sites within a CpG island at the 5′ end of the gene, eventually focusing on two of the CGs, testing hippocampal tissue from four outbred rats per group. The effect of centrally infused trichostatin A, a histone deacetylase inhibitor, to reverse the molecular and physiological phenotypes was taken to be strong evidence for the causality of the molecular regulatory processes.

Although the lessons subsequently learned about epigenetic association studies and described in Chapter 7 would now cause us to be more cautious in interpreting these results, there is no doubt that this was a landmark study and prompted substantial follow-up work by many groups who saw this as an excellent model for mechanistic mediation of DOHaD. Almost 20 years later we now have DNA methylation data for this *Nr3c1* promoter locus in multiple species, including rat hippocampus (Agba et al. 2017). These authors studied the same regions of 17 CGs as shown in Figure 9.3 (the region they described as "Nr–00") and found the locus to be almost completely without DNA methylation, the highest levels at CGs 9 and 11 never exceeding 3% and 5%, respectively. Their data support this CpG island being constitutively unmethylated in rats in both sexes at 3, 9, and 24 months of age (Agba et al. 2017). The original Weaver et al. study performed bisulfite

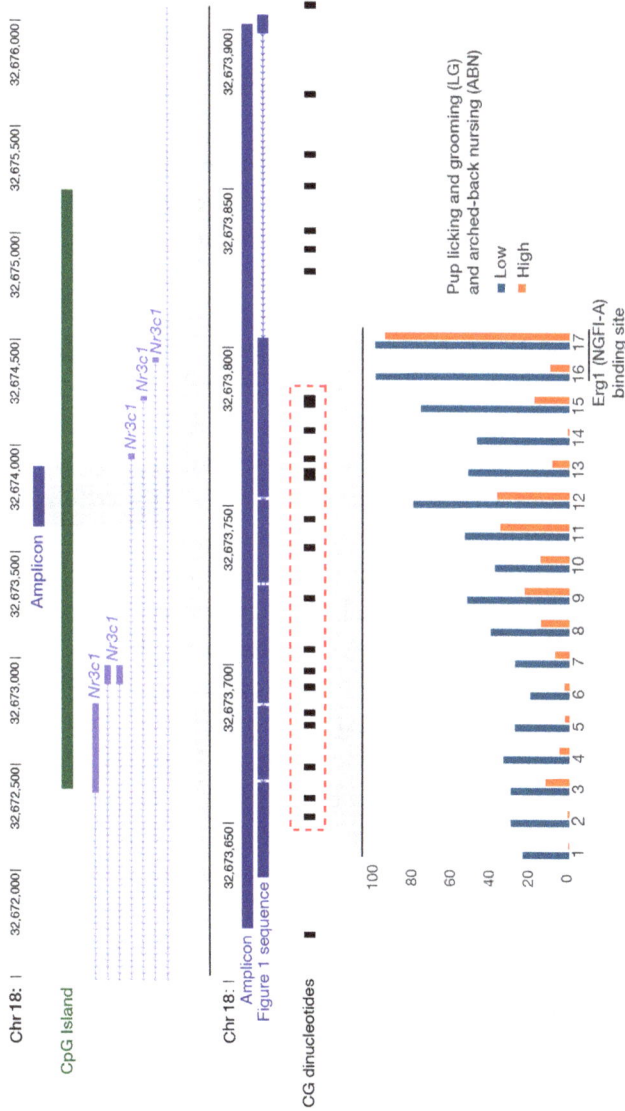

Figure 9.3. Looking back at the *Nr3c1* locus with current Rat Jul. 2014 (RGSC 6.0/rn6) annotations. The published PCR primer sequences should amplify a 285-bp product, but the sequence described in Figure 1 of Weaver et al. (2004) indicates polymorphism in the rats studied (a deleted area plus several small variants). The locus tested exists within a canonical CpG island, within 200 bp of one of many alternative transcription start sites annotated for this gene. Plotting the approximate DNA methylation values published (using http://www .graphreader.com) illustrates the substantial variability across this locus and between conditions. It would be potentially informative if today's more comprehensive DNA methylation and other transcriptional regulatory assays could be applied in a replicated experiment, allowing more quantitative studies, with single-cell assays providing insights into possible cell subtype changes occurring in the hippocampus in these conditions.

conversion of DNA, PCR, and sequencing of the "6–10 clones sequenced/animal; $n = 4$ animals/group," finding levels of DNA methylation at most CGs exceeding 10% and including measurements from ~40% to 95% in the control animals. Today's experience would suggest at least some of these measurements to be unrealistically high, raising the question whether the even higher DNA methylation values in the test animals are as reliable as we would need in order to make conclusions about transcriptional regulatory changes occurring in these hippocampal cells.

If these DNA methylation changes were found to be reproducible, they could also reflect changes in the cellular architecture of the hippocampus and not in reprogramming of the cells within the brain region. An intriguing model of prenatal exposure of glucocorticoid to male mice while in utero revealed the later life phenotype of increased anxiety (Tsiarli et al. 2017). This might sound like a great model for cellular reprogramming—an agent that works through a transcription factor (glucocorticoid receptor) leading to an adult phenotype. However, this group performed very detailed anatomical studies of the brains of these mice and found the cerebral cortex to be grossly thinner in the exposed animals, with histological alterations of the cellular distributions and neuronal architecture within the tissue and an increase in the number of neural progenitor cells. If these investigators had sampled a brain region like the cerebral cortex for molecular genomic assays, it is likely that they would have found differences in DNA methylation, attributable to these cell subtype proportion changes. A study focused on the hippocampus in postmortem samples of human neonates exposed to glucocorticoids in utero revealed similar change in cell subtype composition (Tijsseling et al. 2012). These results highlight two issues. One is that when an epigenomic assay is performed on a tissue, the cell subtype composition of the tissue needs to be tested at the same time. Second, a great way of maintaining a memory of an early-life exposure is to cause a remodeling of a developing tissue in a way that is maintained through life, as noted in Chapter 5.

Birth weight is a readily obtained phenotype from early life that has been associated in epidemiological studies with later life diseases. Low birth weight is associated with increased blood pressure in adulthood, with rodent and human studies further associating low birth weight with reduced numbers of nephrons in the kidneys, a plausible mediator of the hypertension phenotype (Luyckx and Brenner 2005). Our own work on intrauterine growth restriction in rats demonstrated changes in DNA methylation and gene expression in pancreatic islet cells from adult animals (Thompson et al.

2010), but we failed to take into account how the β-cell mass in the islets of this rodent model decreases significantly with age compared with controls (Simmons et al. 2001). We have to assume that our molecular genomic findings were at least in part due to changes in the subtype composition of the islet cells we studied.

DOHaD and Metastable Epialleles

Accounting for cell subtype changes in DOHaD models should be a part of how we understand the effects of early-life events on later life health and disease, but this does not exclude the possibility that cells also undergo reprogramming, reflected by changes in molecular transcriptional regulatory patterns in cells. An intriguing possibility is that the loci mediating these long-term effects are "metastable"—that is, capable of being influenced to adopt different binary states in terms of transcriptional regulation, the so-called "metastable epialleles."

Chapter 5 included the description of examples of metastable epialleles in mice, focusing on the viable yellow (A^{vy}) mice and their variable phenotype depending on whether an IAP transposable element at the *Nonagouti* locus was silenced or not. What was not described in that chapter was a further aspect of the phenotype, the preferential transmission of the coat color of the mother, and not the father, to the offspring, an effect studied in two generations (Morgan et al. 1999). Mechanistically, this is attributable to the silencing state of the IAP element being completely reversible in the male germline but only partially reversible in the female germline, so that a repressed state is allowed to maintain itself when inherited from the mother. A comparable mutation by a retrotransposon at the *Axin1* locus to cause the Fused phenotype (Axin1^{Fu}) was discovered by Shirley Tilghman and colleagues in 1997 (Vasicek et al. 1997). It was already known that the penetrance of the kinked tail phenotype of Axin1^{Fu} mice was dependent on its parent of origin (Ruvinsky and Agulnik 1990), subsequently found to be associated with the DNA methylation at the IAP transposable element at the *Axin1* locus and influenced by the genetic background of the mouse strain (Rakyan et al. 2003). When Rob Waterland and colleagues fed female Axin1^{Fu} mice methyl donor supplementation before and during pregnancy, they linked this environmental influence with increased DNA methylation at the IAP element and reduced penetrance of the kinked tail phenotype (Waterland et al. 2006).

Famines and Human Diseases

Another well-studied DOHaD model is the influence of early-life diet on adult health. As long ago as 1968 it was being observed that early-life nutrition in rats altered the number and size of adipocytes in adult tissue (Knittle and Hirsch 1968). René Dubos, mentioned earlier, had a multifaceted career that involved some of the earliest discoveries of antibiotics, but in his later career he was a prolific author who wrote about topics as varied as environmentalism and social determinants of health (Honigsbaum 2016). In 1966 he published a study that looks strikingly contemporary, entitled *Biological Freudianism: Lasting Effects of Early Environmental Influences.* Although this was overtly addressing the social determinants of health, Dubos focused on early-life events as critical mediators of adult health (and associated "socio-medical problems"). He presented a study of pathogen-free (gnotobiotic) mice, noting that different lines had different adult growth potential, attributed to genetic differences between the lines. This observation prompted him and his co-investigators to cross-foster 240 pups from different strains to 30 different mothers, leading to the realization that the growth of the pups was more dependent on the "nursing effectiveness of the foster mother" than their genetic background. They then tested the effects of modification of the lactating mother and infection of the offspring early in life, revealing lifelong effects of both interventions on offspring growth. Dubos ends by quoting Wordsworth, "*The child is the father of the man,*" as a poetic way of expressing his conclusion that early-life events involving nutrition and infection have strong influences on adult health.

It may be astonishing that questions still central to the DOHaD field were being tested so carefully more than 50 years ago, but it should be noted that Dubos was not proposing anything conceptually new. The British nutritionists Elsie Widdowson (1906–2000) and Robert McCance (1989–1993) were cited by Dubos for their earlier work on rat growth. They performed a simple experiment, "varying the number of rats to be suckled by a single mother from the first day of their lives," finding that animals from the smaller litters, who had increased availability to maternal lactation, grew more quickly and attained sexual maturity earlier (Widdowson and McCance 1960). Even prior to Widdowson and McCance were the studies of Clement Smith (1901–1988), one of the first pediatricians to focus on the care of newborns, which became known as today's field of neonatology. As part of his interest in the effects of maternal nutrition during pregnancy on the health of the offspring, Smith obtained funding to visit the Netherlands at the end of World War II in Europe. The Netherlands had just suffered a period of famine that had been meticulously recorded with what Smith later called the "painstaking ways of administrators and civil servants in the Netherlands" (Smith 1976).

What had recently happened in the Netherlands was recounted by Smith in 1947 (Smith 1947a). In September 1944, three months after the D-Day Normandy landings, British troops attacked Arnhem in the Southern Netherlands, hoping to capture a bridgehead across the Rhine. The Dutch government in exile in London simultaneously broadcast on Radio Orange an appeal to railroad workers to strike, to prevent the mass transportation of German soldiers and stolen goods back to Germany. The workers responded with a very effective strike that started in September 1944. What was not anticipated by the Dutch government was the failure of the Allies to take the Arnhem bridgehead, coupled with the severe winter that froze the barges in canals. The strike that they requested persisted until liberation in May 1945, with the unintended consequence that food and fuel transportation to major Dutch cities in the Northwestern Netherlands almost completely ceased, which was compounded by the German troops adding road and canal blockades to prevent other modes of transportation being used. That winter of 1944–1945 became known as the *Hongerwinter* (or Hunger Winter) and resulted in an estimated 91,000 excess deaths (Ekamper et al. 2017).

The pattern of nutritional deprivation was one of a progressive decline from October through December 1944, with a sustained low point from January through April 1945, and restoration of normal nutrition with liberation in May 1945. Smith noted the high prevalence of amenorrhea (50%) during the hunger months, and that timing irregularities in those still menstruating caused estimation of gestational age "even less trustworthy than is commonly the case" (Smith 1947a). Although birth weights dropped 8%–9% overall, the potential that some of the effect could be due to inexact calculation of gestational age and increased rates of prematurity could not be excluded. Another observation was that the overall rate of pregnancies dropped drastically. Records from the National School for Midwives in Rotterdam and the Obstetrical Service of the Zuidwal Hospital at The Hague show the pregnancy rate dropping to one-third of the expected number of births during the undernutrition period (Smith 1947b).

Epidemiologists looking at this tragedy saw how it could offer the opportunity to gain insights into the long-term consequences of intrauterine starvation within a lifespan, by studying the babies that were born during this period of Dutch history. The challenge was taken up by Zena Stein and Mervyn Susser (1921–2014). If there is ever a film made with epidemiologists as the heroes, it should be about this couple, whose lives were genuinely extraordinary (Neugebauer and Paneth 1992). Their scientific rigor and intellectual honesty were described by Neugebauer and Paneth:

> [The book] conveyed well Mervyn and Zena's analytical principles and aggressive intellectual stance. It said in effect: you are the master of your epidemiological fate; you, not chance or convenience, must take charge of the variables under study. Choose to control your variables either in the design or analysis of the study. If you opt for the latter, accept no association at face value. Rather, your data analysis, conducted warily at every turn, must be a relentless effort to refine each variable and to identify its position and possible contribution to the causal sequence under investigation. Beware of confusing variables antecedent to the exposure of interest with those that are intervening; never be deceived by a suppressor; above all else, drive confounders out into the open.

Zena Stein, while on maternity leave in 1952 in South Africa, read Clement Smith's 1947 publications describing the Dutch Hunger Winter and realized that this was an opportunity to understand whether nutritional deprivation during pregnancy was associated with later life health issues. Her attention turned to the question whether cognitive abilities in later life were affected by this early-life deprivation, a motivation later described by Zena Stein to have been prompted by the civil war in Nigeria of the 1960s. This civil war was accompanied by mass starvation, and the fear was that the next generation of Nigerians would grow up with diminished cognitive abilities because of poor nutrition during development (Wilcox 2003).

Stein and Susser embarked on a study of 125,000 Dutch males born in the calendar years 1944–1946, spanning the period of famine, and including 20,000 exposed to the famine by maternal starvation during intrauterine development. Their data were from the Dutch military induction examinations that included physical, psychological, and educational tests. In their 1972 publication, they devote as much text to the description of "Independent Variables" as they do to their results. This meticulousness allowed them to understand a paradoxical finding from the study, that the intelligence metrics were increased in those exposed to famine in utero, attributable to the increased fertility of people in higher socioeconomic classes (Stein et al. 1972).

The Dutch military also put the young men onto weighing scales. In a subsequent 1976 publication, Stein and Susser reported that exposure to nutritional deprivation in the first half of pregnancy was associated with increased rates of obesity (rising from a baseline of about 1.5% in controls to a peak of 2.8%), but that those exposed in the last trimester of pregnancy and first few months of life had significantly lower rates of obesity (Ravelli et al. 1976). This time they found no effect of socioeconomic class and linked their findings to existing animal model studies to propose that the effects could be due to alterations in the number of adipocytes generated during development.

It is principally the obesity phenotype that has captured our attention in the DOHaD field, but if we remain with the Stein–Susser family a bit longer, another consideration emerges. Zena Stein's 1952 maternity leave produced a son, Ezra Susser, who followed the family tradition and became an epidemiologist, also pursuing questions in his own career about the Dutch Hunger Winter. At the time of his studies, the cohorts being studied included women and a broader range of phenotypic outcomes. Ezra Susser found increased risks of schizophrenia in females but not males with maternal starvation during the first trimester and of affective disorder in males but not females when the exposure was in the second trimester (Brown et al. 1995). Ezra Susser's particular focus was on the reproducibility of his findings in other populations who had undergone famine. He noted that in studies of those exposed to the multiyear Chinese famine of 1959–1961 there was evidence for an increased rate of schizophrenia in people exposed prenatally to starvation (Susser and St Clair 2013). A meta-analysis by Li and Lumey of adults who were exposed to the Chinese famine in early development showed that the choice of controls is important; if you compare those born during the famine years with those born afterward, you are introducing age as a source of variation, associated with the increased rates of being overweight or having type 2 diabetes, hyperglycemia, or the metabolic syndrome. These associations disappear when controls selected from both before and after the famine are used (Li and Lumey 2017). After driving the confounders out into the open, as Zena Stein and Mervyn Susser would have put it (Neugebauer and Paneth 1992), the only association that survives is schizophrenia.

David Barker and Discordance among Epidemiologists

Much of the contemporary interest in DOHaD can be traced to the work of David Barker (1938–2013). Described as the "founding father of DOHaD" (Fall and Cooper 2015), his publication in 1986 linking poor nutrition in early life with ischemic heart disease in adulthood set the course he would continue to pursue throughout his career (Barker and Osmond 1986). His life's work is described in vivid detail in an admiring 2019 tribute (Barker et al. 2019). His breakthrough observation was based on studying maps of the United Kingdom showing increased rates of death of heart disease in poor regions in the 1970s. Barker was not convinced that adult cardiovascular disease was solely attributable to adult lifestyle, reasoning that there was no reason for people from poorer regions to have a worse lifestyle than those from affluent areas (Pearson 2016). When he found that the same regions were associated with increased infant mortality 50 years previously, this allowed him to propose the alternative idea that adult cardiovascular disease was attributable to poor nutrition early in development (Barker and Osmond 1986). What distinguished Barker from others who had been working on similar ideas was his intriguing model—that nutritional deprivation in the womb and early life causes an adaptation by the baby to reprogram its metabolism permanently, even when nutrition became plentiful later in life, which he described as the "thrifty phenotype hypothesis" (Hales and Barker 2001). Early adaptive responses that later in life become maladaptive is a powerful idea and became popular, not least because of the "evangelical zeal with which he promoted his ideas" (Pearson 2016).

The exceptional ability of Barker to communicate his ideas is undeniable. What is not as often recognized is that his epidemiological methods were not always felt by his epidemiology colleagues to be appropriately rigorous. In 1995 Mervyn Susser wrote a commentary with University of Michigan epidemiologist Nigel Paneth in which they raised major concerns about the methods used by David Barker and his University of Southampton colleagues (Paneth and Susser 1995). Susser was being consistent with his guiding philosophy, described above, that confounding influences need to be sought in epidemiological studies. Susser and Paneth described concerns about recently published studies by the Southampton group of a cohort from Hertfordshire in the South of England (Fall et al. 1995). They raised specific concerns about the potential confounding effect of migration, of body mass index as a variable intervening between birth weight and cardiovascular disease, and of the use of weight as a surrogate for nutrition in early life. However, they reserved their major concern for the breadth of the starting hypothesis of the Southampton group that "...a baby's nourishment...influences the diseases it will experience in later life"

(Barker 1994). Their concern was that this was just too all-encompassing (Paneth & Susser, 1995):

> With so broad a hypothesis, researchers are free to test the relation between a whole range of possible markers of a baby's nourishment and any diseases it will experience and to pronounce important those relations that are confirmed. In this work, researchers faced with findings that fail to support the hypothesis seem not to be treating them as threats to the integrity of the hypothesis.

Their criticisms then get more pointed,

> The notion of induction, that knowledge is gained by the summarizing of facts and experiences, has fallen on hard times as a credible approach to research. Indeed, it is easy to see the barrage of papers from Barker's group as an inductionist's delight. Example is piled on example, each somewhat consistent with the hypothesis but none seriously testing it.

Paneth and Susser were not alone in publishing their concerns. Peter Gluckman (Gluckman et al. 2005) put the concern about the use of birth weight as a surrogate for developmental conditions succinctly:

> Using fetal growth as a marker of an adverse intrauterine environment is a convenient but imprecise marker, namely because not all fetuses exposed to an abnormal fetal environment have altered growth, and not all altered fetal growth is a function of responses to environmental stimuli.

Another concern about Barker's approach was voiced by L.H. (Bertie) Lumey (Lumey 2001) in published correspondence in which he expressed concern about Barker's "failure to note that the study findings did not agree with the original study hypothesis":

> The original study hypothesis was not specified in the report, and therefore it was not obvious that the findings provided no support for the a priori hypothesis but rather for a hypothesis formulated later. This presentation of the findings is disappointing and misleading, since I set up the study to test several explicit hypotheses about the relation between fetal nutrition and CVD risk.

Although this discussion may appear arcane, it flags a major potential issue given the way that the studies of prenatal malnutrition break down outcomes by exposure timing and sex of the offspring. The multiple testing being performed can be looked on in two ways—this is a valuable exploratory approach that can generate interesting hypotheses, but it also creates a problem in understanding the significance of findings when multiple tests are being performed. When the focus is on statistical testing for significance, measures like preregistration (Nosek et al. 2018) allow the researchers to state in advance what outcomes they are seeking, permitting testing focused on those outcomes to be used more powerfully. What Lumey was raising as a concern was the possibility that Barker's group was claiming, only after seeing the results, that they had been looking from the outset for the outcomes found to be significant. This kind of analysis, described as "data torturing" (Mills 1993), does not represent best practices but can be difficult to recognize in published work.

The reason for stressing this cautionary note is that the early claims about the developmental origins of health and disease, so influential in the DOHaD field, may have suffered from methodological flaws and overenthusiastic interpretation. One interesting insight after looking back at these early studies of the Dutch Hunger Winter and other cohorts is how these scientists revealed the human and emotional pressures at play. In rebuking Bertie Lumey for the published correspondence quoted above (Lumey 2001), Jan van der Meulen (van der Meulen 2001) concluded with a caution:

> The publication of Lumey's allegations, as well as the implied support by *The Lancet*, may be damaging for the future of the Dutch Famine study. Such damage is regrettable, since we have a unique opportunity to investigate the effects of maternal starvation during gestation on health later in life.

The wording is unfortunate, as it suggests that public criticism could have the undesirable effect to compromise the funding of an ongoing study in which the author is involved, while failing to address the more substantive issue raised by Lumey about the flaws he perceived in the methodology and interpretation of results. In a fascinating interview (Wilcox 2003), Zena Stein recounted the response from colleagues when she published the first results of her studies of the Dutch Famine:

> In the Dutch Famine Study, we found no evidence that 6 months of starvation in pregnancy, among a reasonably well-nourished population, had any effect on the intelligence of the offspring. Many of our colleagues were furious with us because it would have been much more satisfactory (in terms of social justice) to have found that food did matter.

It is extremely uncommon to gain these kinds of insights in the published literature and reveals human nature to be more of a contributor to science than dry, impersonal manuscripts would normally indicate. Probably the worst accusations that could be made in this instance remain relatively benign, that personal investment in these studies and their overenthusiastic interpretation have led to weak foundations for the DOHaD field. This flaw does not need to be irreversible, and it is encouraging that more contemporary studies, exemplified by work on the Dutch famine birth cohort led by Susanne de Rooij (Bleker et al. 2021), pay substantially more attention to the confounding influences that could be affecting the outcomes they observe, resembling the epidemiological stringency championed by Zena Stein and Mervyn Susser. For example, the de Rooij group acknowledges the potential confounding influence of selective conception described earlier and adds some new potential reasons for concern, including the selection for survival of more robust individuals and "information on the time period between discharge from the hospital after birth and age 50 when the cohort was first traced, making it impossible to adjust for potential confounders that may have had an influence during this period" (Bleker et al. 2021). If this is a sign of the increased care now being taken with the interpretation of observations of DOHaD cohort studies, we should be able to have more confidence in the interpretation of future results of these human experiments.

Testing Molecular Memory: Epigenetic Association Studies in Human Cohorts

The potential that molecular genomic regulators maintain memories of early-life exposures was of obvious interest, prompting a study of DNA methylation testing same-sex siblings, only one of whom was exposed to the Dutch famine prenatally (Heijmans et al. 2008). Five CG dinucleotides at the differentially methylated region of the imprinted *IGF2* gene were tested in 60 sib pairs. Whole-blood samples were used, and a statistically significant decrease in DNA methylation was found for those exposed around the time of conception but not later in gestation (Heijmans et al. 2008). Although this result was interpreted to support the authors' conclusion that this represented "empirical support for the hypothesis that early-life environmental conditions can cause epigenetic changes in humans that persist throughout life," the study was published before we were aware of many of the confounding influences in these studies, including cell subtype proportion changes. As the changes found in the group showing statistical significance represented ~5% changes in DNA methylation, they could easily be explained by cell subtype proportional changes, as could later sib pair studies from the same group that looked at further genomic loci, finding similar modest changes in DNA methylation (Tobi et al. 2014). Empirical support for early-life experiences causing epigenetic (molecular genomic regulatory) changes is certainly suggested by these studies but remains to be formally proved. Although there have now been numerous studies of DNA methylation in human DOHaD cohorts, they universally suffer from the same issues of interpretability as these Heijmans and Tobi report, for the reasons described in Chapter 7.

One of the ways that we can fail to interpret an epigenetic association study correctly is by ignoring the potential for genetic differences between individuals to affect outcomes. With DOHaD studies, the assumption is made that we're looking for nongenetic effects—that environmental influences such as those affecting birth weight can affect any individual. To perform an interpretable epigenetic association study, we need to understand how genetic variation may be influencing the results. An even more intriguing possibility arises when we consider the role of genetic variation in studies linking low birthweight with later-life diseases. In a genome-wide association study of birth weights of more than 500,000 individuals, a Mendelian randomization approach (Smith et al. 2007) was used to dissect the causality of relationships between birth weight, maternal and offspring genotypes, and adult type 2 diabetes and systolic blood pressure (Warrington et al. 2019). They looked specifically at the relationships between maternal blood pressure and the birth weight and later-life blood pressure in the offspring. The typical DOHaD supposition would probably be that increased maternal blood pressure, by leading to decreased intrauterine growth and birth weight, would cause a reprogramming in the offspring leading to increased blood pressure later in life. Instead their causality studies found that the critical factor in the offspring having increased blood pressure later in life was their inheritance of maternal risk alleles for hypertension (Warrington et al. 2019).

A genetic contribution to the phenotypes studied in DOHaD projects should be a cause for concern if the possibility of a genetic influence were not considered and not sought in the execution of the study. When we look back at the original data supporting obesity as a consequence of famine-induced intrauterine growth restriction, the kind of observation found to be significant is a rate of obesity in the famine-exposed individuals of 2.77% compared with 1.44% in control individuals (Ravelli et al. 1976). By dichotomizing (splitting into two groups) the participants in the study in terms of a definition of obesity, for which there are well-known genetic influences, the potential exists for a nonrandom segregation of genetic risk alleles for obesity into one group with this kind of result as a consequence. Studying genetic influences as part of the interpretation of DOHaD studies represents another area of potential improvement of and better insights into the results of studies linking early-life exposures with later-life phenotypes.

How Extraordinary Is the Evidence for Intergenerational Memory in Humans?

Animal models show persuasive evidence for developmental or early-life events affecting later-life health and behaviors, making it likely that similar processes occur in humans. The tragedy of famines has allowed later research to be performed to study the effects of prenatal nutritional deprivation, with some interesting findings but some reasons for concern that weaken our confidence in the generalizability of the findings and the strength of these effects in human health. Laplace and Sagan would probably raise their collective eyebrows and send us back to work things out a bit further. We have a lot more to do before we can conclude that intergenerational effects on phenotypes involve the nongenetic model of cellular reprogramming mediated by molecular genomic regulators of transcription.

POTENTIAL MOLECULAR MECHANISMS OF MEMORIES IN HUMANS

Mechanistically, what do we know about how memories could be transmitted from early to later life, from one generation to the next, or across multiple generations? In this section we suspend all skepticism temporarily and assume that nongenetic memories do indeed occur within the human lifespan, from parents to offspring, and across multiple generations, asking how such memories could be propagated mechanistically. The molecular mechanisms of Chapter 4 and cellular memories of Chapter 5 serve as the starting point for discussing these possibilities.

Within the Life Course, the DOHaD Model

Probably the easiest of all the mechanisms to understand is that of early-life events influencing later-life phenotypes. To keep our categories distinct, we have to perform a separation between events occurring during intrauterine life, which have to take into account the potential influence of the mother, falling instead into the intergenerational category.

This discussion of retention of memories from one time in life through later life gets us back to a central question in Chapter 4: How do we propagate a molecular genomic regulatory state through mitotic cell division? As well as DNA methylation, we saw varying strengths of evidence for other molecular genomic regulators to maintain their states from parent to daughter cells, including prior cell signaling states and TF bookmarking. For such a regulatory process to persist through the life course, either (1) stem and progenitor cells need to have acquired these new characteristics or (2) they have affected long-lived postmitotic cells, exemplified by neurons and T lymphocytes. In Chapter 5 we were presented with the alternative possibility that memory could be mediated by changes in the cell subtype composition of the tissue. Probably the best-understood model of early-life exposures influencing long-term health is the development of the immune system repertoire (Nielsen et al. 2019). This process involves the generation of a wider repertoire of cells than is needed, including cells that could cause harm by recognizing self-antigens, followed by the selective loss of these cell clones and the expansion of clones capable of recognizing pathogens. Memory in this context is a function of the retention of certain cell clones, which does not sound like nongenetic inheritance of molecular regulatory processes and is not widely considered to be epigenetic. However, this issue raises a valuable point for consideration—the mechanism of cell fate. For the immune system, the fate of cells is determined by the interplay of TFs and molecular regulators of transcription (Busslinger and Tarakhovsky 2014). In a search for an epigenetic (molecular genomic regulatory) mechanism mediating a perturbation in the development of the immune system repertoire, the focus could be on changes in the activities of these regulatory mediators. For a comparable example outside the immune system, the prenatal glucocorticoid model described earlier also found an effect of the prenatal steroids on neuronal stem and progenitor cell proliferation (Tsiarli et al. 2017), providing another model for studying the molecular genomic regulatory mediators of this response.

It is, however, difficult to study molecular influences on cell fate in humans, as it is very challenging to access the cells in which cell fate decisions are being made during development. We are left with the derived, mature cells of later life, whose proportions within a tissue may reflect their prior cell fate decisions, but, if their differentiation has occurred through typical mechanisms, may not have any innate molecular changes reflecting their prior exposures. These points highlight some of the practical difficulties in molecular genomic studies when the effect of perturbation is on cell fate.

Intergenerational Transmission of Memories to Offspring

The biological mechanisms become more mysterious (and, not coincidentally, more likely to be described as epigenetic) when we start to talk about a parent's "experience" being transmitted to the offspring. This intergenerational model has to be looked at separately when thinking about maternal and paternal effects.

Maternal effects: Maternal effects can be usefully separated into three groups: those that precede pregnancy, those during pregnancy, and those after pregnancy. Of these, probably the least studied are the effects of events prior to pregnancy, which have been described as "oocyte imprinting" by Jacquetta Trasler (Lucifero et al. 2004). The molecular genomic regulators of transcription in mammalian oocytes have been extensively characterized (Sendžikaitė and Kelsey 2019; Stäubli and Peters 2021). It is challenging, even in rodents, to collect enough oocytes to perform genome-wide studies of these molecular epigenetic regulators, especially when looking for differences from a defined normal state in these cells. Direct evidence for molecular perturbations in oocytes associated with preconceptual exposures or events is therefore sparse.

Effects during pregnancy are mechanistically relatively easy to conceptualize because of the ability of the mother to influence the baby through the transmission of molecules across the placenta to the fetal blood circulation. There are many well-known examples of maternal diseases causing phenotypes in offspring, mediated by high maternal glucose levels causing the production of excess fetal insulin, with consequent growth enhancement of the fetus, or neonatal lupus and heart block in offspring of mothers with systemic lupus erythematosus who transfer autoantibodies across the placenta. The effects of prenatal glucocorticoid exposures in mice and the later-life phenotype of increased anxiety were described earlier in terms of anatomical changes in the brains of these animals (Tsiarli et al. 2017). Each of these examples is associated with cellular changes in the developing fetus; for example, the heart block in neonatal lupus is due to the death of cells in the atrioventricular node, with their replacement by fibrotic tissue and calcifications (Llanos et al. 2012).

Possibly the best studied example of a defined maternal exposure during pregnancy associated with adverse outcomes for the offspring is maternal smoking, despite some caveats (Knopik 2009). Maternal smoking fits a mechanistic model of toxins being transferred across the placenta to the developing fetus. A careful study to date aimed toward understanding the association of adverse offspring outcomes with maternal smoking was performed on blood leukocytes from the Avon Longitudinal Study of Parents and Children (ALSPAC) cohort, testing DNA methylation longitudinally, accounting for confounding influences when possible, and exploring critical windows of exposure and testing causality (Richmond et al. 2015). Despite what must have been the significant temptation to jump to conclusions about causation when interpreting some of the strong associations they found, the authors did not speculate whether the DNA methylation changes mediated the phenotypes associated with smoking and acknowledged that cell subtype composition could mediate at least some of the effects observed. The restraint shown by these authors is remarkable for its contrast with other comparable studies of maternal (and fetal) exposures during pregnancy, which frequently drift into assumptions of causality despite being poorly interpretable. In general, it can be said that we have little evidence to support molecular mechanisms of cellular reprogramming mediating the adverse outcomes of maternal–fetal exposures during pregnancy, but in theory we can continue to invoke the potential effects of an altered environment on epigenetic regulators, bearing in mind the associated caveats outlined in Chapter 7.

The last category, that of maternal effects exerted after pregnancy, could be mediated in mammals by the direct transfer of molecules through breast feeding, which has clear longer-term benefits for the neonate (Ho et al. 2018). Focusing on the postweaning period, we are posed with a substantially more complex mechanistic model of transmission of an influence from a mother to an offspring when there is no physical means of transfer of a message. The example earlier in the chapter of the rat maternal nurturing model is in this category. The Meaney group focused on DNA methylation as the potential mediator of these postnatal effects during a plastic period of development (Weaver et al. 2004). Ignoring the methodological concerns described earlier for that study, an example of a DNA methylation change at a locus should prompt the question why that locus was chosen for alteration, which ultimately involves discussions about TFs and their upstream regulation either as nuclear receptors or by cell signaling pathways

(Chapter 4). What could be imagined mechanistically is that the activation of the hypothalamic pituitary adrenal (HPA) axis proposed by the Meaney group could lead to increased glucocorticoid exposure to the hippocampal neurons, with associated transcriptional regulatory changes, that for some reason become stable at the molecular genomic level. This is exactly what the Meaney group is proposing, and it could indeed be occurring, bearing in mind the experimental concerns described earlier in the chapter. A question becomes why these glucocorticoid exposures resulted in fixed reprogramming of these neurons when the essence of the hippocampus is that it is functionally plastic (Horner and Doeller 2017) and can change both dendritogenesis and neurogenesis in response to steroid hormones (Sheppard et al. 2019). To lock in a transcriptional regulatory change long term as a cellular reprogramming event would probably require a strong silencing signal such as those from Polycomb or classical heterochromatin formation. Were such an observation to be made in hippocampal neurons at specific loci following post-pregnancy adverse events, that would make a more convincing case for genuine cellular reprogramming.

Paternal effects: Post-pregnancy effects by fathers on offspring are not only theoretically possible, it has been found (again a result of studying the ALSPAC cohort) that paternal depression in the postnatal period is associated with depression in the offspring at age 18 years (Gutierrez-Galve et al. 2019). This is not, however, the focus of studies of nongenetic inheritance from fathers, typically studying diet, stress, or exposures (Galan et al. 2020). Paternal effects used to be dismissed mechanistically based on the idea that the sperm contributed almost nothing apart from a haploid genome in the process of fertilizing an egg. A 2018 study of 17 individuals from three families generated a lot of attention by suggesting that paternal mitochondrial DNA could be found in offspring, implying transmission by sperm (Luo et al. 2018), but this was not replicated in a larger study of 41 trios (patients and parents) (Rius et al. 2019). We do not have clear evidence for transmission of nonnuclear DNA to offspring by sperm at present.

What we focus on instead are three mechanisms described in Chapter 4: DNA methylation, nucleosomes, and small RNAs. As a molecular genomic regulator with a mechanism for propagating itself through cell division, DNA methylation was the first and obvious choice for studies. The idea was that an exposure to the father would somehow result in a signal to spermatogenic cells to change DNA methylation at specific loci that would maintain this pattern through embryogenesis to lead to functional consequences for differentiated cells in the offspring. There were two main problems that led to DNA methylation falling out of favor as a potential mediator of paternal effects. One is the dramatic erasure of DNA methylation in early embryogenesis—somehow the specific DNA methylation signal would have to be established in such a way at target loci that it would be resistant to this physiological erasure. The second problem was that the degree of change of DNA methylation at specific loci in sperm cells in test mice tended to be small, no greater than 10%–20%. For haploid cells, this means that only 10%–20% of cells were undergoing any changes, assuming these changes did not represent experimental noise or were confounded by the influences described in Chapter 7. As Oliver (Ollie) Rando and his colleagues put it (Galan et al. 2020):

> Thus, modest methylation changes should only affect the *penetrance* of a phenotype within a litter, rather than affecting the majority of offspring across litters as observed in many paternal effect paradigms.

Nucleosomes represent a second, intriguing potential mediator of messages from sperm to embryo. In Chapter 4 we were introduced to the dramatic phenomenon of replacement of nucleosomes during spermatogenesis by histone variants. Initially during the condensing spermatid stage of spermatogenesis, the nucleosomes are replaced by transition proteins 1 and 2 (TP1, TP2) (Bao and Bedford 2016), whereas later, at the elongating spermatid stage, further replacement by protamines takes place (Hao et al. 2019). It has been estimated that 10%–15% of the human genome avoids replacement by these histone variants in contrast to mouse, for which 1% of the genome remains nucleosomally organized, although this appears to reflect a technical issue of inclusion of human sperm with incomplete histone replacement (Yoshida et al. 2018). As the embryo restores nucleosomal organization to the haploid paternal genome, a preexisting nucleosome with a specific pattern of post-translational modifications could conceivably mediate signaling from the paternal germline to the embryo.

Stephen Krawetz noted in a 2005 review that sperm also contribute longer, protein-coding RNAs as well as small RNAs to the zygote (Krawetz 2005). In 2011 he followed up with a sequencing study of short RNAs in human sperm, revealing that the majority was transcribed from repetitive elements, with added representation from piRNAs, miRNAs, and short transcripts that mapped to promoter sequences (Krawetz et al. 2011). Déborah Bourc'his and Olivier Voinnet proposed that small RNAs in sperm could have a role to silence chromatin, which they described as the confrontation–consolidation pathway (Bourc'his and Voinnet 2010). This, they suggested, was a mechanism for variation in the complement of small RNAs in sperm to silence chromatin in a variable manner and lead to long-term silencing. Although it is difficult to imagine how small RNAs derived from repetitive elements could have such an effect, certainly the small RNAs homologous with promoter sequences mapped by Krawetz would be candidates for such locus-specific targeting.

It gets more complex than DNA methylation, nucleosomes, and RNA and their potential roles transmitting information from the sperm to the embryo. Krawetz noted that sperm also delivers signaling molecules and transcription factors to the zygote (Krawetz et al. 2011), which could also have at least short-term effects on transcriptional regulation. Ollie Rando has highlighted the heterogeneity of sperm, showing that the more mature sperm from the distal epididymis have a different repertoire of small RNAs, and that this mature RNA repertoire is needed for appropriate gene expression during early embryogenesis (Conine et al. 2018).

Modern-Day Gemmules: Getting a Message from the Soma to the Germline

Probably the most challenging question is whether an experience, acquired trait, or disease state in a parent can transmit a signal to the germline that causes a similar phenotype to be induced in subsequent generations. Darwin's idea of the inheritance of variation involving gemmules turned out not to be necessary when DNA was found to do the job, but it illustrates the intuitiveness of a model involving molecules from the soma influencing the germline.

Some studies involve inducing a phenotype (often "stress") in an adult male mouse, careful selection of mature sperm from the distal epididymis, and purifying them from contaminating somatic cells by having them swim away from the remainder of the cells, followed by injection into zygotes (Gapp et al. 2020). Rando has shown that small RNAs can be transferred from the epididymal epithelium to maturing sperm, representing a somatic to germline transfer of potentially regulatory molecules. The elements are therefore in place to test whether small RNAs are generated by a cell or tissue mediating a parental phenotype, transferred to the developing germline cells, targeted to specific loci, and involved in the development of a similar phenotype in the offspring. With a focus on inheritance of metabolic states, Rando has noted that the mechanism might be mediated by effects on the placenta, an intriguing and testable model (Galan et al. 2020).

Multigenerational Nongenetic Heritability

In theory, if the process above can self-propagate with a germline-mediated susceptibility to a phenotype transmitted to the offspring in a way that causes gemmule-like small RNAs to get trafficked again to the germline in the next generation, the phenotype could become genuinely transgenerational. Also, in theory, another way to propagate the memory would be for the phenotype to become replicated in the offspring and de novo send a message to the germline to set up susceptibility for the next generation.

It's worth emphasizing once again that we already have a great candidate for mediating transgenerational phenotypes—DNA sequence variability—so this needs to be excluded as an influence in any study of transgenerational heritability of traits believed to be mediated by transcriptional regulatory processes. That requires deep sequencing that is rarely performed as part of these studies, even when using outbred animals (such as humans).

We can't just invoke the well-studied systems in *C. elegans* and plants and say that this is a paradigm for mammals, when we mammals lack RNA-dependent RNA polymerases, and we have a major erasure of DNA methylation and chromatin states during gametogenesis and embryogenesis that does not occur in plants.

It's probably fair to say that we have interesting early observations worth pursuing further when considering the idea of nongenetic transgenerational inheritance. It is also fair to say that we do not yet have the evidence base to satisfy Laplace or Sagan, and we need to move beyond correlative studies to those that are more definitive one way or the other.

The Problem of Using Outbred Animals

Have you noticed the common theme to some of the above rodent models of nongenetic heritability—they focused on the use of outbred strains? This may have its roots in the field of toxicology, which has a traditional preference for using outbred strains, with the vague justification that they are representing human genetic variation, but incurring significant problems of reducing test sensitivity and obscuring the influences of genetic variation, as critically reviewed by Festing (2010).

The problem with outbred strains is that you are not sampling from a highly diverse population: You are using the limited number of animals you purchased to create a colony and then performing matings within that group. Recognizing that this can lead to the development of subgroups of animals with distinctive genotypes, the practice of genetic monitoring (GeMo) is a requirement for maintenance of these colonies (Benavides et al. 2020), with the Jackson Laboratory recommending that colonies maintain 25 unrelated males for breeding (Lambert 2009).

Looking back at the studies cited earlier showing evidence for nongenetic heritability of traits in outbred animals, it is not at all clear that these best practices were followed in the animal breeding. When you have a great mechanism for heritability of a trait—DNA sequence variation—and you are potentially selecting animals that have undergone genetic drift, population bottlenecks, and founder effects, and you don't directly look for the influences of genetic variation on your phenotype of interest, you could be misled into thinking that the effects you observe are nongenetic. Large, genetically monitored outbred colonies could be a solution, but the better approach is probably to test whether the phenotype is replicated in large cohorts of inbred animals. If not, any nongenetic heritability hypothesis is highly questionable. Right now, we cannot assume any of the models based on outbred animals that have not been replicated with inbred strains are valid.

Implications of Multigenerational Nongenetic Heritability

To have locked in a pattern of transcriptional regulation in one generation that cannot be overcome in later generations means that normal cell signaling to the genome is being blocked. We are preventing acclimation, the adaptive changes in a plastic phenotype within a generation in response to the individual's environment (Baugh and Day 2020). It is difficult to understand how multigenerational nongenetic heritability offers any advantages for a species apart from amplifying the effects of adverse events in your progenitors. Eva Jablonka was an early leader in exploring the evolutionary implications of transgenerational epigenetic inheritance (Jablonka and Raz 2009), a question that remains actively explored (Lind and Spagopoulou 2018).

The neuroscientist Kevin Mitchell posted on social media the summary shown in Figure 9.4 that represented his frustration with the idea that a message emanates from neurons to the gametes and finds its way back to those specific neurons in the next generation. A valuable point within the

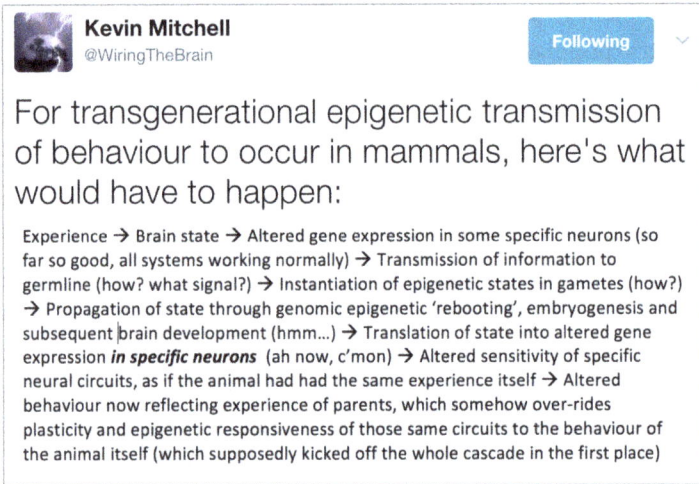

Kevin Mitchell
@WiringTheBrain
Following

For transgenerational epigenetic transmission of behaviour to occur in mammals, here's what would have to happen:

Experience → Brain state → Altered gene expression in some specific neurons (so far so good, all systems working normally) → Transmission of information to germline (how? what signal?) → Instantiation of epigenetic states in gametes (how?) → Propagation of state through genomic epigenetic 'rebooting', embryogenesis and subsequent brain development (hmm...) → Translation of state into altered gene expression *in specific neurons* (ah now, c'mon) → Altered sensitivity of specific neural circuits, as if the animal had had the same experience itself → Altered behaviour now reflecting experience of parents, which somehow over-rides plasticity and epigenetic responsiveness of those same circuits to the behaviour of the animal itself (which supposedly kicked off the whole cascade in the first place)

Figure 9.4. A social media posting from Kevin Mitchell describing succinctly his concerns about a model of transgenerational transmission in mammals of behavior through epigenetic mechanisms (reproduced with permission from Dr. Mitchell, Trinity College, Dublin).

entertaining rant is that we are assuming that neural plasticity in one generation can cause plasticity to be abolished in the next. What could be added to Mitchell's concerns is the fact that neuroplasticity is not solely determined by the way that neurons express their genes but also by the connections and organization of neurons in the brain and the ways that synapses form and function (Mateos-Aparicio and Rodríguez-Moreno 2019). A perspective that proposes that all plasticity is abolished by the transcriptional regulation of a limited number of genes is challenging.

Focusing on the Germ Cell Cycle

It is all very well to say that something is mechanistically implausible, but you're left with the difficulty of dismissing observations like the multiple generations of mice with oligospermia following vinclozolin exposure (Guerrero-Bosagna et al. 2012) (bearing in mind the caveat earlier in this chapter about the irreproducibility of these studies). If DNA mutations potentially caused by the vinclozolin can be excluded as a cause, how do you explain the persistence of this kind of phenotype?

Germ cells give rise to germ cells in mammals, with a short intermediate stage of zygotic and early embryonic development. If we focus on this "germ cell cycle," a term that seems to have been first used more than a century ago

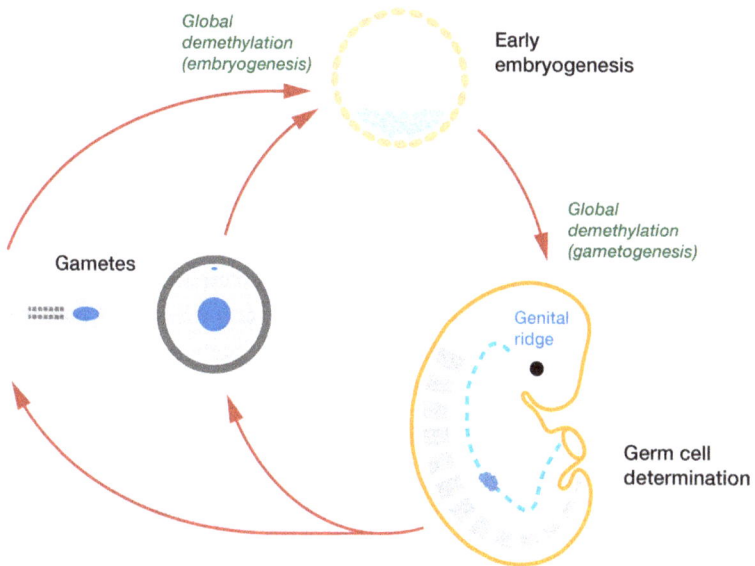

Figure 9.5. An illustration of the germ cell cycle, to show (a) how germ cell commitment in embryogenesis and fetal development create the gametes of the next generation and (b) how there is global loss of DNA methylation at two stages in this process, early embryogenesis and gametogenesis.

(Hegner 1914) (Fig. 9.5), the early embryo quickly specifies within the inner cell mass a subset of epiblast cells that become primordial germ cells (PGCs) at the time of gastrulation. These PGCs then populate the developing gonad to become the germ cells for the next generation.

As a process of cell differentiation, it is a reasonable possibility that a molecular process established in a progenitor cell could persist to derived cells. What makes this germ cell–germ cell differentiation process distinctive are two things: The progenitor cells end up as progenitor cells at the end of the process, and the process involves two waves of global demethylation of DNA—one during gametogenesis, the other during early embryogenesis. In addition, during mammalian spermatogenesis, most of the chromatin is replaced by histone variants. No other differentiation process involves such a radical removal of global regulators of transcription.

The global demethylation is not, however, complete. Azim Surani studied the dynamic DNA methylation patterns in human and mouse PGCs using whole-genome bisulfite sequencing. By identifying the loci where DNA methylation persisted during this stage of gametogenesis, he hoped to be able to find the loci that could mediate transgenerational inheritance through transcriptional regulatory mechanisms. Although the level of DNA methylation genome-wide fell to ~3%–4%, some regions retained ≥30% DNA methylation. These were almost all repetitive sequences—in particular, the evolutionarily youngest and most active retrotransposons, and the targets of the ZFP57 (ZNF698) transcription factor described in Chapter 4—selecting loci at imprinted domains for targeting by DNA methylation (Tang et al. 2015).

Although this suggests that there are indeed loci that could be resistant to erasure of DNA methylation and mediate a message across generations, the quantities of DNA methylation at these loci were still relatively modest. As discussed in Chapter 5, 30% DNA methylation means that 30% of alleles are methylated, and if this represents both alleles on homologous chromosomes being concordantly methylated, then 30% of cells are methylated at the locus and 70% are not. If a DNA methylation signal is to be maintained across generations, it is difficult to imagine how this can occur robustly when only a subset of cells retains the molecular mark mediating such a memory.

This pattern of active retrotransposons retaining a distinctive pattern of DNA methylation brings us back to the viable yellow A^{vy} mouse described in Chapter 7. That variable phenotype arose because of an active retrotransposon that acquires DNA methylation differently in different animals. George Wolff studied breeding data from his viable yellow mice from the 1970s and 1990s, noting that if a mother (dam) had a black or pseudoagouti coat color, indicating DNA methylation of the IAP retrotransposon, they had a higher proportion of pups with similar coat colors than those of yellow (unmethylated) dams (Wolff et al. 1998). He also observed that the pseudoagouti coat color was more likely to be manifested in pups of pseudoagouti fathers (sires) than dams, but this was found later by Emma Whitelaw to be strain-dependent, with the opposite parental effect in a C57BL/6J background (Morgan et al. 1999). These observations indicate that a preexisting DNA methylation state in the gamete can persist in the offspring, and that there are differences in the ability to alter these preexisting DNA methylation states in spermatogenesis and oogenesis—differences that are influenced by genetic background.

This was a very intriguing observation that prompted Anne Ferguson-Smith to ask whether the striking examples of viable yellow A^{vy} and Axin1^{Fu} mice represented a broader set of ERV retrotransposons acting as metastable epialleles in the mammalian genome (Vasicek et al. 1997). Her group found that IAP elements with variable DNA methylation retained no memory of the parental DNA methylation state and reset every generation (Kazachenka et al. 2018). In a more recent study, she asked whether these variably methylated IAPs were responsive to putative influences on DNA methylation such as the endocrine disruptor bisphenol A, an obesogenic diet, or methyl donor supplementation. The group found no differences in DNA methylation at these loci (Bertozzi et al. 2021). It is therefore likely that the viable yellow A^{vy} and Axin1^{Fu} examples represent unusual examples of DNA methylation variability at retrotransposons linked to somatic phenotypes and not the tip of the iceberg of a broader group of metastable epialleles.

Studying IAP retrotransposons does not exclude there being other loci that could be metastable epialleles and mediate environmental exposures and the heritability of DNA methylation patterns, but the Ferguson-Smith studies make it much less likely that the viable yellow and axin-fused examples of IAP-mediated effects can be generalized. We still lack evidence for heritability of DNA methylation at non-retrotransposon sequences in the mammalian genome. We are led to the conclusion that the mammalian genome is designed for erasure of transcriptional regulatory signals and that only rarely can retrotransposed DNA sequences overcome this erasure.

The Problem of Small Changes of DNA Methylation in Sperm

Because human sperm samples are routinely collected in fertility studies, representing one of the cell types accessible for molecular studies without an invasive procedure, they have been tested in many studies for changes of DNA methylation as an indicator of altered transcriptional regulatory programming. For example, a very consistent pattern of DNA methylation changes with age has been found by Douglas Carrell (Jenkins et al. 2014), who subsequently found that these changes do not persist in grandchildren (Jenkins et al. 2019). This should not be surprising when the degree of change of DNA methylation is considered: The values, although statistically significant, are generally in the 2%–3% range (Jenkins et al. 2014). In Chapter 6 we noted how a 3% change of DNA methylation, assuming concordantly methylated alleles in a diploid cell, reflects 3% of the cells changing DNA methylation. In haploid sperm you don't need to include the caution about both alleles having to be concordantly methylated, a 3% change of DNA methylation means 3% of the sperm have a distinctive DNA methylation pattern. Whether this is due to an alteration of cell subtypes within the sperm population remains to be tested.

As mentioned earlier, these small changes in DNA methylation in sperm being the general finding rather than the exception has been noted by Ollie Rando (Galan et al. 2020). It follows that the sperm with the molecular regulatory change are very underrepresented within the population of sperm competing to fertilize the ovum. Carrell's finding that DNA methylation differences do not persist across generations (Jenkins et al. 2019) may have an explanation distinct from the more fundamental mechanistic question about the tendency of these DNA methylation marks to survive erasure; they may simply be outcompeted by the majority of sperm present that have not undergone molecular regulatory changes. We have no insights yet into the fraction of sperm that have unusual small RNA payloads, but this may be testable using single-cell transcriptomic approaches, helping to understand the likelihood that these RNAs mediate transmission of information across generations. It would not be surprising to find that the small RNA complement of sperm is relatively uniform. Since 1955 we have known, through electron microscopy studies, that "spermatids derived from the same spermatogonium are joined together by intercellular bridges" in multiple species (Burgos and Fawcett 1955), which is thought to homogenize the contents of these cells.

How can we explain a situation in which the DNA methylation changes in sperm are modest (10%–20%), but DNA methylation changes persist to the next generation (Radford et al. 2014)? With only 10%–20% of the sperm reprogrammed, it would be expected that the 80%–90% of the sperm with no alteration would effectively outcompete during fertilization. Otherwise, we must invoke an advantage for the reprogrammed sperm in competition for fertilization.

In theory, this is possible. Geoff Parker in 1970 defined sperm competition in terms of sperm from two fathers competing to fertilize the same eggs in the polyandrous matings of insects (Parker 1970). Polyandry is not required, however, as each haploid sperm from a single father has a different genetic makeup, potentially allowing different means of competition and cooperation between "sibling" sperm, as reviewed by Simone Immler (Immler 2008). The outcome of selective advantages for certain genotypes violates Mendel's First Law of independent segregation and random union of gametes at fertilization and manifests as a transmission ratio distortion, with 12 genes identified that appear to mediate these effects in mammals, as reviewed by Joseph (Joe) Nadeau (Nadeau 2017).

Studies of intergenerational transmission of nongenetic signals have, like most other studies of nongenetic influences, tended to focus on molecular assays with less attention to cellular properties and mechanisms. There is a lot to gain by applying the lessons of epigenome-wide association studies from Chapter 7 to the area of multigenerational heritability, using some of the ideas above as a starting point.

The Problem of Invoking Epigenetic Determinism

The remaining concern about nongenetic heritability is the degree of determinism that appears to be attributed to these processes in the phenotypes of subsequent generations. We would hesitate to attribute such determinism to all genetic causes of Mendelian disorders, recognizing that even very damaged genes can inconsistently have an associated phenotype (penetrance) or severity (expressivity).

The viable yellow A^{vy} mice represent the problem nicely. A C57BL/6J dam with a pseudoagouti phenotype, and therefore a less methylated IAP retrotransposon, is not destined to have offspring that are all unmethylated at this locus. Instead, she herself has a variegated population of cells with different DNA methylation states at this locus and has offspring that start with a single methylated state (fully methylated or fully unmethylated) in the ovum but can end up pseudoagouti because of switching of the DNA methylation state within the developing embryo. A litter of multiple offspring can have the full range of black to pseudoagouti to yellow pigmentation phenotypes. What the maternal state influences is the likelihood of ending up preferentially within this spectrum, but it does not determine this outcome.

This in turn highlights a major problem in the published mouse studies testing multigenerational nongenetic heritability—the very small number of mice used, typically in single digits. If an exposure of a parent is being tested for effects in offspring and later generations, and the expected outcome is an influence on the tendency to develop a phenotype rather than a uniform outcome, it is difficult to imagine how small numbers of animals will reveal what might be very modest influences on phenotypes.

♦ ♦ ♦

Waddington's epigenetic landscape model did not define where a specific ball would roll within the choices of creodes at each bifurcation. Instead, a population of balls rolling down the landscape sequentially reveals the relative likelihood of each ending up in a specific creode. As a model, it captures nicely the nondeterminism of many aspects of development. This idea of determinism in development has been dissected nicely by Magdalena Żernicka-Goetz and Sui Huang, who note the temptation to regard transcriptional nonuniformity ("noise") as a manifestation of randomness and stochasticity (Żernicka-Goetz and Huang 2010). Like the bifurcation of a creode, they describe how a balance between mutually repressive TFs creates a metastable state that resolves into two separate differentiation outcomes and can be perturbed by influences to tilt the balance in favor of one of the cellular outcomes. These influences are not only environmental and extrinsic to the cell but can also be encoded in the genome, as mentioned in Chapter 7. What Żernicka-Goetz and Huang also note is the extension of this model encompasses a perturbation leading to an alteration of the properties of a fluctuating system that is maintained over multiple cell generations. Although this remains largely a theoretical possibility at present and is difficult to study, the concept can legitimately be brought into the discussion of nongenetic transmission of properties across generations of cells, which can potentially extend to gametes and their propagation of cellular characteristics to embryos.

Getting It Right: How to Study Intergenerational Nongenetic Heritability

A goal for this chapter has been to raise the possibility that some of the assumptions in the fields of DOHaD and intergenerational and transgenerational inheritance research should be revisited now that we have updated techniques and new insights about the potential for being misled by prior study designs. To balance what may come across as a pessimistic viewpoint, we can look to a more recent study that is impressive in its design and execution.

Jamie Hackett had trained with Azim Surani as a postdoctoral fellow. When he started his laboratory in Rome, he asked himself what the most significant question was that he could test experimentally with his training, and how he could answer this question in a robust way. Coming from the Surani laboratory, he was familiar with imprinted genes representing an exception to the global reprogramming of molecular regulators of transcription during development, and how other loci also appeared to escape complete loss of DNA methylation (Tang et al. 2015). Hackett's question was whether he could find evidence that these escaping loci could be influenced by the environment. But first he wanted to identify an environmental perturbation that resulted in a strong phenotypic response in offspring. He identified three pillars he'd need to put in place:

- A big hammer. Some sort of environmental perturbation likely to have major disruptive effects.

- An order of magnitude more experimental animals than the typical 10–20 used in comparable studies, providing enough statistical power to detect consequences when the result of the environmental perturbation may have small effect sizes and be partially penetrant.

- Reversibility. To support nongenetic mechanisms of heritability, the removal of the environmental perturbation in the parent should lead to the reversion of the phenotype in subsequent offspring.

He chose a system involving a paternal exposure, as the heritability could be more focused on information carried in sperm. At the time, it looked like disruptions to the microbiome were having substantial physiological consequences, prompting his decision to treat the animals with oral, nonabsorbable antibiotics to modify gut flora.

Key to the success of the project was new postdoctoral fellow Ayele Argaw-Denboba (now a group leader at the Max Planck Institute of Immunobiology and Epigenetics in Freiburg, Germany). Jamie Hackett (2024, pers. comm.) describes him as "[n]ot only a wonderful scientist, but the most motivated hard worker I'd ever seen":

> I remember it was his first lab meeting. He had about 30 offspring from dysbiotic fathers, and about 30 from controls. He was reporting something that was completely unbelievable, that about 30-40% of the offspring were sick, but not a single offspring from the controls.

The phenotype they were observing was a proportion of offspring who were unusually small, with some failing to survive. They didn't believe their data and responded by increasing the number of offspring tested to more than 350 in total. Sure enough, the proportion of offspring with the growth phenotype dropped to 5%–10%, but with only ~1% of controls affected, they now had strong evidence that the difference was real (Argaw-Denboba et al. 2024).

They then performed different ways of altering paternal gut microbiomes, replicating their phenotype. To test reversibility, they allowed the male mice to recover a normal gut microbiome and mated them again; they no longer saw the effects on offspring growth.

Hackett was concerned about being misled by multiple testing; if they looked at too many phenotypic outcomes, some would emerge as significant by chance. That led to the restricted focus on the growth and related survival manifestations.

The obvious next question was the mechanism involved, involving three questions:

- How is the microbiome perturbation signaling its disruption to the germline?
- What in the germline is altered to reflect this signaling?
- What is the etiology of the phenotypic manifestation in the offspring?

They were transparent in their report that they did not find any answers for the first two questions, but implicated abnormal placental vascularity as the cause of the poor growth in the affected offspring. As part of the testing of the possible mediators of the germline signaling, they tested DNA methylation, finding no differences in sperm from the treated and control animals. They also found no evidence for DNA sequence changes mediating these effects in offspring, as part of the process of ruling out what would be confounding influences.

This represents a valuable paradigm for a study of nongenetic heritability that can guide more rigorous studies in this field. The use of adequate numbers of experimental animals for high-confidence results would probably have appealed to René Dubos, but Hackett recognized that his study, involving more than 2000 mice, was potentially troubling because of the ethics of animal research. His justification was that it is more responsible to perform a definitive study with greater numbers of animals, publishing the positive or negative outcomes, than to perform hundreds of studies that are underpowered to deliver reliable results.

Hackett (2024, pers. comm.) recognizes the continuity of his research with that of his former mentor:

> It's only subsequent to our completing that study that I've realized that there's some nice symmetry to this in terms of Azim's discovery of genomic imprinting in mice, arguably the first example of inherited epigenetic information in the mammals.

> We're standing on the shoulders of giants like Azim, adding an extra layer that there might be other inherited regulatory marks that can be environmentally sensitive on top of genomic imprints.

> That's nice retrospectively, it's a really nice kind of continuity.

TRANSGENERATIONAL INHERITANCE AND THE POTENTIAL TO DO HARM

The intuitiveness of the idea of a parent's experience being transmissible to offspring, through mechanisms other than DNA sequence changes, is very attractive. Without knowing about genetic mechanisms, Jeremiah, Aristotle, Lamarck, and Darwin all embraced the idea. In the field of epigenetics, the idea of nongenetic heritability of experiences and exposures of parents and prior generations has become popular beyond the scientific literature, especially when seeking answers for the detrimental effects of trauma across generations, with the types of trauma including famine (Stein et al. 1975), the Holocaust (Yehuda et al. 2016), and racism (Aroke et al. 2019). The attractive idea that life events can cause adverse health outcomes in subsequent generations mediated by transcriptional regulatory changes has given rise to a whole new subfield of epigenetics, social epigenomics.

What should be emerging from the discussion in this chapter is that we cannot yet be certain that there exist nongenetic mechanisms for heritability of acquired traits across generations. As scientists, we are free to speculate when we have interesting preliminary data, thinking aloud and publicly about next steps and the implications of our work. Such speculation is a part of science and is valuable.

A problem worth considering, especially in the field of multigenerational nongenetic heritability, is that this speculation could have unintended negative consequences. The intuitive and relatable nature of the focus of these studies engages the public, but the complexity of the advanced genomic technologies is difficult to evaluate by the general public and is likely to be regarded by the nonexpert as more definitive and conclusive than it actually is.

Gemma Sharp has written about one of these unintended negative outcomes in DOHaD research—the emphasis on the mother's role in a child's future health. This has the consequence of neglecting how paternal influences and exposures in later life may be contributing to the phenotypes being studied (Sharp et al. 2018). In a follow-up study (Sharp et al. 2019), her group raised a valuable point about a negative effect of this biased focus:

> In the media, DOHaD findings are often reported using alarmist, inflammatory language, with pregnant mothers presented as individually responsible for a host of specific harms to future generations, ignoring the societal systems that influence health behaviours. This public discourse can have coercive and autonomy-limiting effects for women.

When we talk about the inheritance of trauma from prior generations, those most invested in learning about the scientific advances are those who are already at risk because of their family's negative experiences. To hear discussions about marks added to DNA transmitted through gametes that propagate memories of adverse events in prior generations could leave the impression that the person in such a family is fundamentally and irreversibly damaged.

This is harmful. We should not be giving the impression that we have proof that people have been marked at the molecular level by experiences of past generations; we simply do not have the evidence for this currently. Furthermore, it is a fundamental principle in clinical genetics that when you don't have an intervention, you don't want to thoughtlessly burden the individual with knowledge of their future risks of devastating diseases (like Alzheimer's or Huntington's). By being more aware of the potential to do harm to the vulnerable among us, we can do better in communicating our scientific uncertainties, especially when dealing with research questions involving families that have undergone or continue to undergo trauma.

RECOMMENDED FURTHER READING

Heard E, Martienssen RA. 2014. Transgenerational epigenetic inheritance: myths and mechanisms. *Cell* **157**: 95–109. doi:10.1016/j.cell.2014.02.045

A review from Edith Heard and Rob Martienssen that remains relevant today as a caution against overinterpreting studies indicating that transcriptional regulatory mechanisms may mediate nongenetic heritability across generations in mammals.

Horsthemke B. 2018. A critical view on transgenerational epigenetic inheritance in humans. *Nat Commun* **9**: 2973. doi:10.1038/s41467-018-05445-5

Bernhard Horsthemke continuing the message of concern about uncritical acceptance of the idea that epigenetic (nongenetic, transcription regulator–mediated) inheritance happens in humans, while adding some clear, constructive guidelines for those wishing to test this idea.

Jung YH, Wang H-LV, Ali S, Corces VG, Kremsky I. 2023. Characterization of a strain-specific CD-1 reference genome reveals potential inter- and intra-strain functional variability. *BMC Genomics* **24**: 437. doi:10.1186/s12864-023-09523-x

A report highlighting the reasons for caution when using outbred mice for studies involving epigenomic assays, as mapping sequencing reads to the standard mouse reference genome is misleading and creates false positive results.

Wilcox AJ. 2003. A conversation with Zena Stein. *Epidemiology* **14**: 498–501. doi:10.1097/01.ede.0000071471.35756.96

Although the published conversation between Allen Wilcox, the then Editor-in-Chief of the journal Epidemiology, and Zena Stein is by itself a fascinating read, as a bonus the interview is available on YouTube; search for "A Conversation with Zena Stein" or at this link: https://youtu.be/G5ECc-59Ax4.

CHAPTER 10

The Landscape Ahead

Having rolled down the landscape of the preceding nine chapters, we are now terminally differentiated epigeneticists. What was the point? Why spend all this time delving into the arcane details of epigenetics from its historical roots to the complexity of its molecular mechanisms, what it tells us about disease, and even the multiple meanings of the word?

As it turns out, a book has some advantages over alternative ways of describing a topic. By covering the multiple facets of epigenetics, a book can become akin to a series of linked reviews in one location, allowing some central themes to be maintained while exploring subtopics as diverse as cancer, transgenerational inheritance, and genomic biochemistry. The goal all along was clarity: We don't have to agree about our personal vision of the meaning or value of epigenetics as a scientific field, but if we are at least aware that this fascinating topic is distinctively beautiful in the eyes of different beholders, we have achieved a valuable realization.

REVISITING OUR TERRIBLE EXPERIMENT FROM CHAPTER 1

Let's return to the made-up experiment of Chapter 1 armed with our new insights into epigenetics:

> **HYPOTHESIS: SUNLIGHT CAUSES OBESITY IN GRANDCHILDREN THROUGH "EPIGENETICS"**
>
> - *Foundational observation:* Today's grandparents were more sun-exposed as children, and today's grandchildren are more obese, so sunlight is a candidate for nongenetic (epigenetic) effects on the health of these kids.
>
> - *Primary mediator:* Sunlight helps to create the active form of vitamin D in human skin, and vitamin D binds to transcription factors to regulate gene expression (epigenetic).
>
> *Continued*

- *Secondary mediator:* Changes result in methyl groups added to DNA (epigenetic).
  - – The vitamin D response to sunlight is therefore changing DNA methylation (epigenetic).
- *Inheritance across generations:* DNA methylation can be passed from parent to daughter cells and from generation to generation (epigenetic).
  - – The sunlight response can therefore be passed on to the next couple of generations in a non-DNA-mediated way (epigenetic).
- *Intervention:* By eating a superfood, you can reverse the DNA methylation changes and change your health, overcoming the epigenetic curse of your ancestors (epigenetic).

With the benefit of the insights gained in the prior chapters, we can think through the logic above. DNA demethylation does occur at vitamin D receptor binding sites (Català-Moll et al. 2022). The propagation of DNA methylation through cell division is a biochemical fact, but Chapter 4 should have pushed us out of our comfort zone by noting that DNA methylation in mammalian genomes is the default state, and it's where we don't see DNA methylation that interesting things are happening (e.g., binding of sequence-specific transcription factors [TFs]). For this reason, the information content of the loci where there is no DNA methylation is typically greater than where there is DNA methylation, and we note that TFs may be able to linger at the replicated DNA and pass on their information to daughter cells. The major exception was the situation of heterochromatinization, involving Polycomb and other mediators acting on histones, in which DNA methylation can be part of the repressive process.

But we're not talking about the acquisition of new DNA methylation or heterochromatin in response to sunlight and vitamin D but instead about the loss of DNA methylation at vitamin D receptor binding sites. How this state could be passed on through the gametes would probably require continued binding of the vitamin D receptor in the chromatin of gametes (for which there is no current evidence) and would be a direct effect by the primary mediator (vitamin D) and not a secondary epigenetic (transcriptional regulatory) mediator. Any such change would have to survive the global reprogramming in gametogenesis and early embryogenesis, for which the best examples (like imprinted loci) involve heterochromatin formation. It is very difficult to believe that this state can be transmitted across generations at one or more loci in the genome. We know that any change of DNA methylation or another transcriptional regulatory mediator at a locus could be due to a change in the DNA sequence locally, including causing the loss of DNA methylation (Stefansson et al. 2024). Should this be the case, an intervention like consuming a dietary supplement is unlikely to have any effect, if such supplements have effects to start with.

It should be apparent from the above how we can always bring mechanistic questions back to cells. The entirety of Chapter 5 was devoted to making this point, which is so easily overlooked when we view biology exclusively through the lens of molecular events. If sunlight and vitamin D are passing on their influences onto the next generation, they have to be influencing gametes. We can consider the possibility raised in Chapter 9 that the effects may involve the alteration of the types of gametes produced, noting that sperm are inherently heterogeneous.

We then get to the idea that epigenetic events are reversible. Setting aside both the conflict with the your-grandmother-ate-the-wrong-food-so-your-health-is-irreparably-ruined transgenerational heritability advocates and the taint of commercialism for epigenetic consumer products based on weak evidence, the idea of reversibility sounds plausible when you think about the molecular properties of transcriptional regulators, which have to be able to reprogram all loci, even those where the long-lasting mediators of genomic imprinting act. In practice, the human genome is typically a lot more complicated than a single gene like the retrotransposon upstream from the *Nonagouti* locus in viable yellow mice. Even for these mice, feeding them diets supplemented with single carbon donors was complicated by the need to introduce these diets early in maternal pregnancy.

So, if you are going to eat a so-called superfood like kale to reverse your epigenetic problem, you may need to go back in time and ask your grandmother to be a very early adopter to include this forbidding-looking brassica in her diet, when its contemporary use at the time was probably exclusively as cattle feed.

The bottom line is you should avoid saying "epigenetic" when there's a more accurate way of describing what you're trying to say, so that these artificial and misleading linkages can be avoided in trains of logic. Also, think of what's happening with cells and not just molecules when trying to figure out nongenetic mechanisms of disease.

WHAT WAS THE POINT OF ALL THAT HISTORY, ANYWAY?

A lot of the book has been based on history. In part because it was fun to dig into primary sources, getting old books delivered, opening them and getting that distinctive scent of printed paper from another era, reading while imagining the world in which the author was living at the time, marveling at the hand-drawn illustrations, and appreciating the use of the medium of books, which lend themselves to broader, more thoughtful musings about a topic than today's typically terse formats.

History is not all remote and from another era. During the writing of this book, it was sad news to hear of the passings of Art Riggs, after having had him provide his perspective to help understand the development of the field since the 1970s, and of David Allis, who shared his personal insights into the genesis of the term "the histone code." Having a record of the contributions of great scientists like them honors their memory. The discussions with contemporary scientists about their work from decades ago were fascinating, in part because of the humanity that came through (such as the technical mistakes that proved to be illuminating), choosing career directions because they had just met their future partner and wanted to be with them, or emigrating to avoid the possibility of fighting their heavily armed ex-mentor in a revolutionary war. A goal of recounting these stories is to embolden today's early-stage colleagues with the realization that today's senior luminaries made mistakes and faced challenges but still succeeded in making major scientific advances.

The history of epigenetics has been told in terms of the underpinnings of theories about fertilization, embryogenesis, and sex determination. The recurring theme is one of conflict, with deep schisms dividing those passionate about preformationism versus epigenesis, ovism versus spermism or animalculism, Lamarckism versus the Weismann germ cell cycle model, and cytoplasmic versus nuclear regulation of development. The advocates on both sides were all leading scientists of the day and vulnerable to the human failings that persist today. The young upstart Wolff was correct about his theories of epigenesis, but von Haller and the rest of the German academic establishment felt the need to put the politely insubordinate Wolff in his place. Those who dissented about the nuclear regulation of development have since had their cytoplasmic theories validated by the recognition of siRNAs produced by Dicer in the cytoplasm, and there is a growing interest today in the rehabilitation of Lamarckism, or nongenetic inheritance of acquired characteristics. Conflict is not always a problem in science, may be inevitable, and is probably productive most of the time, as colleagues challenge each other to defend their assumptions. If this is the major lesson from the recounting of history in this book, a valuable goal will have been achieved.

◆ ◆ ◆

Waddington was unusual for being a broker of ideas across the divide of a twentieth-century conflict—that between embryologists and geneticists. He was also unafraid to propose grand unifying ideas and had a skill in communication, making his relatively unexciting epigenetic landscape model more compelling than it probably deserved at the time. Oddly enough, the era of large, population-scale genotype–phenotype correlation studies and single-cell genomics have made this kind of model relevant again. A great example of a large population study came from a study of the UK Biobank in 2016, in which 173,480 participants were studied, linking hundreds of common and rare variants with cell subtype proportions in peripheral blood (Astle et al. 2016). Although there is more to the steady state proportions of blood cell subtypes than influences on lineage commitment during differentiation, it is tempting to assume that the model of genetic variation changing cell fate is responsible for at least some of these changes. It is also noteworthy that the informative sequence variants were enriched at loci with distinctive "epigenetic" (usually histone modification) properties (Astle et al. 2016). When the group of Dan Landau wanted to portray how differentiation of hematopoietic cells was influenced by mutations of genes encoding enzymes involved in DNA methylation maintenance, they generated "differentiation topologies" based on their single-cell transcriptional data, eerily recreating Waddington's hand-drawn landscape but based on molecular genomic data. As example of the Landau laboratory's unpublished work is shown in Figure 10.1. They reanalyzed data from a study of chromatin accessibility in hematopoietic cell differentiation (Izzo et al. 2024) using techniques developed for DNA methylation data from the same cell types (Izzo et al. 2020). This is an example of beautiful data visualization—esthetically pleasing while communicating data effectively. John Piper (1903–1992), the artist friend of Waddington who drew the original depiction of the epigenetic landscape (Waddington 1940) (Fig. 2.3), would have been gratified.

EPIGENETICS IS A CONSTANTLY EVOLVING, EXCITING FIELD OF RESEARCH—EMBRACE THE CHURN

If it is not obvious at this stage, a goal in the preceding chapters was to probe the foundations of this world of science living under the umbrella of epigenetics. Don't be put off by any perceived doubts about the field; as with any fast-moving, technologically driven area of science, we have some legacy areas of research that could do with revisiting. What we can define as

A

Data-driven construction of 3D space

B

Density projection into 3D space

C

Differentiation landscape by genotype

○ HSC   --- Monocytic   --- Erythroid

Wild type   JAK2^V617F

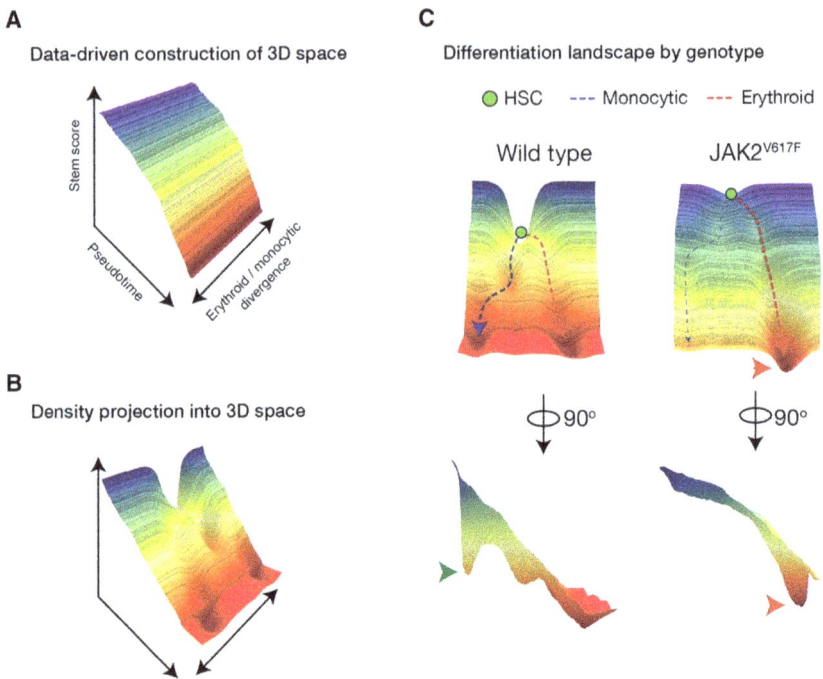

Figure 10.1. Data generated and visualized by Franco Izzo and Dan Landau (see Izzo et al. 2020). They used single-cell assays of chromatin accessibility to identify the cells as they differentiated through hematopoiesis, quantifying the relative numbers committing to each lineage. This quantitative commitment was then represented visually as the topologies shown. Waddington's predicted effect of genetic mutation (Fig. 2.6) comes to life vividly in *C*, the JAK2^V617F missense variant skewing differentiation strongly toward erythroid cell fate. Reprinted from Izzo et al. 2020. *Nat Genet* **52:** 378–387, with permission from SpringerNature.

unquestionable strengths are the areas of biochemical genomics, model organism research, genomic assay development, bioinformatic analytical progress, and significant drug discoveries. A major driver of insights in these and other areas of "epigenetics" has been the advances in technologies available to us (in particular, genomic sequencing assays). It is notable that when these technologies are not central to progress (e.g., in studies of epidemiological cohorts or animal models of multigenerational nongenetic heritability), questions linger about how to interpret where we are today, but that uncertainty indicates opportunities for new, incisive projects.

In our world of environmental pollutants, climate change, and population health challenges, we seek a measurable molecular mechanism for the environment on phenotypes. The idea of epigenetic mechanisms as those that mediate environmental influences on the cell is appealing and may not be misplaced; this does indeed look like a valuable research avenue to pursue. As stressed many times, we don't need to invoke the instinctive idea of reprogramming (i.e., a cell changing its inherent characteristics in response to the environment through molecular genomic changes). A great way of developing a memory of a past exposure is instead to change the composition of cell subtypes in the organ. Both outcomes are now readily explored, especially with the maturation of single-cell genomics approaches.

We are probably guilty in the world of epigenetics of not paying sufficient attention to the world of genetics. The geneticists exploring the influence of the environment on the genome work with DNA sequence information that varies minimally between different cell and tissue types. The epigenetics researcher sampling the cell type(s) likely to be influenced by the environmental exposure can test for both cellular reprogramming and cell subtype changes. If they use a genome-wide assay like ATAC-seq, they will probably narrow their search to the tiny fraction of the genome where changes in chromatin states are occurring. At that point, the epigenetics and genetics researchers need to start talking, because the overlap of loci where genetic and chromatin states vary becomes extremely informative, potentially combining to mediate the condition. The epigenetics researcher, used to studying very limited numbers of samples, may not appreciate nor look for influences of genetic sequence variability on their genome-wide assays. The geneticist may still be treating haplotypes as uniformly informative as opposed to focusing on their embedded loci with distinctive regulatory functions. In the example of multigenerational inheritance of traits, we need to understand whether the powerful mediator of heritability, DNA sequence, is covertly influencing the phenotype being tested. Part of the churn of modern epigenetics is being challenged to think not just as an epigenetics researcher, but also as a geneticist, which should improve our ability to deliver insights into these kinds of research avenues.

◆ ◆ ◆

Equity in research is another challenge being faced in the field of genetics as a whole. In epigenetics, our first point of contact with genetics is the influence of DNA sequence variants on molecular genomic states, the molecular quantitative trait loci (molQTLs) described in Chapters 5 and 7. By understanding DNA sequence variants that influence how genes are regulated, we have narrowed down on variants that could be mediating disease phenotypes. A concern in human genetics research is that the continued focus on genomes and cell resources from individuals of Northern European white ancestry is failing to define sequence variants that could be causing diseases in the majority of the world's populations (Martin et al. 2019). A striking observation in a 2020 review was the degree to which "Caucasian" or "European" donors were overrepresented in U.S. and European induced pluripotent stem cell (iPSC) collections (Nehme and Barrett 2020). Scientists will see two self-serving but positive opportunities in these kinds of situations: the opportunity to build new resources that are more representative of world populations and the opportunity to discover new functional variants in the human genome. The derived opportunity then becomes the understanding of genetic variants that could be involved in ancestry-related differences in disease susceptibility or severity, recognizing that the genetic contribution to such differences will often be intertwined with and dwarfed by socioeconomic status, societal choices about access to health care, environmental and cultural differences, and the malign effects of racism. Who wouldn't want to be at this nexus of research interests, bringing epigenetics expertise to help unlock discoveries about human diseases for all humankind?

This is skimming the surface; there are clearly plenty of other exciting possibilities for this broad field of epigenetics. One concept worth stressing is hopefully obvious from the preceding chapters—the idea that thinking in terms of epigenetics should make us more comfortable with the idea of chance in development and disease. We saw this with the viable yellow mice of Chapters 5 and 7, in which the presence of the genetic mutation (the retrotransposon insertion) and dietary supplementation increase the chances of phenotypic outcomes. In these mice, just as in Muller's *Drosophila* with position-effect variegation affecting the *white* gene (Chapter 5), the decisions about whether to activate or silence a gene operates on a cell-by-cell basis. The overall coat color of the viable yellow mice or eye color of Muller's flies depends on many individual cell decisions that can be influenced by genetic variation acting in *trans*, which is the basis for identifying activating and silencing influences in *Drosophila* and Emma Whitelaw's mouse *Momme* system (Chapter 5). This all comes back conceptually to the balls rolling down Waddington's landscape; the trajectory of any individual ball is unpredictable, but we can predict the relative likelihood of a ball ending up in a specific creode. Cell fate is not predetermined and can vary like any probabilistic event, which is something we see when we measure X-chromosome inactivation in individuals with two X chromosomes—not every person has a 50:50 ratio of paternal and maternal inactivation; the standard deviation is 15.4% in newborns and 19.3% in adults (Amos-Landgraf et al. 2006).

A fascinating example of stochasticity in phenotypic outcomes comes from one of the *Momme* loci, *Trim28* (Whitelaw et al. 2010). Genetically identical mice, all haploinsufficient for *Trim28*, have two phenotypic outcomes: obese or normal weight. Removing the gene selectively in differentiated tissues has no comparable effects, indicating that the obesity phenotype is related to events during early development (Dalgaard et al. 2016). The resulting model is very Waddingtonian: The mice can proceed down one developmental fate or another, a bifurcation that involves the presence or absence of an obesity phenotype. This illustrates a practical problem that is emerging in clinical diagnostics, which is the idea that we can look at the entire DNA sequence of an individual and only end up with a probability of developing a disease—something difficult for a patient or family to process when we are used to diagnosing diseases and their risks in more binary present or absent categories.

Kevin Mitchell, in his book *Innate* (Mitchell 2018), makes the point that the same genes run through their developmental program twice in an individual, once on each side, and inevitably end up associated with subtle differences in morphology. Any ideas that we have that DNA sequence variation should always end up with the same outcomes is at odds with the realization that development is modestly stochastic, that X-chromosome inactivation outcomes can vary like coin tosses, that heterochromatin formation can be variable between cells, or that a ball can end up rolling down different creodes on Waddington's hillside. This does not make our lives easier in interpreting genomic information, but it does allow us a better perspective on how phenotypes are related to our DNA sequences. So, add uncertainty about phenotypic outcomes involving molecular genomic processes to the churn that we can be embracing as the field of epigenetics progresses over the next several years.

EPIGENETICS AND HUMAN DISEASE, RECONSIDERED

In Chapter 7 the discussion was focused on the use of molecular genomic assays to understand human disease and mostly on the deficiencies of the standard epigenome-wide association study (EWAS), which remains the default approach for trying to understand whether altered cellular programming is contributing to the disease phenotype. The optimistic and constructive outcome of the chapter is that we can learn a lot from these investigations if we design and execute these studies optimally and include outcomes other than just cellular reprogramming as potential insights into the disease being studied.

Using EWASs is only the start of the use of all things epigenetic in human diseases, however. In Chapter 8 we tracked how cancer therapeutics have involved insights into alterations of biochemical regulators of the cancer genome all the way to drug interventions targeting these regulators. This sets a high bar for other areas of epigenetic studies in human disease, in which pharmacological intervention with the hope of curing the condition is not likely to happen for some time. How else can the broad field of epigenetics be of likely value in years to come?

One way to think about the road ahead is to divide diseases into those that are rare and those that are common while recognizing how these categories can blur when rare variants contribute to common diseases, as thoughtfully discussed by Gregory (Greg) Gibson (Gibson 2012).

Starting with rare diseases, one of the sources of poor interpretability of epigenomic assays is nicely primed to help us to understand rare variants. In Chapter 7 we highlighted how molecular genomic organization can be influenced by DNA sequence variation, revealing so-called functional variants, prime candidates for mediating disease phenotypes within the large haplotypes implicated in genome-wide association studies (GWASs). There are two ways that these functional variants can be identified. One is to study the molecular genomic outcome (transcription, DNA methylation, etc.) in a large number of individuals, correlating their genotypes with the range of molecular genomic values obtained. This allows the presence of, for example, a nearby position in the genome with a C compared with an A to be associated with higher or lower levels of gene expression, DNA methylation, or other molecular genomic outcomes. This approach typically involved only common variants in the genome but has more recently started to include very rare variants found with whole-genome sequencing, revealing much stronger effects for these ultrarare variants, which are likely to be deleterious and strongly subject to purifying selection (Hernandez et al. 2019). The other approach is to use the results of the molecular genomic assay to identify loci where there is "allelic skewing"—that is, where one of the two homologous alleles is overrepresented. Practically, the only way you would observe such an event is when the two alleles are distinguishable by the presence of a sequence variant on one allele. This kind of outcome has been identified in studies using ATAC-seq (Atak et al. 2021), DNase-seq (Vierstra et al. 2020), ChIP-seq (Tehranchi et al. 2016), and bisulfite sequencing testing DNA methylation (Onuchic et al. 2018). Ultrarare variants that Ryan Hernandez identified to have the strongest effects on gene expression (Hernandez et al. 2019) are almost never going to be homozygous in an individual, allowing them to be tested using allelic skewing approaches. Not yet tested but a likely outcome is that distinct ultrarare functional variants at the same *cis*-regulatory locus will have similar effects on genomic regulation. This would, in effect, collapse the effects of multiple distinct variants down to a common outcome on chromatin, DNA methylation, or other states, increasing the power to attribute an association of the individual rare variants with an organismal phenotype. Our epigenomic assays may end up being essential tools for revealing why an individual has a disease, even if their rare variant has never been seen previously.

The other way that rare variants can be understood using epigenomic assays is through the power of these assays as biomarkers. We really don't understand why DNA methylation of peripheral blood leukocytes is so strongly associated with aging, cigarette smoking, or other phenotypes and exposures (Chapter 7). Potentially the many sources of variability of DNA methylation combine to make these assays unusually sensitive to exposures. Rosanna Weksberg proposed that when a gene encoding an "epigenetic" regulator is mutated, this should result in distinctive patterns of DNA methylation in peripheral blood leukocytes. Her group developed a model for DNA methylation changes in individuals with Sotos syndrome, due to pathogenic variants in the *NSD1* histone methyltransferase, that is the basis for improved diagnostic testing when a patient has an equivocal variant in the gene (Choufani et al. 2015). This approach has been expanded by Bekim Sadikovic to develop "episignatures" for several genes and the implementation of the test in clinical practice (Sadikovic et al. 2021). One intriguing possibility is that there may be DNA methylation signatures in peripheral blood leukocytes for genes other than those obviously associated with transcriptional regulation, potentially expanding the scope and value of this kind of approach.

The same property of DNA methylation in peripheral blood leukocytes acting as a biomarker is now being applied to common diseases. It should probably not come as a shock to us after reading Chapter 7, in which we learned about the association of obesity-related phenotypes with DNA methylation in peripheral blood leukocytes, that a DNA "methylation risk score" is predictive of patient laboratory test phenotypes (Thompson et al. 2022). In fact, these authors showed that the association based on DNA methylation outperformed the polygenic risk scores on the same patients, driven substantially by variability due to ancestry and cell type composition estimates derived from DNA methylation data. These results suggest that we may be seeing DNA methylation in action as a molecular sentinel reflecting many biological responses by host cells. Many of the laboratory results studied reflected age-associated conditions (e.g., hemoglobin A1C, cholesterol, indices of renal function), raising the intriguing possibility that aging clocks based on DNA methylation may be reflective of the sum of many age-associated biomarkers. As there is a lot of current discussion about the value of implementing polygenic risk scores as part of population health (Lewis and Vassos 2020), we can anticipate a subsequent wave of interest in the use of methylation risk scores instead, if they continue to show better potential as instruments allowing personalized, precision medicine.

WHAT IS THE PATH TO EPIGENETIC THERAPIES?

It was in Chapter 8 that we saw how cancer therapeutics have successfully used drugs targeting molecular regulators of transcription. The logical progression appears to have been that "epigenetic" regulation of the genome is dysregulated in cancer, so that a drug targeting the mediators of the DNA methylation or histone states should be targeting the mediator of the dysregulation and could have the ability to reverse the problem. The targeting of the BET proteins that bind to acetylated histones reflects a more selective choice based not only on their tendency to be overexpressed in cancers but also on their known roles in oncogenesis (Shorstova et al. 2021).

The mindset in each case is that the therapeutic agent is reversing some sort of detrimental cellular reprogramming. If we return to the ideas from earlier in the book about how epigenetic changes can occur, cellular reprogramming is one clear means of causing and therapeutically reversing a damaged cellular state. However, we should be as conscious of the possibility that the disease state reflects influences on cell differentiation, which points to another avenue for therapeutic intervention. This is already in the crosshairs of the cancer therapeutics community, with a spectacular example of a successful differentiation therapy represented by the use of all-*trans* retinoic acid (ATRA) in the treatment of acute promyelocytic leukemia (PML), which typically involves a reciprocal translocation t(15;17)(q24;q21) that fuses the amino-terminal part of the PML protein with the carboxy-terminal part of the RARA protein (Fig. 10.2). ATRA is the ligand for the RARA protein and causes the fusion protein to drive the cells to differentiate to mature cells that do not divide, which is part of the curative process. The history of how differentiation agents began to be used in cancer therapeutics has been comprehensively reviewed by Zhen-Yi Wang and Zhu Chen (Wang and Chen 2008). The broader topic of intervening to change cell fate as a way of treating diseases has been the subject of a review by Jayaraj Rajagopal and colleagues, in which they highlight the potential to target the epithelial–mesenchymal transition (EMT) that underlies treatment-refractory cancer, but they also include the potential for differentiation, trans-differentiation, and dedifferentiation strategies to be applied in other diseases including pulmonary, cardiovascular, ocular, and endocrinological diseases (Lin et al. 2018). They also discuss how this might involve the targeting of regulators of chromatin and DNA methylation.

It should be no surprise that CRISPR is being considered as a potential therapeutic approach for altering the epigenome (Katti et al. 2022). By using a Cas9 that lacks the ability to digest DNA and instead using it to tether an enzyme that can alter molecular genomic regulation, the guide RNAs can be designed to bring the modified CRISPR complex to targets in the genome to cause increased or decreased transcription of genes. There remain obvious challenges for this approach to be successful, including the problem that sending a guide RNA to a locus generates an R-loop and disrupts the double-stranded DNA bound by TFs, a potential mechanism for the observed loss of binding of TFs in vitro (Shariati et al. 2019). One great advantage of a CRISPR-based system is that multiple sites in the genome can be targeted simultaneously using a set of guide RNAs, so that if the reprogramming or cell fate change requires such multilocal targeting, CRISPR could allow such flexibility.

Of course, by the time you are talking about clamping a transcriptional regulator onto something that binds to specific DNA sequences, you have effectively reinvented the TF. TFs have been considered "undruggable" because they are intrinsically disordered (Liu et al. 2006), forming highly dynamic interactions with DNA and proteins, and lack domains with concave, positively charged pockets where small molecules bind predictably (Henley and Koehler 2021). Advances in computer-aided drug discovery (CADD) appear to be helping to solve this problem (Radaeva et al. 2021). It is a bit

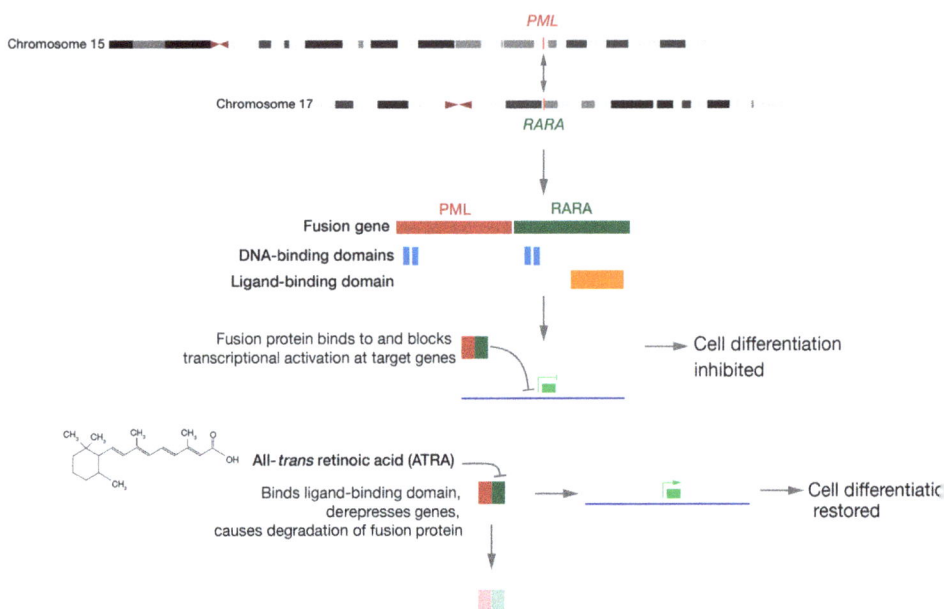

Figure 10.2. The *PML-RARA* fusion gene. A translocation between Chromosomes 15 and 17 juxtaposes the first part of the *PML* gene with the downstream part of *RARA*. The fusion protein is now a combination of a transcription factor (PML) and a transcription factor-like nuclear receptor (RARA). Each brings its zinc finger DNA-binding domains, and RARA adds its ligand-binding domain. The outcome includes repression of many genes associated with a block to cell differentiation, accumulating the immature, malignant promyelocytes. When the RARA ligand all-*trans* retinoic acid (ATRA) is exposed to these cells, it binds to the fusion gene, relieving its gene regulatory effects and causing degradation of the fusion protein, leading to a renewed differentiation capacity of the leukemic cells.

different for the subgroup of TFs that act as nuclear receptors. By definition, they bind to an effector of some type to become activated, typically hormones of different types, although the peroxisome proliferator activated receptors (PPARs) are bound by fatty acids and eicosanoids and have been targeted by drugs in the thiazolidinedione family for amelioration of type II diabetes mellitus (Kroker and Bruning 2015).

Targeting TFs using small molecules looks like an attractive strategy when you consider the in vivo delivery issues for CRISPR-based systems or any alternative that requires getting constructs to be expressed in human tissues. If the tissue is composed mostly of postmitotic, long-lived cells, the challenge is to get as many cells as possible to take up and express the construct you have introduced. If the tissue turns over rapidly, the focus turns to getting the system introduced in its stem cells, so that progeny cells will be producing the introduced protein long-term. One positive outcome from the otherwise grim COVID-19 era was the extraordinary success of RNA-based vaccines, which give us hope that innovation in delivery of expression constructs will continue to help with these therapeutic strategies.

The arc of research in this field of epigenetic therapies is bending toward thoughtful, innovative strategies that encompass both the ideas of cellular reprogramming and lineage determination as outcomes and of TFs as mediators, providing reasons for substantial optimism that we can progress from mechanistic insights to interventions in the near future.

HOW WILL EMERGING TECHNOLOGIES DRIVE ADVANCES IN EPIGENETICS?

Technology has consistently and powerfully driven advances in the broad world of epigenetics. In Chapter 6 the historical advances that brought us to today's insights were reviewed. The clever use of methylation-sensitive restriction enzymes allowed early insights, but it took the development of oligonucleotide microarrays (in the early 2000s) before we could look at the genome at scale, with the subsequent transformative introduction of sequencing-based assays.

Today, the development of robust single-cell platforms is probably the greatest driver of insights. The laboratory of Azim Surani is credited with the first report of single-cell RNA-seq in 2009 (Tang et al. 2009). Since then, there have been further developments in single-cell RNA-seq (scRNA-seq), as well as single-cell chromatin studies (ATAC-seq, ChIP-seq, CUT&Tag, CUT&RUN, chromatin conformation), DNA methylation, and genomic sequencing studies. The vague term "multi-omics" is being used to describe any study of more than one assay per single cell, with scRNA-seq combined with scATAC-seq a relatively common combination currently in use.

The result of these technological advances is that we now think in terms of single cells; the Chapter 6 maxim *The unit of organization of molecular genomic states is the cell* is now fully in our consciousness when we study the data from these studies. Ideas about reprogramming and cell subtype proportional changes when comparing samples are no longer abstract; they are vividly demonstrated in single-cell data. By looking at these multiple events in individual cells, we can address the problem raised in Chapter 6 that events that colocalize in the genome do not necessarily occur in the same cell or chromatid.

Another revelation from single-cell studies is the relationships between the cell subtypes found in a mixed cell sample. Different analytical techniques can be applied that reveal the differentiation trajectories within the sample, showing how one cell subtype is likely to give rise to one or more other cell subtypes present. The resulting mental image of Waddington's epigenetic landscape has been turned into a reality using a Markov-chain entropy (MCE) approach (Shi et al. 2020), a vivid illustration of the decrease in cell potency as differentiation proceeds.

What is certain is that we are about to see the cost of short-read sequencing plummet in the mid-2020s, as patents expire and more competitors enter the marketplace. This will allow deeper sequencing, more replicates, and more experimental conditions for assays that are based on counting reads at loci in the genome, the basis for most gene expression, DNA methylation, and chromatin state assays. Of even greater value will be the potential to sequence concurrently the genomes of the cells being tested, so that we can relate DNA sequence with molecular genomic variability, revealing functional variants at unprecedented rates.

Possibly the most fascinating potential for discovery will come from the long-read platforms, both nanopore-based (currently typified by Oxford Nanopore Technologies) and the zero-mode waveguide (ZMW)-based approach (offered by Pacific Biosciences of California, Inc. [PacBio]). In each case, the individual reads for whole-genome sequencing are of the order of 15–30 kb, allowing multiple sequence variants to be identified for each allele, which in turn permits the construction of contiguous assemblies of DNA sequences of hundreds of kilobases for each chromosome.

It starts to get interesting for the epigenetics enthusiasts when each platform is used to distinguish cytosine from a cytosine modified by methylation. Now you begin to see DNA methylation "haplotypes"—how the presence or absence of DNA methylation at one site on a chromatid is associated with DNA methylation states in *cis*. Figure 10.3 shows examples of why this offers powerful new insights, including more obvious ideas like the detection of differentially methylated regions in imprinted domains, but also the ability to detect skewed X-chromosome inactivation, the kind of repeat expansion and DNA methylation that causes Fragile X syndrome, cell subtype deconvolution, and the detection of functional variants. The development of assays that expose 6mA methyltransferases to nuclei to deposit this modification at loci of open chromatin (Abdulhay et al. 2020; Stergachis et al. 2020) can use the capability of long-read sequencing platforms to detect 6mA, allowing haplotypes of chromatin accessibility to be constructed concurrently. It is conceivable that the CUT&Tag approach that sends an antibody to bind to a specific histone type or modification and uses an attached methyltransferase to decorate the nearby DNA with 6mA could expand the range of molecular genomic events detectable using these haplotypic approaches.

Now we have the possibility of using all these techniques in the setting of the study of functional variants. What happens when one allele contains a functional variant, and you perform one of these long-read sequencing approaches? You should be able to see in DNA from untreated cells (no 6mA methyltransferase use) effects on 5mC and can see whether it is functional in only some or all the cell subtypes from which DNA was extracted. You can also pursue the observation by Mike Snyder that functional variants cause changes over hundreds of kilobases (Grubert et al. 2015), something we cannot explore with short-read approaches.

With longer reads comes the direct detection of all kinds of structural variants, including translocations, inversions, amplification, and deletions. You're looking at a genome browser depiction of your long-read sequencing and see the breakpoint plus large regions of DNA on either side. Now you channel Hermann Muller from Chapter 4, thinking of the silencing of the *Drosophila white* gene because it had the misfortune to land beside a chunk of heterochromatin. What do you see when you look at the DNA methylation haplotypes on both sides of the breakpoint? Is there a position-effect variegation–like phenomenon of altered DNA methylation on one or both sides of the breakpoint? Are any promoters or other *cis*-regulatory loci contracting or entirely losing their hypomethylated states? Is it in all chromatids and cells, or is it variable and variegated? Right now, in clinical diagnostics we lack a means of understanding anything about the regulatory effects of structural or copy number variants, so this would be potentially transformative.

All these possibilities serve to illustrate how scientific questions often follow technological advances; when you can't study something, you often don't even consider it a possibility. The development of single-cell approaches and long-read sequencing is going to open possibilities to address most of the concerns raised in Chapter 7's discussions of human diseases. With the right samples, we will be able to consider cellular reprogramming and cell fate models for disease, the influences of functional variants, and how

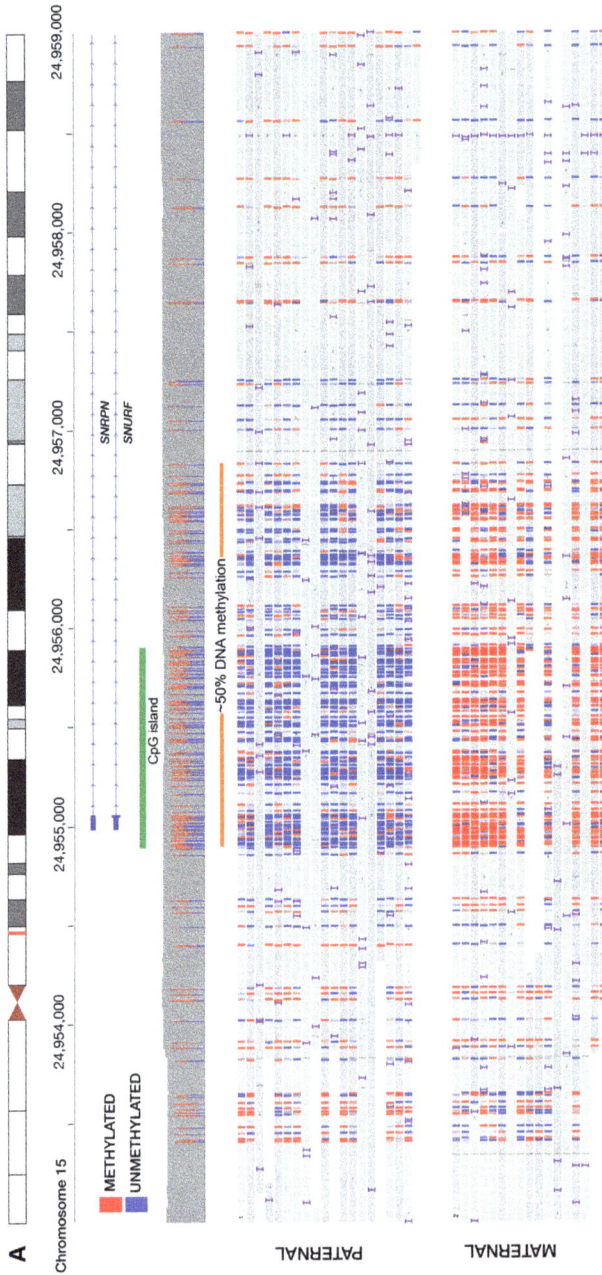

Figure 10.3. Long-read DNA sequencing allows more insights into DNA methylation patterns. Because each long read is likely to contain a number of variants distinguishing the two chromosomes, distinct haplotypes can be constructed, revealing patterns of DNA methylation on each chromosome. Two classical examples of epigenetic regulation are portrayed, genomic imprinting (A) and X-chromosome inactivation (B). (*Figure and legend continue on the following page.*)

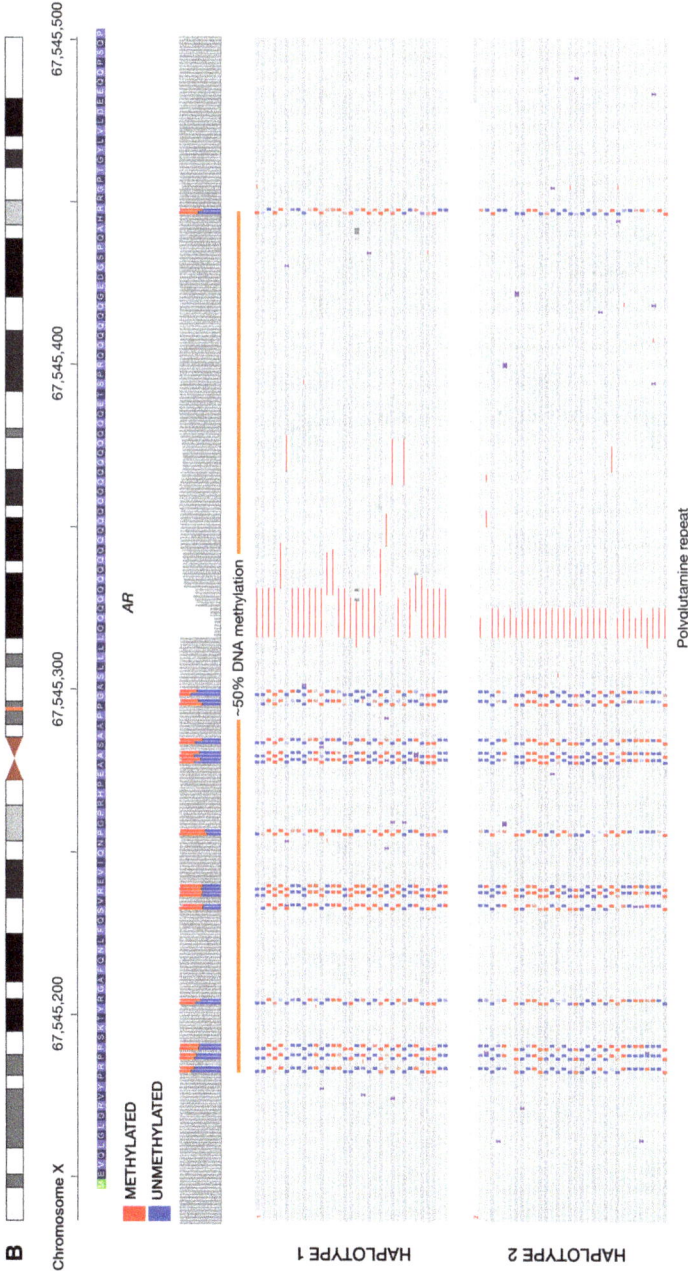

Figure 10.3. (*Continued*) Although each regulatory process results in half the reads being methylated and half unmethylated, the parent of origin–specific pattern of genomic imprinting means that all the methylated CGs are on one haplotype, whereas for X-chromosome inactivation, the maternal and paternal chromosomes are typically equally likely to undergo inactivation, causing each haplotype to be 50% methylated. The imprinted locus shown is the *SNURF-SNRPN* promoter, whereas the X-chromosome locus is at the *Androgen Receptor* (*AR*) gene, showing the locus typically tested by bisulfite PCR to test for X-chromosome inactivation skewing. Thanks to David Yang for generating these data and figures.

TFs mediate targeted regulatory events that go awry in diseases. Technology moves quickly and will be contributing to a lot of the churn we should be anticipating over the next several years.

WHAT ARE OUR LIKELY CHALLENGES?

As some challenges recede, others take their place. When the cost of short-read sequencing diminishes substantially, as it should by the middle of the 2020s, there will be fewer excuses for underpowered studies, greater demands on data storage and analysis, and an even larger deluge of information than currently. The real challenge is going to involve moving from small-scale, exploratory studies generating preliminary findings to the larger-scale studies that answer experimental questions as definitively as possible. We can learn from René Dubos and his approach to mouse experiments (Schaedler 2006):

> He [Dubos] did not like statistics and avoided using them by repeating his experiments over and over, deleting parameters that seemed fruitless and adding others that seemed promising. This repetition gave his experiments numbers that were large enough not to need statistical analysis, and none of his published articles ever had to be retracted. His conclusions were based on the data at hand; he never overstepped this boundary.

How does this approach look more than 50 years later? Dubos would probably have issues with the typically small numbers (no more than low double digits) of rodents used in the groundbreaking trans- and intergenerational models of epigenetic (nongenetic) heritability of traits (Weaver et al. 2004; Crews et al. 2007; Skinner et al. 2008; Franklin et al. 2010; Dietz et al. 2011; Rodgers et al. 2013; Gapp et al. 2014). The problem of data torturing that was highlighted in the discussions of epidemiological studies in Chapter 9 would probably violate Dubos' approach of basing conclusions only on the data at hand. Our tendency as scientists and humans to overinterpret early, intriguing studies allied with the passage of time has ended up creating a canon of beliefs in the broad world of epigenetics, necessitated in part by cost considerations, but which could benefit from renewed scrutiny and definitive experiments. To allow such adequately powered, controlled, definitive studies to be performed, funding agencies must have the courage to support this research, and journals must be prepared to publish negative outcomes. To avoid the temptation to torture data, the study preregistration approach described in Chapter 9 will be a key component to study design.

Most of all, we have to acknowledge that human instincts exist in science, which combine with the practical problems of funding larger studies to keep us in a state of constant preliminary phase research. Nobody in a field has an incentive to take the risk that a definitive study will yield negative results and arrest their chosen research path. Some of the biggest challenges throughout the history recounted in this book have had to do with changing the minds of the scientists of the time, often in the face of strong evidence that their long-held beliefs were unsupported. Nothing has changed today.

Another looming challenge is the understanding of how noncoding variants influence genomic function. Because we know the "rules" of protein-coding sequences—how triplets encode amino acids, how missense variants probably affect protein function, how splice sites can be altered, and so on—we do not need to have already seen every prior variant to predict how each affects a protein-coding sequence. We do not yet have rules available to us that allow us to predict the effects of variants in the noncoding majority of the genome. The development of the Combined Annotation-Dependent Depletion (CADD) score for noncoding loci (Rentzsch et al. 2019) and the comparable de novo risk score for DNA sequence variants in individuals with autism (An et al. 2018) represent early, valuable steps toward this goal, but much remains to be done to develop these approaches to the point that they are as predictive as protein-coding algorithms. Complicating matters further is the observation by Christopher (Chris) Glass that a substantial proportion of variability of TF binding at mice occurs at loci with no sequence variability (Link et al. 2018), implying that functional testing of molecular genomic regulators may always be needed in understanding phenotypes.

Even as the cost of short-read sequencing drops, with the increased recognition of the value of long-read sequencing platforms that also generate DNA methylation information, we can anticipate that sequencing investments will shift to these more costly studies. The ability to identify DNA methylation haplotypes will starkly reveal the kinds of expansion and contraction on unmethylated loci shown in the work of Emily Hodges (Hodges et al. 2011) and described in Chapter 6. We will have to develop new ways of describing DNA methylation patterns in the genome, like "core constitutive unmethylated locus," reflecting our use of this new kind of information. No longer will a CG dinucleotide be informative on its own; it will represent part of a more informative DNA methylation haplotype.

Anything that increases the dimensionality of the molecular regulatory information of the genome is going to be a challenge to address. An

understudied area that warrants attention is the post-translational modifications (PTMs) of TFs (Filtz et al. 2014). Given the number of different
modifications, the number of amino acids that can be targeted, and the likely functional implications of these PTMs, TF PTMs have been described
as "molecular barcodes," using the same logic as described in Chapter 4 to
describe the "histone code." TF PTMs are implicated in just about every
property attributed to TFs—"subcellular localization, stability, interactions
with cofactors, other post-translational modifications, and transcriptional
activities" (Tootle and Rebay 2005). Studying how TF PTMs differentially
affect molecular genomic regulation is a daunting task, but it will be needed
if we are to understand more fully the regulation of the genome and may
offer a way of being precise in targeting specific loci for reprogramming in
cancer (Williams et al. 2020) and potentially other diseases.

♦ ♦ ♦

Now it's time to violate the maxim that *The unit of organization of molecular genomic states is the cell* in highlighting another challenge. Alain
Prochiantz has been pursuing for many years the fascinating model that
homeoprotein TFs are secreted from their cell of origin and influence the
genomic regulation of nearby cells (Di Nardo et al. 2018). Cell to adjacent
cell communication is also mediated by tunneling nanotubes, structures that
allow transfer of everything from molecules to organelles between nearby
cells (Rustom et al. 2004). In a cell's microenvironment there exist means of
transfer of regulatory information, as well as evidence that TFs are actively exchanged. Our reliance on dissociated cells for single-cell technologies
will cause us to fail to recognize such local intercellular communications.
As the world of spatially resolved transcriptomics (Marx 2021) generates
increasing resolution with the ultimate goal of testing individual cells in a
tissue section, we will gain insights into how cellular niches develop transcriptional regulatory activities that are both cell autonomous and the result
of intercellular regulation of transcription.

The challenges for those seeking nongenetic mechanisms of inheritance
that propagate acquired characteristics, as reviewed in Chapter 9, are extraordinary. Without rehashing the discussions of that chapter, it is worth focusing
on a central issue: Propagation of a signal from a somatic tissue to the germline and back out to the somatic tissue again cannot use a single molecular
genomic mechanism (like plants or *C. elegans*); one or more *signal transfers*
are needed. This is meant to describe how a potential small RNA (signal 1)

generated by the somatic tissue finds its way into exosomes to target a high proportion of gametes and load them with enough small RNAs to have an effect in a fertilized ovum. At that stage the RNA will be lost through dilution as cells divide, so it would need to induce heterochromatinization (signal 2) of one or (probably) more critical loci very early, potentially through a siRNA-like process, and that heterochromatin needs to persist in the somatic cells that form the tissue in the next generation, at which time the heterochromatinization needs to regenerate the RNAs and exosomes for the next generation, and so on. A daunting but testable challenge is to search for these signal transfers in mammalian systems, excluding selection for functional genomic variants that could mimic these kinds of events.

Finally, it is worth reiterating how Chapter 9 closed—with an explanation of how we can unwittingly do harm in our enthusiasm to make and tell the world about our new scientific findings. That discussion focused on unintentionally reinforcing the impression in vulnerable individuals that the negative experiences of multiple generations in their family has irreversibly damaged them, and how childbearing individuals end up disproportionately targeted for blame for the adverse health of their offspring. To this we can add another potential challenge—how to make sure that when we use molecular genomic assays as biomarkers, they perform equally well in people of all genetic ancestries. Continental (Yuan et al. 2019) and even country of origin (Rahmani et al. 2017) ancestries can be predicted from the data generated by the DNA methylation microarrays typically used for biomarker studies. This raises the specter of variability of these assays affecting their biomarker use, making them of greater value in the ancestral group in whom the biomarker was initially generated (typically European white individuals) and thus needlessly contributing to health disparities. This is a surprisingly underexplored area of research and would benefit from more attention.

FUNDAMENTAL QUESTIONS IN EPIGENETICS

Finally, we can close with a few basic questions that verge on the philosophical. The first has to do with the idea that a central idea in epigenetics is that it represents how a cell retains a memory of a prior event. What we never seem to discuss is the question whether this involves a time threshold. For example, when you expose *Arabidopsis thaliana* to a heat shock for 5 minutes, there is an immediate transcriptional response which has mostly dissipated after 3 hours (Oyoshi et al. 2020). Is this an epigenetic memory? If not, why not, if it uses cell signaling, TF-mediated gene regulation, and histone PTMs (Lämke et al. 2016) and results in altered gene expression? We note how Adrian Bird in Chapter 4 wanted a definition of epigenetics and cellular memory to encompass nondividing cells, so we can't just wriggle out of this by requiring that the state has to be transmitted through mitosis or meiosis, which would exclude a lot of cells in the body from having "epigenetic" memories. Instead, we probably need to acknowledge the commonality between the biochemical mechanisms involved with acute and long-term transcriptional regulation and not place any arbitrary limits that define "epigenetic" memory as in some way distinctive, because it is difficult to see how such limits can be defined based on objective information.

The second question that comes up is one shared by the community now performing single-cell genomics assays: How do you define a cell type? This question has been thoughtfully explored by Hongkui Zeng, focusing on the complexity of cells in the mammalian brain (Zeng 2022). She not only gives an overview of the many techniques that can be used to address this issue, but she also draws a distinction between cell types and cell states—in other words, how can a canonical cell type exist in several states? Once again, we are brought back to the concept of the epigenetic landscape and a delta of creodes that split off and recombine around the canonical path down the hillside. This reinforces the idea that the original Waddingtonian idea of epigenetics represents a concept that is more relevant today than ever, as we struggle to deal with the complexity of information about cell types and states.

A third question is prompted by the extraordinary ability of quantitative testing of peripheral blood cells to associate with a diversity of human phenotypes. We have discussed how the DNA methylation profile robustly associates with age (Hannum et al. 2013; Horvath 2013), cancers (Houseman et al. 2012), and smoking (Bauer et al. 2015), but we now appreciate that even the relatively crude quantification of blood cell types in the complete blood count (CBC) is associated with a number of common diseases (Foy et al. 2024). It may be that the interconnectedness of physiology in an individual causes a disease or other phenotype to set off ripples across multiple organs and tissues, which will be revealed if you look carefully and quantitatively enough. The early promise of methylation risk scores (Thompson et al. 2022) and the ability to categorize patients with Mendelian diseases using DNA methylation signatures (episignatures) indicate that we may be able to correlate blood cell DNA methylation profiles with both common and rare diseases, helping with risk stratification for the former and diagnostics for the latter.

The final question is the most fundamental of all: In this fast-moving field of epigenetics, how soon will a book like this will need to be updated? The reality is that by the time this is available to read, it is already out of date. This is as it should be: We should hope that the field continues to advance at such a pace that revisions will be needed sooner rather than later, a welcome problem.

RECOMMENDED FURTHER READING

Argentieri MA, Baccarelli AA, Shields AE. 2021. Special focus issue—epigenomics and health disparities. *Epigenomics* **13**: 1673–1676. doi:10.2217/epi-2021-0359

An introduction to an issue of the journal Epigenomics *that contains several papers exploring the intersection of epigenomics and health disparities.*

Morris JA, Caragine C, Daniloski Z, Domingo J, Barry T, Lu L, Davis K, Ziosi M, Glinos DA, Hao S, et al. 2023. Discovery of target genes and pathways at GWAS loci by pooled single-cell CRISPR screens. *Science* **380**: eadh7699. doi:10.1126/science.adh7699

The STING-seq assay is a paradigm for the use of CRISPR to screen how regulatory loci in the genome are affected by sequence polymorphism and interact with nearby and distant genes, using single-cell sequencing assays and exhibiting a convergence of technologies revealing new insights into the regulatory genome.

Nabais MF, Gadd DA, Hannon E, Mill J, McRae AF, Wray NR. 2023. An overview of DNA methylation–derived trait score methods and applications. *Genome Biol* **24**: 28. doi:10.1186/s13059-023-02855-7

A review of the use of methylation risk scores in disease risk prediction.

References

Abdulhay NJ, McNally CP, Hsieh LJ, Kasinathan S, Keith A, Estes LS, Karimzadeh M, Underwood JG, Goodarzi H, Narlikar GJ, Ramani V. 2020. Massively multiplex single-molecule oligonucleosome footprinting. *eLife* **9**: e59504. doi:10.7554/eLife.59404

Abercrombie M. 1967. General review of the nature of differentiation. In *Ciba Foundation Symposium—Cell Differentiation* (eds., de Reuck AVS, Knight JE), pp. 3–17. Wiley, Hoboken, NJ. doi:10.1002/9780470719589.ch2

Achour C, Aguilo F. 2018. Long non-coding RNA and Polycomb: an intricate partnership in cancer biology. *Front Biosci (Landmark Edition)* **23**: 2106–2132. doi:10.2741/4693

Adam AP, George A, Schewe D, Bragado P, Iglesias BV, Ranganathan AC, Kourtidis A, Conklin DS, Aguirre-Ghiso JA. 2009. Computational identification of a p38^{SAPK}-jregulated transcription factor network required for tumor cell quiescence. *Cancer Res* **69**: 5664–5672. doi:10.1158/0008-5472.CAN-08-3820

Adorján P, Distler J, Lipscher E, Model F, Müller J, Pelet C, Braun A, Florl AR, Gütig D, Grabs G, et al. 2002. Tumour class prediction and discovery by microarray-based DNA methylation analysis. *Nucl Acids Res* **30**: e21. doi:10.1093/nar/30.5.e21

Agba OB, Lausser L, Huse K, Bergmeier C, Jahn N, Groth M, Bens M, Sahm A, Gall M, Witte OW, et al. 2017. Tissue-, sex-, and age-specific DNA methylation of rat glucocorticoid receptor gene promoter and insulin-like growth factor 2 imprinting control region. *Physiol Gen* **49**: 690–702. doi:10.1152/physiolgenomics.00009.2017

Aguirre-Gamboa R, Joosten I, Urbano PCM, van der Molen RG, van Rijssen E, van Cranenbroek B, Oosting M, Smeekens S, Jaeger M, Zorro M, et al. 2016. Differential effects of environmental and genetic factors on T and B cell immune traits. *Cell Rep* **17**: 2474–2487. doi:10.1016/j.celrep.2016.10.053

Aguirre-Ghiso JA. 2007. Models, mechanisms and clinical evidence for cancer dormancy. *Nat Rev Cancer* **7**: 834–846. doi:10.1038/nrc2256

Ahrens M, Ammerpohl O, von Schönfels W, Kolarova J, Bens S, Itzel T, Teufel A, Herrmann A, Brosch M, Hinrichsen H, et al. 2013. DNA methylation analysis in nonalcoholic fatty liver disease suggests distinct disease-specific and remodeling signatures after bariatric surgery. *Cell Metab* **18**: 296–302. doi:10.1016/j.cmet.2013.07.004

Ali A, Han K, Liang P. 2021. Role of transposable elements in gene regulation in the human genome. *Life (Basel, Switzerland)* **11**: 118. doi:10.3390/life11020118

Allis CD, Caparros M-L, Jenuwein T, Reinberg D. 2015. *Epigenetics*, 2nd ed., p. 984. Cold Spring Harbor Laboratory Press, Cold Spring Harbor, NY.

Allshire RC, Madhani HD. 2018. Ten principles of heterochromatin formation and function. *Nat Rev Mol Cell Biol* **19**: 229–244. doi:10.1038/nrm.2017.119

Amos-Landgraf JM, Cottle A, Plenge RM, Friez M, Schwartz CE, Longshore J, Willard HF. 2006. X chromosome-inactivation patterns of 1,005 phenotypically unaffected females. *Am J Human Genet* **79:** 493–499. doi:10.1086/507565

An J-Y, Lin K, Zhu L, Werling DM, Dong S, Brand H, Wang HZ, Zhao X, Schwartz GB, Collins RL, et al. 2018. Genome-wide de novo risk score implicates promoter variation in autism spectrum disorder. *Science* **362:** eaat6576. doi:10.1126/science.aat6576

Andersen J, Delihas N, Ikenaka K, Green PJ, Pines O, Ilercil O, Inouye M. 1987. The isolation and characterization of RNA coded by the *micF* gene in *Escherichia coli*. *Nucl Acids Res* **15:** 2089–2101. doi:10.1093/nar/15.5.2089

Anderson ES, Felix A. 1952. Variation in Vi-phage II of *Salmonella typhi*. *Nature* **170:** 492–494.

Anon. 1974. Obituary. L.C. Dunn. *Nature* **250:** 451–452.

Anway MD, Cupp AS, Uzumcu M, Skinner MK. 2005. Epigenetic transgenerational actions of endocrine disruptors and male fertility. *Science* **308:** 1466–1469. doi:10.1126/science.1108190

Aquilino M, Sánchez-Argüello P, Martínez-Guitarte J-L. 2018. Genotoxic effects of vinclozolin on the aquatic insect *Chironomus riparius* (Diptera, Chironomidae). *Environ Pollut* **232:** 563–570. doi:10.1016/j.envpol.2017.09.088

Aranda S, Mas G, Di Croce L. 2015. Regulation of gene transcription by Polycomb proteins. *Sci Adv* **1:** e1500737. doi:10.1126/sciadv.1500737

Argaw-Denboba A, Schmidt TSB, Di Giacomo M, Ranjan B, Devendran S, Mastrorilli E, Lloyd CT, Pugliese D, Paribeni V, Dabin J, et al. 2024. Paternal microbiome perturbations impact offspring fitness. *Nature* **629:** 652–659. doi:10.1038/s41586-024-07336-w

Argentieri MA, Baccarelli AA, Shields AE. 2021. Special focus issue—epigenomics and health disparities. *Epigenomics* **13:** 1673–1676. doi:10.2217/epi-2021-0359

Aroke EN, Joseph PV, Roy A, Overstreet DS, Tollefsbol TO, Vance DE, Goodin BR. 2019. Could epigenetics help explain racial disparities in chronic pain? *J Pain Res* **12:** 701–710. doi:10.2147/JPR.S191848

Arteaga-Vazquez MA, Chandler VL. 2010. Paramutation in maize: RNA mediated trans-generational gene silencing. *Curr Opin Genet Dev* **20:** 156–163. doi:10.1016/j.gde.2010.01.008

Aryee MJ, Jaffe AE, Corrada-Bravo H, Ladd-Acosta C, Feinberg AP, Hansen KD, Irizarry RA. 2014. Minfi: a flexible and comprehensive Bioconductor package for the analysis of Infinium DNA methylation microarrays. *Bioinformatics* **30:** 1363–1369. doi:10.1093/bioinformatics/btu049

Astle WJ, Elding H, Jiang T, Allen D, Ruklisa D, Mann AL, Mead D, Bouman H, Riveros-Mckay F, Kostadima MA, et al. 2016. The allelic landscape of human blood cell trait variation and links to common complex disease. *Cell* **167:** 1415–1429.e19. doi:10.1016/j.cell.2016.10.042

Atak ZK, Taskiran II, Demeulemeester J, Flerin C, Mauduit D, Minnoye L, Hulselmans G, Christiaens V, Ghanem G-E, Wouters J, Aerts S. 2021. Interpretation of allele-specific chromatin accessibility using cell state–aware deep learning. *Genome Res* **31:** 1082–1096. doi:10.1101/gr.260851.120

Atallah-Yunes SA, Ready A, Newburger PE. 2019. Benign ethnic neutropenia. *Blood Rev* **37**: 100586. doi:10.1016/j.blre.2019.06.003

Babenko VN, Chadaeva IV, Orlov YL. 2017. Genomic landscape of CpG rich elements in human. *BMC Evol Biol* **17**: 19. doi:10.1186/s12862-016-0864-0

Bakulski KM, Dou J, Lin N, London SJ, Colacino JA. 2019. DNA methylation signature of smoking in lung cancer is enriched for exposure signatures in newborn and adult blood. *Sci Rep* **9**: 4576. doi:10.1038/s41598-019-40963-2

Baldwin JM. 1896. A new factor in evolution. *Am Nat* **30**: 441–451. doi:10.1086/276408

Bao J, Bedford MT. 2016. Epigenetic regulation of the histone-to-protamine transition during spermiogenesis. *Reproduction* **151**: R55–R70. doi:10.1530/REP-15-0562

Bardou M, Barkun AN, Martel M. 2013. Obesity and colorectal cancer. *Gut* **62**: 933–947. doi:10.1136/gutjnl-2013-304701

Barker DJ. 1992. Fetal growth and adult disease. *Br J Obstet Gyn* **99**: 275–276. doi:10.1111/j.1471-0528.1992.tb13719.x

Barker DJP. 1994. *Mothers, babies, and disease in later life*, p. 192. BMJ Publishing Group, London.

Barker DJ, Osmond C. 1986. Infant mortality, childhood nutrition, and ischaemic heart disease in England and Wales. *Lancet* **1**: 1077–1081. doi:10.1016/s0140-6736(86)91340-1

Barker M, Fall CH, Osmond C, Cooper C, Fleming TP, Thornburg KL, Burton GJ. 2019. David James Purslove Barker. 29 June 1938–27 August 2013. *Biographical Memoirs of Fellows of the Royal Society* **67**: 29–57. doi:10.1098/rsbm.2019.0021

Barlow DP, Stöger R, Herrmann BG, Saito K, Schweifer N. 1991. The mouse insulin-like growth factor type-2 receptor is imprinted and closely linked to the Tme locus. *Nature* **349**: 84–87. doi:10.1038/349084a0

Baron M, Veres A, Wolock SL, Faust AL, Gaujoux R, Vetere A, Ryu JH, Wagner BK, Shen-Orr SS, Klein AM, et al. 2016. A single-cell transcriptomic map of the human and mouse pancreas reveals inter- and intra-cell population structure. *Cell Systems* **3**: 346–360.e4. doi:10.1016/j.cels.2016.08.011

Barr ML. 1988. Human cytogenetics: some reminiscences. *Bioessays* **9**: 79–82. doi:10.1002/bies.950090210

Barr ML, Bertram EG. 1949. A morphological distinction between neurones of the male and female, and the behaviour of the nucleolar satellite during accelerated nucleoprotein synthesis. *Nature* **163**: 676. doi:10.1038/163676a0

Barreiro LB, Tailleux L, Pai AA, Gicquel B, Marioni JC, Gilad Y. 2012. Deciphering the genetic architecture of variation in the immune response to *Mycobacterium tuberculosis* infection. *Proc Natl Acad Sci* **109**: 1204–1209. doi:10.1073/pnas.1115761109

Batie M, Frost J, Frost M, Wilson JW, Schofield P, Rocha S. 2019. Hypoxia induces rapid changes to histone methylation and reprograms chromatin. *Science* **363**: 1222–1226. doi:10.1126/science.aau5870

Baubec T, Colombo DF, Wirbelauer C, Schmidt J, Burger L, Krebs AR, Akalin A, Schübeler D. 2015. Genomic profiling of DNA methyltransferases reveals a role for DNMT3B in genic methylation. *Nature* **520**: 243–247. doi:10.1038/nature14176

Bauer M, Linsel G, Fink B, Offenberg K, Hahn AM, Sack U, Knaack H, Eszlinger M, Herberth G. 2015. A varying T cell subtype explains apparent tobacco smoking induced single CpG hypomethylation in whole blood. *Clin Epigenet* **7**: 81. doi:10.1186/s13148-015-0113-1

Bauer M, Hackermüller J, Schor J, Schreiber S, Fink B, Pierzchalski A, Herberth G. 2019. Specific induction of the unique GPR15 expression in heterogeneous blood lymphocytes by tobacco smoking. *Biomarkers* **24**: 217–224. doi:10.1080/1354750X.2018.1539769

Baugh LR, Day T. 2020. Nongenetic inheritance and multigenerational plasticity in the nematode *C. elegans*. *eLife* **9**: e58498. doi:10.7554/eLife.58498

Baugh EH, Ke H, Levine AJ, Bonneau RA, Chan CS. 2018. Why are there hotspot mutations in the *TP53* gene in human cancers? *Cell Death Differ* **25**: 154–160. doi:10.1038/cdd.2017.180

Baylin SB, Jones PA. 2011. A decade of exploring the cancer epigenome—biological and translational implications. *Nat Rev Cancer* **11**: 726–734. doi:10.1038/nrc3130

Baylin SB, Jones PA. 2016. Epigenetic determinants of cancer. *Cold Spring Harb Perspect Biol* **8**: a019505. doi:10.1101/cshperspect.a019505

Baylin SB, Mendelsohn G. 1980. Ectopic (inappropriate) hormone production by tumors: mechanisms involved and the biological and clinical implications. *Endocrine Rev* **1**: 45–77. doi:10.1210/edrv-1-1-45

Baylin SB, Höppener JW, de Bustros A, Steenbergh PH, Lips CJ, Nelkin BD. 1986. DNA methylation patterns of the calcitonin gene in human lung cancers and lymphomas. *Cancer Res* **46**: 2917–2922.

Beagrie RA, Scialdone A, Schueler M, Kraemer DCA, Chotalia M, Xie SQ, Barbieri M, de Santiago I, Lavitas L-M, Branco MR, et al. 2017. Complex multi-enhancer contacts captured by genome architecture mapping. *Nature* **543**: 519–524. doi:10.1038/nature21411

Becker JS, McCarthy RL, Sidoli S, Donahue G, Kaeding KE, He Z, Lin S, Garcia BA, Zaret KS. 2017. Genomic and proteomic resolution of heterochromatin and its restriction of alternate fate genes. *Mol Cell* **68**: 1023–1037.e15. doi:10.1016/j.molcel.2017.11.030

Behera V, Evans P, Face CJ, Hamagami N, Sankaranarayanan L, Keller CA, Giardine B, Tan K, Hardison RC, Shi J, Blobel GA. 2018. Exploiting genetic variation to uncover rules of transcription factor binding and chromatin accessibility. *Nat Commun* **9**: 782. doi:10.1038/s41467-018-03082-6

Bell AC, Felsenfeld G. 2000. Methylation of a CTCF-dependent boundary controls imprinted expression of the *Igf2* gene. *Nature* **405**: 482–485. doi:10.1038/35013100

Bell JT, Pai AA, Pickrell JK, Gaffney DJ, Pique-Regi R, Degner JF, Gilad Y, Pritchard JK. 2011. DNA methylation patterns associate with genetic and gene expression variation in HapMap cell lines. *Gen Biol* **12**: R10. doi:10.1186/gb-2011-12-1-r10

Bell JT, Tsai P-C, Yang T-P, Pidsley R, Nisbet J, Glass D, Mangino M, Zhai G, Zhang F, Valdes A, et al. 2012. Epigenome-wide scans identify differentially methylated regions for age and age-related phenotypes in a healthy ageing population. *PLoS Genet* **8**: e1002629. doi:10.1371/journal.pgen.1002629

Bell JC, Jukam D, Teran NA, Risca VI, Smith OK, Johnson WL, Skotheim JM, Greenleaf WJ, Straight AF. 2018. Chromatin-associated RNA sequencing (ChAR-seq) maps genome-wide RNA-to-DNA contacts. *eLife* **7**: e27024. doi:10.7554/eLife.27024

Belmont AS. 2014. Large-scale chromatin organization: the good, the surprising, and the still perplexing. *Curr Opin Cell Biol* **26**: 69–78. doi:10.1016/j.ceb.2013.10.002

Benavides F, Rülicke T, Prins J-B, Bussell J, Scavizzi F, Cinelli P, Herault Y, Wedekind D. 2020. Genetic quality assurance and genetic monitoring of laboratory mice and rats: FELASA Working Group Report. *Lab Animals* **54**: 135–148. doi:10.1177/0023677219867719

Benson KR. 2001. T.H. Morgan's resistance to the chromosome theory. *Nat Rev Genet* **2**: 469–474. doi:10.1038/35076532

Benyajati C, Worcel A. 1976. Isolation, characterization, and structure of the folded interphase genome of *Drosophila melanogaster*. *Cell* **9**: 393–407. doi:10.1016/0092-8674(76)90084-2

Berman BP, Weisenberger DJ, Aman JF, Hinoue T, Ramjan Z, Liu Y, Noushmehr H, Lange CPE, van Dijk CM, Tollenaar RAEM, et al. 2012. Regions of focal DNA hypermethylation and long-range hypomethylation in colorectal cancer coincide with nuclear lamina-associated domains. *Nat Genet* **44**: 40–46. doi:10.1038/ng.969

Bernstein BE, Mikkelsen TS, Xie X, Kamal M, Huebert DJ, Cuff J, Fry B, Meissner A, Wernig M, Plath K, et al. 2006. A bivalent chromatin structure marks key developmental genes in embryonic stem cells. *Cell* **125**: 315–326. doi:10.1016/j.cell.2006.02.041

Berry JL, Polski A, Cavenee WK, Dryja TP, Murphree AL, Gallie BL. 2019. The *RB1* story: characterization and cloning of the first tumor suppressor gene. *Genes* **10**: 879. doi:10.3390/genes10110879

Bertani G, Weigle JJ. 1953. Host controlled variation in bacterial viruses. *J Bacteriol* **65**: 113–121. doi:10.1128/jb.65.2.113-121.1953

Bertozzi TM, Becker JL, Blake GET, Bansal A, Nguyen DK, Fernandez-Twinn DS, Ozanne SE, Bartolomei MS, Simmons RA, Watson ED, Ferguson-Smith AC. 2021. Variably methylated retrotransposons are refractory to a range of environmental perturbations. *Nat Genet* **53**: 1233–1242. doi:1038/s41588-021-00898-9

Bestor TH. 1988. Cloning of a mammalian DNA methyltransferase. *Gene* **74**: 9–12.

Beutler E. 1998. Susumu Ohno: the father of X-inactivation. *Cytogenet Cell Genet* **80**: 16–17. doi:10.1159/000014948

Beyaz S, Mana MD, Roper J, Kedrin D, Saadatpour A, Hong S-J, Bauer-Rowe KE, Xifaras ME, Akkad A, Arias E, et al. 2016. High-fat diet enhances stemness and tumorigenicity of intestinal progenitors. *Nature* **531**: 53–58. doi:10.1038/nature17173

Bhattacharya S, Levy MJ, Zhang N, Li H, Florens L, Washburn MP, Workman JL. 2021. The methyltransferase SETD2 couples transcription and splicing by engaging mRNA processing factors through its SHI domain. *Nat Commun* **12**: 1443. doi:10.1038/s41467-021-21663-w

Bhopal R. 2007. The beautiful skull and Blumenbach's errors: the birth of the scientific concept of race. *Br Med J* **335**: 1308–1309. doi:10.1136/bmj.39413.463958.80

Bibikova M, Lin Z, Zhou L, Chudin E, Garcia EW, Wu B, Doucet D, Thomas NJ, Wang Y, Vollmer E, et al. 2006. High-throughput DNA methylation profiling using universal bead arrays. *Genome Res* **16**: 383–393. doi:10.1101/gr.4410706

Bird AP. 1986. CpG-rich islands and the function of DNA methylation. *Nature* **321**: 209–213. doi:10.1038/321209a0

Bird A. 2007. Perceptions of epigenetics. *Nature* **447**: 396–398. doi:10.1038/nature05913

Bird AP, Southern EM. 1978. Use of restriction enzymes to study eukaryotic DNA methylation: I. The methylation pattern in ribosomal DNA from *Xenopus laevis*. *J Mol Biol* **118**: 27–47. doi:10.1016/0022-2836(78)90242-5

Birney E, Smith GD, Greally JM. 2016. Epigenome-wide association studies and the interpretation of disease -omics. *PLoS Genet* **12**: e1006105. doi:10.1371/journal.pgen.1006105

Bischoff A, Albers J, Kharboush I, Stelzer E, Cremer T, Cremer C. 1993. Differences of size and shape of active and inactive X-chromosome domains in human amniotic fluid cell nuclei. *Microsc Res Tech* **25**: 68–77. doi:10.1002/jemt.1070250110

Blaschke K, Ebata KT, Karimi MM, Zepeda-Martínez JA, Goyal P, Mahapatra S, Tam A, Laird DJ, Hirst M, Rao A, et al. 2013. Vitamin C induces Tet-dependent DNA demethylation and a blastocyst-like state in ES cells. *Nature* **500**: 222–226. doi:10.1038/nature12362

Bleker LS, de Rooij SR, Painter RC, Ravelli AC, Roseboom TJ. 2021. Cohort profile: the Dutch famine birth cohort (DFBC)—a prospective birth cohort study in the Netherlands. *BMJ Open* **11**: e042078. doi:10.1136/bmjopen-2020-042078

Blewitt M, Whitelaw E. 2013. The use of mouse models to study epigenetics. *Cold Spring Harb Perspect Biol* **5**: a017939. doi:10.1101/cshperspect.a017939

Blewitt ME, Vickaryous NK, Hemley SJ, Ashe A, Bruxner TJ, Preis JI, Arkell R, Whitelaw E. 2005. An *N*-ethyl-*N*-nitrosourea screen for genes involved in variegation in the mouse. *Proc Natl Acad Sci* **102**: 7629–7634. doi:0.1073/pnas.0409375102

Bönisch C, Hake SB. 2012. Histone H2A variants in nucleosomes and chromatin: more or less stable? *Nucl Acids Res* **40**: 10719–10741. doi:10.1093/nar/gks865

Bormann F, Rodríguez-Paredes M, Lasitschka F, Edelmann D, Musch T, Benner A, Bergman Y, Dieter SM, Ball CR, Glimm H, et al. 2018. Cell-of-origin DNA methylation signatures are maintained during colorectal carcinogenesis. *Cell Rep* **23**: 3407–3418. doi:10.1016/j.celrep.2018.05.045

Bornelöv S, Reynolds N, Xenophontos M, Gharbi S, Johnstone E, Floyd R, Ralser M, Signolet J, Loos R, Dietmann S, et al. 2018. The nucleosome remodeling and deacetylation complex modulates chromatin structure at sites of active transcription to fine-tune gene expression. *Mol Cell* **71**: 56–72.e4. doi:10.1016/j.molcel.2018.06.003

Boshnjaku V, Shim K-W, Tsurubuchi T, Ichi S, Szany EV, Xi G, Mania-Farnell B, McLone DG, Tomita T, Mayanil CS. 2012. Nuclear localization of folate receptor alpha: a new role as a transcription factor. *Sci Rep* **2**: 980. doi:10.1038/srep00980

Bourc'his D, Voinnet O. 2010. A small-RNA perspective on gametogenesis, fertilization, and early zygotic development. *Science* **330**: 617–622. doi:10.1126/science.1194776

Boyd J, Takahashi H, Waggoner SE, Jones LA, Hajek RA, Wharton JT, Liu FS, Fujino T, Barrett JC, McLachlan JA. 1996. Molecular genetic analysis of clear cell adenocarcinomas of the vagina and cervix associated and unassociated with diethylstilbestrol exposure in utero. *Cancer* **77**: 507–513. doi:10.1002/(SICI)1097-0142(19960201)77:3<5 07::AID-CNCR12>3.0.CO;2-8

Bozler J, Kacsoh BZ, Bosco G. 2019. Transgenerational inheritance of ethanol preference is caused by maternal NPF repression. *eLife* **8**: e45391. doi:10.7554/eLife.45391

Braunschweig U, Hogan GJ, Pagie L, van Steensel B. 2009. Histone H1 binding is inhibited by histone variant H3.3. *EMBO J* **28**: 3635–3645. doi:10.1038/emboj.2009.301

Breitling LP, Yang R, Korn B, Burwinkel B, Brenner H. 2011. Tobacco-smoking-related differential DNA methylation: 27K discovery and replication. *Am J Human Genet* **88**: 450–457. doi:10.1016/j.ajhg.2011.03.003

Brink RA. 1956. A genetic change associated with the R locus in maize which is directed and potentially reversible. *Genetics* **41**: 872–889.

Brinster RL. 1974. The effect of cells transferred into the mouse blastocyst on subsequent development. *J Exp Med* **140**: 1049–1056. doi:10.1084/jem.140.4.1049

Brower-Toland B, Findley SD, Jiang L, Liu L, Yin H, Dus M, Zhou P, Elgin SCR, Lin H. 2007. *Drosophila* PIWI associates with chromatin and interacts directly with HP1a. *Genes Dev* **21**: 2300–2311. doi:10.1101/gad.1564307

Brown CJ, Hendrich BD, Rupert JL, Lafrenière RG, Xing Y, Lawrence J, Willard HF. 1992. The human *XIST* gene: analysis of a 17 kb inactive X-specific RNA that contains conserved repeats and is highly localized within the nucleus. *Cell* **71**: 527–542. doi:10.1016/0092-8674(92)90520-m

Brown AS, Susser ES, Lin SP, Neugebauer R, Gorman JM. 1995. Increased risk of affective disorders in males after second trimester prenatal exposure to the Dutch hunger winter of 1944–45. *Br J Psychiatry* **166**: 601–606. doi:10.1192/bjp.166.5.601

Brown KE, Guest SS, Smale ST, Hahm K, Merkenschlager M, Fisher AG. 1997. Association of transcriptionally silent genes with Ikaros complexes at centromeric heterochromatin. *Cell* **91**: 845–854. doi:10.1016/s0092-8674(00)80472-9

Buenrostro JD, Giresi PG, Zaba LC, Chang HY, Greenleaf WJ. 2013. Transposition of native chromatin for fast and sensitive epigenomic profiling of open chromatin, DNA-binding proteins and nucleosome position. *Nat Meth* **10**: 1213–1218. doi:10.1038/nmeth.2688

Buenrostro JD, Wu B, Litzenburger UM, Ruff D, Gonzales ML, Snyder MP, Chang HY, Greenleaf WJ. 2015. Single-cell chromatin accessibility reveals principles of regulatory variation. *Nature* **523**: 486–490. doi:10.1038/nature14590

Buenrostro JD, Corces MR, Lareau CA, Wu B, Schep AN, Aryee MJ, Majeti R, Chang HY, Greenleaf WJ. 2018. Integrated single-cell analysis maps the continuous regulatory landscape of human hematopoietic differentiation. *Cell* **173**: 1535–1548.e16. doi:10.1016/j.cell.2018.03.074

Bunch H. 2018. Gene regulation of mammalian long non-coding RNA. *Mol Genet Gen* **293**: 1–15. doi:10.1007/s00438-017-1370-9

Burgos MH, Fawcett DW. 1955. Studies on the fine structure of the mammalian testis. I. Differentiation of the spermatids in the cat (*Felis domestica*). *J Biophys Biochem Cytol* **1**: 287–300. doi:10.1083/jcb.1.4.287

Burkhardt RW Jr. 2013. Lamarck, evolution, and the inheritance of acquired characters. *Genetics* **194**: 793–805. doi:10.1534/genetics.113.151852

Burton WG, Grabowy CT, Sager R. 1979. Role of methylation in the modification and restriction of chloroplast DNA in *Chlamydomonas*. *Proc Natl Acad Sci* **76**: 1390–1394.

Busslinger M, Tarakhovsky A. 2014. Epigenetic control of immunity. *Cold Spring Harb Perspect Biol* **6**: doi:10.1101/cshperspect.a019307

Butler M, Pongor L, Su Y-T, Xi L, Raffeld M, Quezado M, Trepel J, Aldape K, Pommier Y, Wu J. 2020. MGMT status as a clinical biomarker in glioblastoma. *Trends Cancer* **6**: 380–391. doi:10.1016/j.trecan.2020.02.010

Cabezón M, Malinverni R, Bargay J, Xicoy B, Marcé S, Garrido A, Tormo M, Arenillas L, Coll R, Borras J, et al. 2021. Different methylation signatures at diagnosis in patients with high-risk myelodysplastic syndromes and secondary acute myeloid leukemia predict azacitidine response and longer survival. *Clin Epigenet* **13**: 9. doi:10.1186/s13148-021-01002-y

Calarco JP, Borges F, Donoghue MTA, Van Ex F, Jullien PE, Lopes T, Gardner R, Berger F, Feijó JA, Becker JD, Martienssen RA. 2012. Reprogramming of DNA methylation in pollen guides epigenetic inheritance via small RNA. *Cell* **151**: 194–205. doi:10.1016/j.cell.2012.09.001

Carr SM, Poppy Roworth A, Chan C, La Thangue NB. 2015. Post-translational control of transcription factors: methylation ranks highly. *FEBS J* **282**: 4450–4465. doi:10.1111/febs.13524

Casciello F, Al-Ejeh F, Kelly G, Brennan DJ, Ngiow SF, Young A, Stoll T, Windloch K, Hill MM, Smyth MJ, et al. 2017. G9a drives hypoxia-mediated gene repression for breast cancer cell survival and tumorigenesis. *Proc Natl Acad Sci* **114**: 7077–7082. doi:10.1073/pnas.1618706114

Català-Moll F, Ferreté-Bonastre AG, Godoy-Tena G, Morante-Palacios O, Ciudad L, Barberà L, Fondelli F, Martínez-Cáceres EM, Rodríguez-Ubreva J, Li T, Ballestar E. 2022. Vitamin D receptor, STAT3, and TET2 cooperate to establish tolerogenesis. *Cell Rep* **38**: 110244. doi:10.1016/j.celrep.2021.110244

Catania S, Dumesic PA, Pimentel H, Nasif A, Stoddard CI, Burke JE, Diedrich JK, Cook S, Shea T, Geinger E, et al. 2020. Evolutionary persistence of DNA methylation for millions of years after ancient loss of a de novo methyltransferase. *Cell* **180**: 263–277. e20. doi:10.1016/j.cell.2019.12.012

Cattanach BM. 1961. A chemically-induced variegated-type position effect in the mouse. *Z Vererbungslehre* **92**: 165–182.

Chakravarthy S, Luger K. 2006. The histone variant macro-H2A preferentially forms "hybrid nucleosomes". *J Biol Chem* **281**: 25522–25531. doi:10.1074/jbc.M602258200

Charlton J, Downing TL, Smith ZD, Gu H, Clement K, Pop R, Akopian V, Klages S, Santos DP, Tsankov AM, et al. 2018. Global delay in nascent strand DNA methylation. *Nat Struct Mol Biol* **25**: 327–332. doi:10.1038/s41594-018-0046-4

Chen F, Marquez H, Kim Y-K, Qian J, Shao F, Fine A, Cruikshank WW, Quadro L, Cardoso WV. 2014. Prenatal retinoid deficiency leads to airway hyperresponsiveness in adult mice. *J Clin Invest* **124**: 801–811. doi:10.1172/JCI70291

Cheung VG, Spielman RS. 2002. The genetics of variation in gene expression. *Nat Genet (Suppl)* **32**: 522–525. doi:10.1038/ng1036

Cheung WA, Shao X, Morin A, Siroux V, Kwan T, Ge B, Aïssi D, Chen L, Vasquez L, Allum F, et al. 2017. Functional variation in allelic methylomes underscores a strong genetic contribution and reveals novel epigenetic alterations in the human epigenome. *Gen Biol* **18:** 50. doi:10.1186/s13059-017-1173-7

Chinnam M, Goodrich DW. 2011. RB1, development, and cancer. *Curr Topics Dev Biol* **94:** 129–169. doi:10.1016/B978-0-12-380916-2.00005-X

Choufani S, Cytrynbaum C, Chung BHY, Turinsky AL, Grafodatskaya D, Chen YA, Cohen ASA, Dupuis L, Butcher DT, Siu MT, et al. 2015. *NSD1* mutations generate a genome-wide DNA methylation signature. *Nat Commun* **6:** 10207. doi:10.1038/ncomms10207

Chubb JR, Trcek T, Shenoy SM, Singer RH. 2006. Transcriptional pulsing of a developmental gene. *Curr Biol* **16:** 1018–1025. doi:10.1016/j.cub.2006.03.092

Chung C-W, Coste H, White JH, Mirguet O, Wilde J, Gosmini RL, Delves C, Magny SM, Woodward R, Hughes SA, et al. 2011. Discovery and characterization of small molecule inhibitors of the BET family bromodomains. *J Med Chem* **54:** 3827–3838. doi:10.1021/jm200108t

Cihák A. 1974. Biological effects of 5-azacytidine in eukaryotes. *Oncology* **30:** 405–422. doi:10.1159/000224981

Clark SJ, Harrison J, Paul CL, Frommer M. 1994. High sensitivity mapping of methylated cytosines. *Nucl Acids Res* **22:** 2990–2997. doi:10.1093/nar/22.15.2990

Clemson CM, Hall LL, Byron M, McNeil J, Lawrence JB. 2006. The X chromosome is organized into a gene-rich outer rim and an internal core containing silenced nongenic sequences. *Proc Natl Acad Sci* **103:** 7688–7693. doi:10.1073/pnas.0601069103

Cobb M. 2000. Reading and writing *The Book of Nature*: Jan Swammerdam (1637–1680). *Endeavour* **24:** 122–128. doi:10.1016/S0160-9327(00)01306-5

Cohen BL. 2000. Guido Pontecorvo ("Ponte"), 1907–1999. *Genetics* **154:** 497–501. doi:10.1093/genetics/154.2.497

Cokus SJ, Feng S, Zhang X, Chen Z, Merriman B, Haudenschild CD, Pradhan S, Nelson SF, Pellegrini M, Jacobsen SE. 2008. Shotgun bisulphite sequencing of the *Arabidopsis* genome reveals DNA methylation patterning. *Nature* **452:** 215–219. doi:10.1038/nature06745

Conine CC, Sun F, Song L, Rivera-Pérez JA, Rando OJ. 2018. Small RNAs gained during epididymal transit of sperm are essential for embryonic development in mice. *Dev Cell* **46:** 470–480.e3. doi:10.1016/j.devcel.2018.06.024

Cook HC. 1997. Origins of … tinctorial methods in histology. *J Clin Pathol* **50:** 716–720. doi:10.1136/jcp.50.9.716

Cooper DN, Taggart MH, Bird AP. 1983. Unmethylated domains in vertebrate DNA. *Nucl Acids Res* **11:** 647–658. doi:10.1093/nar/11.3.647

Corces MR, Granja JM, Shams S, Louie BH, Seoane JA, Zhou W, Silva TC, Groeneveld C, Wong CK, Cho SW, et al. 2018. The chromatin accessibility landscape of primary human cancers. *Science* **362:** eaav1898. doi:10.1126/science.aav1898

Corrigan AM, Tunnacliffe E, Cannon D, Chubb JR. 2016. A continuum model of transcriptional bursting. *eLife* **5:** e13051. doi:10.7554/eLife.13051

Costanzi C, Pehrson JR. 1998. Histone macroH2A1 is concentrated in the inactive X chromosome of female mammals. *Nature* **393:** 599–601. doi:10.1038/31275

Cremer T, Cremer C. 2006. Rise, fall and resurrection of chromosome territories: a historical perspective. Part II. Fall and resurrection of chromosome territories during the 1950s to 1980s. Part III. Chromosome territories and the functional nuclear architecture: experiments and models from the 1990s to the present. *Eur J Histochem* **50**: 223–272.

Crews D, Gore AC, Hsu TS, Dangleben NL, Spinetta M, Schallert T, Anway MD, Skinner MK. 2007. Transgenerational epigenetic imprints on mate preference. *Proc Natl Acad Sci* **104**: 5942–5946. doi:10.1073/pnas.0610410104

Croft JA, Bridger JM, Boyle S, Perry P, Teague P, Bickmore WA. 1999. Differences in the localization and morphology of chromosomes in the human nucleus. *J Cell Biol* **145**: 1119–1131. doi:10.1083/jcb.145.6.1119

Crouse HV. 1960. The controlling element in sex chromosome behavior in sciara. *Genetics* **45**: 1429–1443. doi:10.1093/genetics/45.10.1429

Cubas P, Vincent C, Coen E. 1999. An epigenetic mutation responsible for natural variation in floral symmetry. *Nature* **401**: 157–161. doi:10.1038/43657

Cullen MR, Solomon LR, Pace PE, Buckley P, Duffy TP, McPhedran P, Kelsey KT, Redlich CA. 1992. Morphologic, biochemical, and cytogenetic studies of bone marrow and circulating blood cells in painters exposed to ethylene glycol ethers. *Environ Res* **59**: 250–264.

Cupp AS, Uzumcu M, Suzuki H, Dirks K, Phillips B, Skinner MK. 2003. Effect of transient embryonic in vivo exposure to the endocrine disruptor methoxychlor on embryonic and postnatal testis development. *J Androl* **24**: 736–745. doi:10.1002/j.1939-4640.2003.tb02736.x

Dalgaard K, Landgraf K, Heyne S, Lempradl A, Longinotto J, Gossens K, Ruf M, Orthofer M, Strogantsev R, Selvaraj M, et al. 2016. Trim28 haploinsufficiency triggers bi-stable epigenetic obesity. *Cell* **164**: 353–364. doi:10.1016/j.cell.2015.12.025

Dane DS, Cameron CH, Briggs M. 1970. Virus-like particles in serum of patients with Australia-antigen-associated hepatitis. *Lancet* **1**: 695–698. doi:10.1016/s0140-6736(70)90926-8

Dang L, White DW, Gross S, Bennett BD, Bittinger MA, Driggers EM, Fantin VR, Jang HG, Jin S, Keenan MC, et al. 2009. Cancer-associated *IDH1* mutations produce 2-hydroxyglutarate. *Nature* **462**: 739–744. doi:10.1038/nature08617

Darwin C. 1859. *On the origin of species by means of natural selection, or the preservation of favoured races in the struggle for life.* John Murray, London.

Darwin C. 1871. *The descent of man: and selection in relation to sex.* John Murray, London.

Dawid IB, Brown DD, Reeder RH. 1970. Composition and structure of chromosomal and amplified ribosomal DNA's of *Xenopus laevis. J Mol Biol* **51**: 341–360. doi:10.1016/0022-2836(70)90147-6

da Rocha ST, Boeva V, Escamilla-Del-Arenal M, Ancelin K, Granier C, Matias NR, Sanulli S, Chow J, Schulz E, Picard C, et al. 2014. Jarid2 is implicated in the initial Xist-induced targeting of PRC2 to the inactive X chromosome. *Mol Cell* **53**: 301–316. doi:10.1016/j.molcel.2014.01.002

de Bustros A, Nelkin BD, Silverman A, Ehrlich G, Poiesz B, Baylin SB. 1988. The short arm of chromosome 11 is a "hot spot" for hypermethylation in human neoplasia. *Proc Natl Acad Sci* **85**: 5693–5697. doi:10.1073/pnas.85.15.5693 de Cubas AA, Dunker W, Zaninovich A, Hongo RA, Bhatia A, Panda A, Beckermann KE, Bhanot G, Ganesan S, Karijolich J, Rathmell WK. 2020. DNA hypomethylation promotes transposable element expression and activation of immune signaling in renal cell cancer. *J Clin Invest Insight* **5:** e137569. doi:10.1172/jci.insight.137569

de Mello VD, Matte A, Perfilyev A, Männistö V, Rönn T, Nilsson E, Käkelä P, Ling C, Pihlajamäki J. 2017. Human liver epigenetic alterations in non-alcoholic steatohepatitis are related to insulin action. *Epigenetics* **12:** 287–295. doi:10.1080/15592294.2017.1294305

Degner JF, Pai AA, Pique-Regi R, Veyrieras J-B, Gaffney DJ, Pickrell JK, De Leon S, Michelini K, Lewellen N, Crawford GE, et al. 2012. DNase I sensitivity QTLs are a major determinant of human expression variation. *Nature* **482:** 390–394. doi:10.1038/nature10808

de Graaf R. 1672. *De mulierum organis generationi inservientibus*. Hack, Leiden.

Dekker J, Rippe K, Dekker M, Kleckner N. 2002. Capturing chromosome conformation. *Science* **295:** 1306–1311. doi:0.1126/science.1067799

Dekkers KF, van Iterson M, Slieker RC, Moed MH, Bonder MJ, van Galen M, Mei H, Zhernakova DV, van den Berg LH, Deelen J, et al. 2016. Blood lipids influence DNA methylation in circulating cells. *Gen Biol* **17:** 138. doi:10.1186/s13059-016-1000-6

Demircioğlu D, Cukuroglu E, Kindermans M, Nandi T, Calabrese C, Fonseca NA, Kahles A, Lehmann K-V, Stegle O, Brazma A, et al. 2019. A pan-cancer transcriptome analysis reveals pervasive regulation through alternative promoters. *Cell* **178:** 1465–1477.e17. doi:10.1016/j.cell.2019.08.018

Deng X, Ma W, Ramani V, Hill A, Yang F, Ay F, Berletch JB, Blau CA, Shendure J, Duan Z, et al. 2015. Bipartite structure of the inactive mouse X chromosome. *Gen Biol* **16:** 152. doi:10.1186/s13059-015-0728-8

Deplancke B, Alpern D, Gardeux V. 2016. The genetics of transcription factor DNA binding variation. *Cell* **166:** 538–554. doi:10.1016/j.cell.2016.07.012

de Souza N. 2007. Human chromatin at the FAIRE. *Nat Meth* **4:** 200. doi:10.1038/nmeth0307-200

Dickies MM. 1962. A new viable yellow mutation in the house mouse. *J Hered* **53:** 84–86. doi:10.1093/oxfordjournals.jhered.a107129

Dietz DM, Laplant Q, Watts EL, Hodes GE, Russo SJ, Feng J, Oosting RS, Vialou V, Nestler EJ. 2011. Paternal transmission of stress-induced pathologies. *Biol Psychiatry* **70:** 408–414. doi:10.1016/j.biopsych.2011.05.005

Di Nardo AA, Fuchs J, Joshi RL, Moya KL, Prochiantz A. 2018. The physiology of homeoprotein transduction. *Physiol Rev* **98:** 1943–1982. doi:10.1152/physrev.00018.2017

Domann FE, Rice JC, Hendrix MJ, Futscher BW. 2000. Epigenetic silencing of maspin gene expression in human breast cancers. *Intl J Cancer* **85:** 805–810. doi:/10.1002/(sici)1097-0215(20000315)85:6<805::aid-ijc12>3.0.co;2-5

Doni Jayavelu N, Jajodia A, Mishra A, Hawkins RD. 2020. Candidate silencer elements for the human and mouse genomes. *Nat Commun* **11:** 1061. doi:10.1038/s41467-020-14853-5

Doolittle WF. 2018. We simply cannot go on being so vague about "function". *Gen Biol* **19:** 223. doi:10.1186/s13059-018-1600-4

Down TA, Rakyan VK, Turner DJ, Flicek P, Li H, Kulesha E, Gräf S, Johnson N, Herrero J, Tomazou EM, et al. 2008. A Bayesian deconvolution strategy for immunoprecipitation-based DNA methylome analysis. *Nat Biotechnol* **26:** 779–785. doi:10.1038/nbt1414

Dubos RJ. 1965. *Man adapting.* Yale University Press, New Haven, CT.

Dubos R. 1969. Lasting biological effects of early influences. *Perspect Biol Med* **12:** 479–491. doi:10.1353/pbm.1969.0011

Dulbecco R. 1977. Report on a discussion meeting on the biology of chemical carcinogenesis. *Proc R Soc B: Biol Sci* **196:** 117–130. doi:10.1098/rspb.1977.0033

Dupas T, Lauzier B, McGraw S. 2023. O-GlcNAcylation: the sweet side of epigenetics. *Epigenet Chromatin* **16:** 49. doi:10.1186/s13072-023-00523-5

Edwards SL, Beesley J, French JD, Dunning AM. 2013. Beyond GWASs: illuminating the dark road from association to function. *Am J Human Genet* **93:** 779–797. doi:10.1016/j.ajhg.2013.10.012

Ehrlich M. 2005. The controversial denouement of vertebrate DNA methylation research. *Biochemistry Biokhimiia* **70:** 568–575.

Ehrlich M. 2009. DNA hypomethylation in cancer cells. *Epigenomics* **1:** 239–259. doi:10.2217/epi.09.33

Ehrlich M. 2015. Development-linked changes in DNA methylation and hydroxymethylation in humans: interview with Dr Melanie Ehrlich. *Epigenomics* **7:** 691–694. doi:10.2217/epi.15.44

Ehrlich M, Ehrlich KC. 1981. A novel, highly modified, bacteriophage DNA in which thymine is partly replaced by a phosphoglucuronate moiety covalently bound to 5-(4',5'-dihydroxypentyl)uracil. *J Bioll Chem* **256:** 9966–9972.

Ehrlich M, Wang RY. 1981. 5-Methylcytosine in eukaryotic DNA. *Science* **212:** 1350–1357. doi:10.1126/science.6262918

Ehrlich M, Ehrlich K, Mayo JA. 1975. Unusual properties of the DNA from Xanthomonas phage XP-12 in which 5-methylcytosine completely replaces cytosine. *Biochim Biophys Acta Nucl Acids Protein Synth* **395:** 109–119. doi:10.1016/0005-2787(75)90149-5

Ehrlich M, Gama-Sosa MA, Huang LH, Midgett RM, Kuo KC, McCune RA, Gehrke C. 1982. Amount and distribution of 5-methylcytosine in human DNA from different types of tissues of cells. *Nucl Acids Res* **10:** 2709–2721. doi:10.1093/nar/10.8.2709

Eils R, Dietzel S, Bertin E, Schröck E, Speicher MR, Ried T, Robert-Nicoud M, Cremer C, Cremer T. 1996. Three-dimensional reconstruction of painted human interphase chromosomes: active and inactive X chromosome territories have similar volumes but differ in shape and surface structure. *J Cell Biol* **135:** 1427–1440. doi:10.1083/jcb.135.6.1427

Ekamper P, Bijwaard G, van Poppel F, Lumey LH. 2017. War-related excess mortality in The Netherlands, 1944–45: new estimates of famine- and non-famine-related deaths from national death records. *Hist Meth* **50:** 113–128. doi:10.1080/01615440.2017.1285260

ENCODE Project Consortium. 2012. An integrated encyclopedia of DNA elements in the human genome. *Nature* **489:** 57–74. doi:10.1038/nature11247

Epstein MA, Achong BG, Barr YM. 1964. Virus particles in cultured lymphoblasts from Burkitt's lymphoma. *Lancet* **1**: 702–703. doi:10.1016/S0140-6736(64)91524-7

Erdel F, Rippe K. 2018. Formation of chromatin subcompartments by phase separation. *Biophys J* **114**: 2262–2270. doi:10.1016/j.bpj.2018.03.011

Erdel F, Schubert T, Marth C, Längst G, Rippe K. 2010. Human ISWI chromatin-remodeling complexes sample nucleosomes via transient binding reactions and become immobilized at active sites. *Proc Natl Acad Sci* **107**: 19873–19878. doi:10.1073/pnas.1003438107

Erdelyi I, Levenkova N, Lin EY, Pinto JT, Lipkin M, Quimby FW, Holt PR. 2009. Western-style diets induce oxidative stress and dysregulate immune responses in the colon in a mouse model of sporadic colon cancer. *J Nutr* **139**: 2072–2078. doi:10.3945/jn.108.104125

Esteller M, Garcia-Foncillas J, Andion E, Goodman SN, Hidalgo OF, Vanaclocha V, Baylin SB, Herman JG. 2000. Inactivation of the DNA-repair gene *MGMT* and the clinical response of gliomas to alkylating agents. *New Engl J Med* **343**: 1350–1354. doi:10.1056/NEJM200011093431901

Fairfax BP, Humburg P, Makino S, Naranbhai V, Wong D, Lau E, Jostins L, Plant K, Andrews R, McGee C, Knight JC. 2014. Innate immune activity conditions the effect of regulatory variants upon monocyte gene expression. *Science* **343**: 1246949. doi:10.1126/science.1246949

Fall CHD, Cooper C. 2015. Introduction to special issue on the David Barker commemorative meeting, September 2014; the future of the science he inspired. *J Dev Origins Health Dis* **6**: 364–365. doi:10.1017/s2040174415001464

Fall CH, Osmond C, Barker DJ, Clark PM, Hales CN, Stirling Y, Meade TW. 1995. Fetal and infant growth and cardiovascular risk factors in women. *Br Med J* **310**: 428–432. doi:10.1136/bmj.310.6977.428

Fang M, Ou J, Hutchinson L, Green MR. 2014. The BRAF oncoprotein functions through the transcriptional repressor MAFG to mediate the CpG Island Methylator phenotype. *Mol Cel* **55**: 904–915. doi:10.1016/j.molcel.2014.08.010

Fanti L, Piacentini L, Cappucci U, Casale AM, Pimpinelli S. 2017. Canalization by selection of de novo induced mutations. *Genetics* **206**: 1995–2006. doi:10.1534/genetics.117.201079

Farrelly LA, Thompson RE, Zhao S, Lepack AE, Lyu Y, Bhanu NV, Zhang B, Loh Y-HE, Ramakrishnan A, Vadodaria KC, et al. 2019. Histone serotonylation is a permissive modification that enhances TFIID binding to H3K4me3. *Nature* **567**: 535–539. doi:10.1038/s41586-019-1024-7

Fearnhead NS, Britton MP, Bodmer WF. 2001. The ABC of APC. *Human Mol Genet* **10**: 721–733. doi:10.1093/hmg/10.7.721

Feinberg AP. 2013. The epigenetic basis of common human disease. *Trans Am Clin Climatol Assoc* **124**: 84–93.

Feinberg EH, Hunter CP. 2003. Transport of dsRNA into cells by the transmembrane protein SID-1. *Science* **301**: 1545–1547. doi:10.1126/science.1087117

Feinberg AP, Vogelstein B. 1983a. Hypomethylation distinguishes genes of some human cancers from their normal counterparts. *Nature* **301:** 89–92. doi:10.1038/301089a0

Feinberg AP, Vogelstein B. 1983b. Hypomethylation of ras oncogenes in primary human cancers. *Biochem Biophys Res Commun* **111:** 47–54. doi:10.1016/s0006-291x(83)80115-6

Feinberg AP, Springer WR, Barondes SH. 1979. Segregation of pre-stalk and pre-spore cells of *Dictyostelium discoideum*: observations consistent with selective cell cohesion. *Proc Natl Acad Sci* **76:** 3977–3981. doi:10.1073/pnas.76.8.3977

Feldmann A, Ivanek R, Murr R, Gaidatzis D, Burger L, Schübeler D. 2013. Transcription factor occupancy can mediate active turnover of DNA methylation at regulatory regions. *PLoS Genet* **9:** e1003994. doi:10.1371/journal.pgen.1003994

Fellows R, Denizot J, Stellato C, Cuomo A, Jain P, Stoyanova E, Balázsi S, Hajnády Z, Liebert A, Kazakevych J, et al. 2018. Microbiota derived short chain fatty acids promote histone crotonylation in the colon through histone deacetylases. *Nat Commun* **9:** 105. doi:10.1038/s41467-017-02651-5

Felsenfeld G, McGhee J. 1982. Methylation and gene control. *Nature* **296:** 602–603. doi:10.1038/296602a0

Ferreira MAR, Hottenga J-J, Warrington NM, Medland SE, Willemsen G, Lawrence RW, Gordon S, de Geus EJC, Henders AK, Smit JH, et al. 2009. Sequence variants in three loci influence monocyte counts and erythrocyte volume. *Am J Human Genet* **85:** 745–749. doi:10.1016/j.ajhg.2009.10.005

Festing MFW. 2010. Inbred strains should replace outbred stocks in toxicology, safety testing, and drug development. *Toxicol Pathol* **38:** 681–690. doi:10.1177/0192623310373776

Festuccia N, Gonzalez I, Owens N, Navarro P. 2017. Mitotic bookmarking in development and stem cells. *Development* **144:** 3633–3645. doi:10.1242/dev.146522

Filtz TM, Vogel WK, Leid M. 2014. Regulation of transcription factor activity by interconnected post-translational modifications. *Trends Pharmacol Sci* **35:** 76–85. doi:10.1016/j.tips.2013.11.005

Flavahan WA, Drier Y, Liau BB, Gillespie SM, Venteicher AS, Stemmer-Rachamimov AO, Suvà ML, Bernstein BE. 2016. Insulator dysfunction and oncogene activation in *IDH* mutant gliomas. *Nature* **529:** 110–114. doi:10.1038/nature16490

Foy BH, Petherbridge R, Roth MT, Zhang C, De Souza DC, Mow C, Patel HR, Pate CH, Ho SN, Lam E, et al. 2024. Haematological setpoints are a stable and patient-specific deep phenotype. *Nature* **637:** 430–438. doi:10.1038/s41586-024-08264-5

Franklin TB, Russig H, Weiss IC, Gräff J, Linder N, Michalon A, Vizi S, Mansuy IM. 2010. Epigenetic transmission of the impact of early stress across generations. *Biol Psychiatry* **68:** 408–415. doi:10.1016/j.biopsych.2010.05.036

Frayling TM, Timpson NJ, Weedon MN, Zeggini E, Freathy RM, Lindgren CM, Perry JRB, Elliott K, Lango H, Rayner NW, et al. 2007. A common variant in the *FTO* gene is associated with body mass index and predisposes to childhood and adult obesity. *Science* **316:** 889–894. doi:10.1126/science.1141634

Friend C, Scher W, Holland JG, Sato T. 1971. Hemoglobin synthesis in murine virus-induced leukemic cells in vitro: stimulation of erythroid differentiation by dimethyl sulfoxide. *Proc Natl Acad Sci* **68**: 378–382. doi:10.1073/pnas.68.2.378

Frigola J, Ribas M, Risques R-A, Peinado MA. 2002. Methylome profiling of cancer cells by amplification of inter-methylated sites (AIMS). *Nucl Acids Res* **30**: e28. doi:10.1093/nar/30.7.e28

Frommer M, McDonald LE, Millar DS, Collis CM, Watt F, Grigg GW, Molloy PL, Paul CL. 1992. A genomic sequencing protocol that yields a positive display of 5-methylcytosine residues in individual DNA strands. *Proc Natl Acad Sci* **89**: 1827–1831. doi:10.1073/pnas.89.5.1827

Fudenberg G, Imakaev M, Lu C, Goloborodko A, Abdennur N, Mirny LA. 2016. Formation of chromosomal domains by loop extrusion. *Cell Rep* **15**: 2038–2049. doi:10.1016/j.celrep.2016.04.085

Furuta M, Ueno M, Fujimoto A, Hayami S, Yasukawa S, Kojima F, Arihiro K, Kawakami Y, Wardell CP, Shiraishi Y, et al. 2017. Whole genome sequencing discriminates hepatocellular carcinoma with intrahepatic metastasis from multi-centric tumors. *J Hepatol* **66**: 363–373. doi:10.1016/j.jhep.2016.09.021

Futscher BW, Oshiro MM, Wozniak RJ, Holtan N, Hanigan CL, Duan H, Domann FE. 2002. Role for DNA methylation in the control of cell type specific maspin expression. *Nat Genet* **31**: 175–179. doi:10.1038/ng886

Fyodorov DV, Zhou B-R, Skoultchi AI, Bai Y. 2018. Emerging roles of linker histones in regulating chromatin structure and function. *Nat Rev Mol Cell Biol* **19**: 192–206. doi:10.1038/nrm.2017.94

Galan C, Krykbaeva M, Rando OJ. 2020. Early life lessons: the lasting effects of germline epigenetic information on organismal development. *Mol Metab* **38**: 100924. doi:10.1016/j.molmet.2019.12.004

Galton F. 1871. I. Experiments in pangenesis, by breeding from rabbits of a pure variety, into whose circulation blood taken from other varieties had previously been largely transfused. *Abstr Papers Commun R Soc Lond* **19**: 393–410. doi:10.1098/rspl.1870.0061

Gama-Sosa MA, Slage VA, Trewyn RW, Oxenhandler R, Kuo KC, Gehrke CW, Ehrlich M. 1983. The 5-methylcytosine content of DNA from human tumors. *Nucl Acids Res* **11**: 6883–6894. doi:10.1093/nar/11.19.6883

Ganesan A, Arimondo PB, Rots MG, Jeronimo C, Berdasco M. 2019. The timeline of epigenetic drug discovery: from reality to dreams. *Clin Epigenet* **11**: 174. doi:10.1186/s13148-019-0776-0

Gapp K, Jawaid A, Sarkies P, Bohacek J, Pelczar P, Prados J, Farinelli L, Miska E, Mansuy IM. 2014. Implication of sperm RNAs in transgenerational inheritance of the effects of early trauma in mice. *Nat Neurosci* **17**: 667–669. doi:10.1038/nn.3695

Gapp K, van Steenwyk G, Germain PL, Matsushima W, Rudolph KLM, Manuella F, Roszkowski M, Vernaz G, Ghosh T, Pelczar P, et al. 2020. Alterations in sperm long RNA contribute to the epigenetic inheritance of the effects of postnatal trauma. *Mol Psychiatry* **25**: 2162–2174. doi:10.1038/s41380-018-0271-6

Gardiner-Garden M, Frommer M. 1987. CpG islands in vertebrate genomes. *J Mol Biol* **196:** 261–282. doi:10.1016/0022-2836(87)90689-9

Gaujoux R, Seoighe C. 2012. Semi-supervised nonnegative matrix factorization for gene expression deconvolution: a case study. *Infect Genet Evol* **12:** 913–921. doi:10.1016/j.meegid.2011.08.014

Gaydos LJ, Wang W, Strome S. 2014. Gene repression. H3K27me and PRC2 transmit a memory of repression across generations and during development. *Science* **345:** 1515–1518. doi:10.1126/science.1255023

Gayon J. 2016. From Mendel to epigenetics: history of genetics. *C R Biol* **339:** 225–230. doi:10.1016/j.crvi.2016.05.009

Gebhard C, Schwarzfischer L, Pham T-H, Schilling E, Klug M, Andreesen R, Rehli M. 2006. Genome-wide profiling of CpG methylation identifies novel targets of aberrant hypermethylation in myeloid leukemia. *Cancer Res* **66:** 6118–6128. doi:10.1158/0008-5472.CAN-06-0376

Gelling C, Genetics Society of America. 2017. Treasure your exceptions: an interview with 2017 George Beadle Award Recipient Susan A. Gerbi. *Genes to genomes: a blog from the Genetics Society of America.* http://genestogenomes.org/gerbi/

Genetics and plant breeding in the U.S.S.R. 1937. *Nature* **140:** 296–297. doi:10.1038/140296a0

Georgopoulos K, Moore DD, Derfler B. 1992. Ikaros, an early lymphoid-specific transcription factor and a putative mediator for T cell commitment. *Science* **258:** 808–812. doi:10.1126/science.1439790

Gerhard GS, Malenica I, Llaci L, Chu X, Petrick AT, Still CD, DiStefano JK. 2018. Differentially methylated loci in NAFLD cirrhosis are associated with key signaling pathways. *Clin Epigenet* **10:** 93. doi:10.1186/s13148-018-0525-9

Gertz J, Varley KE, Reddy TE, Bowling KM, Pauli F, Parker SL, Kucera KS, Willard HF, Myers RM. 2011. Analysis of DNA methylation in a three-generation family reveals widespread genetic influence on epigenetic regulation. *PLoS Genet* **7:** e1002228. doi:10.1371/journal.pgen.1002228

Geyer PK, Corces VG. 1992. DNA position–specific repression of transcription by a *Drosophila* zinc finger protein. *Genes Dev* **6:** 1865–1873. doi:10.1101/gad.6.10.1865

Ghantous Y, Schussel JL, Brait M. 2018. Tobacco and alcohol–induced epigenetic changes in oral carcinoma. *Curr Opin Oncol* **30:** 152–158. doi:10.1097/CCO.0000000000000444

Ghavi-Helm Y, Jankowski A, Meiers S, Viales RR, Korbel JO, Furlong EEM. 2019. Highly rearranged chromosomes reveal uncoupling between genome topology and gene expression. *Nat Genet* **51:** 1272–1282. doi:10.1038/s41588-019-0462-3

Ghosh RP, Horowitz-Scherer RA, Nikitina T, Shlyakhtenko LS, Woodcock CL. 2010. MeCP2 binds cooperatively to its substrate and competes with histone H1 for chromatin binding sites. *Mol Cell Biol* **30:** 4656–4670. doi:10.1128/MCB.00379-10

Gibson G. 2012. Rare and common variants: twenty arguments. *Nat Rev Genet* **13:** 135–145. doi:10.1038/nrg3118

Gilbert SF. 2000. *Developmental Biology*, 6th ed. Sinauer Associates, Sunderland, MA.

Gilbert WV, Nachtergaele S. 2023. mRNA regulation by RNA modifications. *Ann Rev Biochem* **92:** 175–198. doi:10.1146/annurev-biochem-052521-035949

Gilly A, Southam L, Suveges D, Kuchenbaecker K, Moore R, Melloni GEM, Hatzikotoulas K, Farmaki A-E, Ritchie G, Schwartzentruber J, et al. 2019. Very low-depth whole-genome sequencing in complex trait association studies. *Bioinformatics* **35:** 2555–2561. doi:10.1093/bioinformatics/bty1032

Giovannucci E, Ogino S. 2005. DNA methylation, field effects, and colorectal cancer. *JNCI: J Natl Cancer Inst* **97:** 1317–1319. doi:10.1093/jnci/dji305

Giresi PG, Kim J, McDaniell RM, Iyer VR, Lieb JD. 2007. FAIRE (formaldehyde-assisted isolation of regulatory elements) isolates active regulatory elements from human chromatin. *Genome Res* **17:** 877–885. doi:10.1101/gr.5533506

Gitan RS, Shi H, Chen C-M, Yan PS, Huang TH-M. 2002. Methylation-specific oligonucleotide microarray: a new potential for high-throughput methylation analysis. *Genome Res* **12:** 158–164. doi:10.1101/gr.202801

Gitschier J. 2009. On the track of DNA methylation: an interview with Adrian Bird. *PLoS Genet* **5:** e1000667. doi:10.1371/journal.pgen.1000667

Glass DJ. 2010. A critique of the hypothesis, and a defense of the question, as a framework for experimentation. *Clin Chem* **56:** 1080–1085. doi:10.1373/clinchem.2010.144477

Glass DJ, Hall N. 2008. A brief history of the hypothesis. *Cell* **134:** 378–381. doi:10.1016/j.cell.2008.07.033

Glass JL, Thompson RF, Khulan B, Figueroa ME, Olivier EN, Oakley EJ, Van Zant G, Bouhassira EE, Melnick A, Golden A, et al. 2007. CG dinucleotide clustering is a species-specific property of the genome. *Nucl Acids Res* **35:** 6798–6807. doi:10.1093/nar/gkm489

Glastonbury CA, Couto Alves A, El-Sayed Moustafa JS, Small KS. 2019. Cell-type heterogeneity in adipose tissue is associated with complex traits and reveals disease-relevant cell-specific eQTLs. *Am J Human Genet* **104:** 1013–1024. doi:10.1016/j.ajhg.2019.03.025

Gluckman PD, Hanson MA, Pinal C. 2005. The developmental origins of adult disease. *Matern Child Nutr* **1:** 130–141. doi:10.1111/j.1740-8709.2005.00020.x

Goldberg AD, Banaszynski LA, Noh K-M, Lewis PW, Elsaesser SJ, Stadler S, Dewell S, Law M, Guo X, Li X, et al. 2010. Distinct factors control histone variant H3.3 localization at specific genomic regions. *Cell* **140:** 678–691. doi:10.1016/j.cell.2010.01.003

Gold M, Hurwitz J, Anders M. 1963. The enzymatic methylation of RNA and DNA. II. On the species specificity of the methylation enzymes. *Proc Natl Acad Sci* **50:** 164–169.

Goll MG, Kirpekar F, Maggert KA, Yoder JA, Hsieh C-L, Zhang X, Golic KG, Jacobsen SE, Bestor TH. 2006. Methylation of tRNA[Asp] by the DNA methyltransferase homolog Dnmt2. *Science* **311:** 395–398. doi:10.1126/science.1120976

Gong T, Hartmann N, Kohane IS, Brinkmann V, Staedtler F, Letzkus M, Bongiovanni S, Szustakowski JD. 2011. Optimal deconvolution of transcriptional profiling data using quadratic programming with application to complex clinical blood samples. *PLoS One* **6:** e27156. doi:10.1371/journal.pone.0027156

Gonzalez-Perez A, Jene-Sanz A, Lopez-Bigas N. 2013. The mutational landscape of chromatin regulatory factors across 4,623 tumor samples. *Gen Biol* **14:** r106. doi:10.1186/gb-2013-14-9-r106

Goriaux C, Théron E, Brasset E, Vaury C. 2014. History of the discovery of a master locus producing piRNAs: the flamenco/COM locus in *Drosophila melanogaster*. *Front Genet* **5:** 257. doi:10.3389/fgene.2014.00257

Gowher H, Jeltsch A. 2018. Mammalian DNA methyltransferases: new discoveries and open questions. *Biochem Soc Trans* **46:** 1191–1202. doi:10.1042/BST20170574

Gray LE, Furr J. 2008. Vinclozolin treatment induces reproductive malformations and infertility in male rats when administered during sexual but not gonadal differentiation; however, the effects are not transmitted to the subsequent generations. *Biol Reprod* **78**(Suppl_1): 227–228. doi:10.1093/biolreprod/78.s1.227c

Greally JM. 2018. A user's guide to the ambiguous word "epigenetics". *Nat Rev Mol Cell Biol* **19:** 207–208. doi:10.1038/nrm.2017.135

Greger V, Passarge E, Höpping W, Messmer E, Horsthemke B. 1989. Epigenetic changes may contribute to the formation and spontaneous regression of retinoblastoma. *Human Genet* **83:** 155–158. doi:10.1007/BF00286709

Greiner D, Bonaldi T, Eskeland R, Roemer E, Imhof A. 2005. Identification of a specific inhibitor of the histone methyltransferase SU(VAR)3-9. *Nat Chem Biol* **1:** 143–145. doi:10.1038/nchembio721

Grishok A, Tabara H, Mello CC. 2000. Genetic requirements for inheritance of RNAi in *C. elegans*. *Science* **287:** 2494–2497. doi:0.1126/science.287.5462.2494

Grosjean H. 2005. Modification and editing of RNA: historical overview and important facts to remember. In *Fine-tuning of RNA functions by modification and editing* (ed. Grosjean H), Vol. 12, pp. 1–22. Springer, Berlin. doi:10.1007/b106848

Grubert F, Zaugg JB, Kasowski M, Ursu O, Spacek DV, Martin AR, Greenside P, Srivas R, Phanstiel DH, Pekowska A, et al. 2015. Genetic control of chromatin states in humans involves local and distal chromosomal interactions. *Cell* **162:** 1051–1065. doi:10.1016/j.cell.2015.07.048

Grützmann R, Molnar B, Pilarsky C, Habermann JK, Schlag PM, Saeger HD, Miehlke S, Stolz T, Model F, Roblick UJ, et al. 2008. Sensitive detection of colorectal cancer in peripheral blood by septin 9 DNA methylation assay. *PLoS One* **3:** e3759. doi:10.1371/journal.pone.0003759

Guan Z, Yu H, Cuk K, Zhang Y, Brenner H. 2019. Whole-blood DNA methylation markers in early detection of breast cancer: a systematic literature review. *Cancer Epidemiol Biomarkers Prev* **28:** 496–505. doi:10.1158/1055-9965.EPI-18-0378

Guelen L, Pagie L, Brasset E, Meuleman W, Faza MB, Talhout W, Eussen BH, de Klein A, Wessels L, de Laat W, van Steensel B. 2008. Domain organization of human chromosomes revealed by mapping of nuclear lamina interactions. *Nature* **453:** 948–951. doi:10.1038/nature06947

Guerrero-Bosagna C, Covert TR, Haque MM, Settles M, Nilsson EE, Anway MD, Skinner MK. 2012. Epigenetic transgenerational inheritance of vinclozolin induced mouse adult onset disease and associated sperm epigenome biomarkers. *Reproduct Toxicol* **34:** 694–707. doi:10.1016/j.reprotox.2012.09.005

Guinney J, Dienstmann R, Wang X, de Reyniès A, Schlicker A, Soneson C, Marisa L, Roepman P, Nyamundanda G, Angelino P, et al. 2015. The consensus molecular subtypes of colorectal cancer. *Nat Med* **21:** 1350–1356. doi:10.1038/nm.3967

Gutierrez-Galve L, Stein A, Hanington L, Heron J, Lewis G, O'Farrelly C, Ramchandani PG. 2019. Association of maternal and paternal depression in the postnatal period with offspring depression at age 18 years. *JAMA Psychiatry* **76:** 290–296. doi:10.1001/jamapsychiatry.2018.3667

Hales CN, Barker DJ. 2001. The thrifty phenotype hypothesis. *Br Med Bull* **60:** 5–20. doi:10.1093/bmb/60.1.5

Han J, Zhang Z, Wang K. 2018. 3C and 3C-based techniques: the powerful tools for spatial genome organization deciphering. *Mol Cytogenet* **11:** 21. doi:0.1186/s13039-018-0368-2

Hannum G, Guinney J, Zhao L, Zhang L, Hughes G, Sadda S, Klotzle B, Bibikova M, Fan J-B, Gao Y, et al. 2013. Genome-wide methylation profiles reveal quantitative views of human aging rates. *Mol Cell* **49:** 359–367. doi:10.1016/j.molcel.2012.10.016

Hänsel-Hertsch R, Spiegel J, Marsico G, Tannahill D, Balasubramanian S. 2018. Genome-wide mapping of endogenous G-quadruplex DNA structures by chromatin immunoprecipitation and high-throughput sequencing. *Nat Prot* **13:** 551–564. doi:10.1038/nprot.2017.150

Hansen RS, Wijmenga C, Luo P, Stanek AM, Canfield TK, Weemaes CM, Gartler SM. 1999. The *DNMT3B* DNA methyltransferase gene is mutated in the ICF immunodeficiency syndrome. *Proc Natl Acad Sci* **96:** 14412–14417. doi:10.1073/pnas.96.25.14412

Hansen KD, Timp W, Bravo HC, Sabunciyan S, Langmead B, McDonald OG, Wen B, Wu H, Liu Y, Diep D, et al. 2011. Increased methylation variation in epigenetic domains across cancer types. *Nat Genet* **43:** 768–775. doi:10.1038/ng.865

Hansen KD, Sabunciyan S, Langmead B, Nagy N, Curley R, Klein G, Klein E, Salamon D, Feinberg AP. 2014. Large-scale hypomethylated blocks associated with Epstein–Barr virus–induced B-cell immortalization. *Genome Res* **24:** 177–184. doi:10.1101/gr.157743.113

Hansen MF, Greibe E, Skovbjerg S, Rohde S, Kristensen ACM, Jensen TR, Stentoft C, Kjær KH, Kronborg CS, Martensen PM. 2015. Folic acid mediates activation of the pro-oncogene STAT3 via the Folate Receptor alpha. *Cell Signal* **27:** 1356–1368. doi:10.1016/j.cellsig.2015.03.020

Hansen JL, Loell KJ, Cohen BA. 2022. A test of the pioneer factor hypothesis using ectopic liver gene activation. *eLife* **11:** 73358. doi:10.7554/eLife.73358

Hao S-L, Ni F-D, Yang W-X. 2019. The dynamics and regulation of chromatin remodeling during spermiogenesis. *Gene* **706:** 201–210. doi:10.1016/j.gene.2019.05.027

Harrison RG. 1937. Embryology and its relations. *Science* **85:** 369–374. doi:10.1126/science.85.2207.369

Hatada I, Hayashizaki Y, Hirotsune S, Komatsubara H, Mukai T. 1991. A genomic scanning method for higher organisms using restriction sites as landmarks. *Proc Natl Acad Sci* **88**: 9523–9527. doi:10.1073/pnas.88.21.9523

Hawkins RD, Larjo A, Tripathi SK, Wagner U, Luu Y, Lönnberg T, Raghav SK, Lee LK, Lund R, Ren B, et al. 2013. Global chromatin state analysis reveals lineage-specific enhancers during the initiation of human T helper 1 and T helper 2 cell polarization. *Immunity* **38**: 1271–1284. doi:10.1016/j.immuni.2013.05.011

Hayashizaki Y, Hirotsune S, Okazaki Y, Hatada I, Shibata H, Kawai J, Hirose K, Watanabe S, Fushiki S, Wada S. 1993. Restriction landmark genomic scanning method and its various applications. *Electrophoresis* **14**: 251–258. doi:10.1002/elps.1150140145

Heard E, Martienssen RA. 2014. Transgenerational epigenetic inheritance: myths and mechanisms. *Cell* **157**: 95–109. doi:10.1016/j.cell.2014.02.045

Hegner RW. 1914. *The germ cell cycle in animals.* Macmillan, New York.

Heijmans BT, Kremer D, Tobi EW, Boomsma DI, Slagboom PE. 2007. Heritable rather than age-related environmental and stochastic factors dominate variation in DNA methylation of the human *IGF2/H19* locus. *Human Mol Genet* **16**: 547–554. doi:10.1093/hmg/ddm010

Heijmans BT, Tobi EW, Stein AD, Putter H, Blauw GJ, Susser ES, Slagboom PE, Lumey LH. 2008. Persistent epigenetic differences associated with prenatal exposure to famine in humans. *Proc Natl Acad Sci* **105**: 17046–17049. doi:10.1073/pnas.0806560105

Hendrich B, Hardeland U, Ng HH, Jiricny J, Bird A. 1999. The thymine glycosylase MBD4 can bind to the product of deamination at methylated CpG sites. *Nature* **401**: 301–304. doi:10.1038/45843

Henley MJ, Koehler AN. 2021. Advances in targeting "undruggable" transcription factors with small molecules. *Nat Rev Drug Discov* **20**: 669–688. doi:10.1038/s41573-021-00199-0

Herman JG, Latif F, Weng Y, Lerman MI, Zbar B, Liu S, Samid D, Duan DS, Gnarra JR, Linehan WM. 1994. Silencing of the VHL tumor-suppressor gene by DNA methylation in renal carcinoma. *Proceedings of the National Academy of Sciences of the United States of America* **91**: 9700–9704. doi:10.1073/pnas.91.21.9700

Herman JG, Graff JR, Myöhänen S, Nelkin BD, Baylin SB. 1996. Methylation-specific PCR: a novel PCR assay for methylation status of CpG islands. *Proc Natl Acad Sci* **93**: 9821–9826. doi:10.1073/pnas.93.18.9821

Hermant C, Torres-Padilla M-E. 2021. TFs for TEs: the transcription factor repertoire of mammalian transposable elements. *Genes Dev* **35**: 22–39. doi:10.1101/gad.344473.120

Hernandez RD, Uricchio LH, Hartman K, Ye C, Dahl A, Zaitlen N. 2019. Ultrarare variants drive substantial *cis* heritability of human gene expression. *Nat Genet* **51**: 1349–1355. doi:10.1038/s41588-019-0487-7

Hill KA. 1985. Hartsoeker's homunculus: a corrective note. *J Hist Behav Sci* **21**: 178–179. doi:10.1002/1520-6696(198504)21:2<178::AID-JHBS2300210208>3.0.CO;2-T

Histopathology is ripe for automation. 2017. *Nat Biomed Eng* **1**: 925. doi:10.1038/s41551-017-0179-5

Hitchins MP, Rapkins RW, Kwok C-T, Srivastava S, Wong JJL, Khachigian LM, Polly P, Goldblatt J, Ward RL. 2011. Dominantly inherited constitutional epigenetic silencing of *MLH1* in a cancer-affected family is linked to a single nucleotide variant within the 5'UTR. *Cancer Cell* **20**: 200–213. doi:10.1016/j.ccr.2011.07.003

Ho NT, Li F, Lee-Sarwar KA, Tun HM, Brown BP, Pannaraj PS, Bender JM, Azad MB, Thompson AL, Weiss ST, et al. 2018. Meta-analysis of effects of exclusive breastfeeding on infant gut microbiota across populations. *Nat Commun* **9**: 4169. doi:10.1038/s41467-018-06473-x

Hodges E, Molaro A, dos Santos CO, Thekkat P, Song Q, Uren PJ, Park J, Butler J, Rafii S, McCombie WR, et al. 2011. Directional DNA methylation changes and complex intermediate states accompany lineage specificity in the adult hematopoietic compartment. *Mol Cell* **44**: 17–28. doi:10.1016/j.molcel.2011.08.026

Hodgkinson A, Eyre-Walker A. 2011. Variation in the mutation rate across mammalian genomes. *Nat Rev Genet* **12**: 756–766. doi:10.1038/nrg3098

Holliday R. 1979. A new theory of carcinogenesis. *Br J Cancer* **40**: 513–522. doi:10.1038/bjc.1979.216

Holliday R. 2006. Epigenetics: a historical overview. *Epigenetics* **1**: 76–80. doi:10.4161/epi.1.2.2762

Holliday R, Pugh JE. 1975. DNA modification mechanisms and gene activity during development. *Science* **187**: 226–232. doi:10.1126/science.187.4173.226

Holloman B. 2014. Robin Holliday: 1932–2014. *Cell* **157**: 1001–1003.

Honigsbaum M. 2016. Antibiotic antagonist: the curious career of René Dubos. *Lancet* **387**: 118–119. doi:10.1016/S0140-6736(15)00840-5

Horak CE, Mahajan MC, Luscombe NM, Gerstein M, Weissman SM, Snyder M. 2002. GATA-1 binding sites mapped in the β-globin locus by using mammalian ChIP-chip analysis. *Proc Natl Acad Sci* **99**: 2924–2929. doi:10.1073/pnas.052706999

Horn WT, Tars K, Grahn E, Helgstrand C, Baron AJ, Lago H, Adams CJ, Peabody DS, Phillips SEV, Stonehouse NJ, et al. 2006. Structural basis of RNA binding discrimination between bacteriophages Qβ and MS2. *Structure* **14**: 487–495. doi:10.1016/j.str.2005.12.006

Horn S, Figl A, Rachakonda PS, Fischer C, Sucker A, Gast A, Kadel S, Moll I, Nagore E, Hemminki K, et al. 2013. TERT promoter mutations in familial and sporadic melanoma. *Science* **339**: 959–961. doi:10.1126/science.1230062

Horner AJ, Doeller CF. 2017. Plasticity of hippocampal memories in humans. *Curr Opin Neurobiol* **43**: 102–109. doi:10.1016/j.conb.2017.02.004

Horvath S. 2013. DNA methylation age of human tissues and cell types. *Gen Biol* **14**: R115. doi:10.1186/gb-2013-14-10-r115

Hoshida Y, Villanueva A, Llovet JM. 2009. Molecular profiling to predict hepatocellular carcinoma outcome. *Expert Rev Gastroenterol Hepatol* **3**: 101–103. doi:10.1586/egh.09.5

Hotta K, Kitamoto T, Kitamoto A, Ogawa Y, Honda Y, Kessoku T, Yoneda M, Imajo K, Tomeno W, Saito S, Nakajima A. 2018. Identification of the genomic region under epigenetic regulation during non-alcoholic fatty liver disease progression. *Hepatol Res* **48**: E320–E334. doi:10.1111/hepr.12992

Houri-Zeevi L, Korem Kohanim Y, Antonova O, Rechavi O. 2020. Three rules explain transgenerational small RNA inheritance in *C. elegans*. *Cell* **182**: 1186–1197.e12. doi:10.1016/j.cell.2020.07.022

Houseley J, Tollervey D. 2009. The many pathways of RNA degradation. *Cell* **136**: 763–776. doi:10.1016/j.cell.2009.01.019

Houseman EA, Accomando WP, Koestler DC, Christensen BC, Marsit CJ, Nelson HH, Wiencke JK, Kelsey KT. 2012. DNA methylation arrays as surrogate measures of cell mixture distribution. *BMC Bioinformat* **13**: 86. doi:10.1186/1471-2105-13-86

Hrelia P, Fimognari C, Maffei F, Vigagni F, Mesirca R, Pozzetti L, Paolini M, Cantelli Forti G. 1996. The genetic and non-genetic toxicity of the fungicide Vinclozolin. *Mutagenesis* **11**: 445–453. doi:10.1093/mutage/11.5.445

Hrit J, Goodrich L, Li C, Wang B-A, Nie J, Cui X, Martin EA, Simental E, Fernandez J, Liu MY, et al. 2018. OGT binds a conserved C-terminal domain of TET1 to regulate TET1 activity and function in development. *eLife* **7**: 34870. doi:10.7554/eLife.34870

Huan T, Rong J, Liu C, Zhang X, Tanriverdi K, Joehanes R, Chen BH, Murabito JM, Yao C, Courchesne P, et al. 2015. Genome-wide identification of microRNA expression quantitative trait loci. *Nat Commun* **6**: 6601. doi:10.1038/ncomms7601

Huang TH, Perry MR, Laux DE. 1999. Methylation profiling of CpG islands in human breast cancer cells. *Human Mol Genet* **8**: 459–470. doi:0.1093/hmg/8.3.459

Huang FW, Hodis E, Xu MJ, Kryukov GV, Chin L, Garraway LA. 2013. Highly recurrent TERT promoter mutations in human melanoma. *Science* **339**: 957–959. doi:10.1126/science.1229259

Huang H, Lin S, Garcia BA, Zhao Y. 2015. Quantitative proteomic analysis of histone modifications. *Chem Rev* **115**: 2376–2418. doi:10.1021/cr500491u

Huang Y, Gu L, Li G-M. 2018. H3K36me3-mediated mismatch repair preferentially protects actively transcribed genes from mutation. *J Biol Chem* **293**: 7811–7823. doi:10.1074/jbc.RA118.002839

Huang D, Petrykowska HM, Miller BF, Elnitski L, Ovcharenko I. 2019. Identification of human silencers by correlating cross-tissue epigenetic profiles and gene expression. *Genome Res* **29**: 657–667. doi:10.1101/gr.247007.118

Huff JT, Plocik AM, Guthrie C, Yamamoto KR. 2010. Reciprocal intronic and exonic histone modification regions in humans. *Nat Struct Mol Biol* **17**: 1495–1499. doi:10.1038/nsmb.1924

Huxley J. 1956. Epigenetics. *Nature* **177**: 807–809. doi:10.1038/177807a0

Huxley J. 1957. Cancer biology: viral and epigenetic. *Biol Rev* **32**: 1–37. doi:10.1111/j.1469-185X.1957.tb01575.x

Huxley J. 1958. *Biological aspects of cancer.* Harcourt, Brace, New York.

Illingworth RS, Gruenewald-Schneider U, Webb S, Kerr ARW, James KD, Turner DJ, Smith C, Harrison DJ, Andrews R, Bird AP. 2010. Orphan CpG islands identify numerous conserved promoters in the mammalian genome. *PLoS Genet* **6**: e1001134. doi:10.1371/journal.pgen.1001134

Imai S-I, Guarente L. 2016. It takes two to tango: NAD⁺ and sirtuins in aging/longevity control. *NPJ Aging Mech Dis* **2:** 16017. doi:10.1038/npjamd.2016.17

Imielinski M, Guo G, Meyerson M. 2017. Insertions and deletions target lineage-defining genes in human cancers. *Cell* **168:** 460–472.e14. doi:10.1016/j.cell.2016.12.025

Immler S. 2008. Sperm competition and sperm cooperation: the potential role of diploid and haploid expression. *Reproduction* **135:** 275–283. doi:10.1530/REP-07-0482

Inawaka K, Kawabe M, Takahashi S, Doi Y, Tomigahara Y, Tarui H, Abe J, Kawamura S, Shirai T. 2009. Maternal exposure to anti-androgenic compounds, vinclozolin, flutamide and procymidone, has no effects on spermatogenesis and DNA methylation in male rats of subsequent generations. *Toxicol Appl Pharmacol* **237:** 178–187. doi:10.1016/j.taap.2009.03.004

Irizarry RA, Ladd-Acosta C, Wen B, Wu Z, Montano C, Onyango P, Cui H, Gabo K, Rongione M, Webster M, et al. 2009. The human colon cancer methylome shows similar hypo- and hypermethylation at conserved tissue-specific CpG island shores. *Nat Genet* **41:** 178–186. doi:10.1038/ng.298

Isaac CE, Francis SM, Martens AL, Julian LM, Seifried LA, Erdmann N, Binné UK, Harrington L, Sicinski P, Bérubé NG, et al. 2006. The retinoblastoma protein regulates pericentric heterochromatin. *Mol Cell Biol* **26:** 3659–3671. doi:10.1128/MCB.26.9.3659-3671.2006

Ito S, Shen L, Dai Q, Wu SC, Collins LB, Swenberg JA, He C, Zhang Y. 2011. Tet proteins can convert 5-methylcytosine to 5-formylcytosine and 5-carboxylcytosine. *Science* **333:** 1300–1303. doi:10.1126/science.1210597

Ito-Ishida A, Yamalanchili HK, Shao Y, Baker SA, Heckman LD, Lavery LA, Kim J-Y, Lombardi LM, Sun Y, Liu Z, Zoghbi HY. 2018. Genome-wide distribution of linker histone H1.0 is independent of MeCP2. *Nat Neurosci* **21:** 794–798. doi:10.1038/s41593-018-0155-8

Iwafuchi-Doi M, Zaret KS. 2014. Pioneer transcription factors in cell reprogramming. *Genes Dev* **28:** 2679–2692. doi:10.1101/gad.253443.114

Izzo F, Lee SC, Poran A, Chaligne R, Gaiti F, Gross B, Murali RR, Deochand SD, Ang C, Jones PW, et al. 2020. DNA methylation disruption reshapes the hematopoietic differentiation landscape. *Nat Genet* **52:** 378–387. doi:10.1038/s41588-020-0595-4

Izzo F, Myers RM, Ganesan S, Mekerishvili L, Kottapalli S, Prieto T, Eton EO, Botella T, Dunbar AJ, Bowman RL, et al. 2024. Mapping genotypes to chromatin accessibility profiles in single cells. *Nature* **629:** 1149–1157. doi:10.1038/s41586-024-07388-y

Jablonka E, Raz G. 2009. Transgenerational epigenetic inheritance: prevalence, mechanisms, and implications for the study of heredity and evolution. *Q Rev Biol* **84:** 131–176. doi:10.1086/598822

Jacob F, Monod J. 1961. Genetic regulatory mechanisms in the synthesis of proteins. *J Mol Biol* **3:** 318–356. doi:10.1016/S0022-2836(61)80072-7

Jacobsen SE, Meyerowitz EM. 1997. Hypermethylated SUPERMAN epigenetic alleles in *Arabidopsis. Science* **277:** 1100–1103. doi:10.1126/science.277.5329.1100

Jaitin DA, Kenigsberg E, Keren-Shaul H, Elefant N, Paul F, Zaretsky I, Mildner A, Cohen N, Jung S, Tanay A, Amit I. 2014. Massively parallel single-cell RNA-seq for marker-free decomposition of tissues into cell types. *Science* **343:** 776–779. doi:10.1126/science.1247651

Jang YE, Jang I, Kim S, Cho S, Kim D, Kim K, Kim J, Hwang J, Kim S, Kim J, et al. 2020. ChimerDB 4.0: an updated and expanded database of fusion genes. *Nucl Acids Res* **48**: D817–D824. doi:0.1093/nar/gkz1013

Jaric I, Rocks D, Greally JM, Suzuki M, Kundakovic M. 2019. Chromatin organization in the female mouse brain fluctuates across the oestrous cycle. *Nat Commun* **10**: 2851. doi:10.1038/s41467-019-10704-0

Jenkins TG, Aston KI, Pflueger C, Cairns BR, Carrell DT. 2014. Age-associated sperm DNA methylation alterations: possible implications in offspring disease susceptibility. *PLoS Genet* **10**: e1004458. doi:10.1371/journal.pgen.1004458

Jenkins TG, James ER, Aston KI, Salas-Huetos A, Pastuszak AW, Smith KR, Hanson HA, Hotaling JM, Carrell DT. 2019. Age-associated sperm DNA methylation patterns do not directly persist trans-generationally. *Epigenet Chromat* **12**: 74. doi:10.1186/s13072-019-0323-4

Jenuwein T, Allis CD. 2001. Translating the histone code. *Science* **293**: 1074–1080. doi:10.1126/science.1063127

Jeon Y, Lee JT. 2011. YY1 tethers Xist RNA to the inactive X nucleation center. *Cell* **146**: 119–133. doi:10.1016/j.cell.2011.06.026

Johannsen W. 1909. *Elemente der exakten Erblichkeitslehre.* Gustav Fischer, Stuttgart.

Johnson DR. 1974. Hairpin-tail: a case of post-reductional gene action in the mouse egg. *Genetics* **76**: 795–805. doi:10.1093/genetics/76.4.795

Johnston AD, Simões-Pire CA, Thompson TV, Suzuki M, Greally JM. 2019. Functional genetic variants can mediate their regulatory effects through alteration of transcription factor binding. *Nat Commun* **10**: 3472. doi:0.1038/s41467-019-11412-5

Jones P. 2011. Out of Africa and into epigenetics: discovering reprogramming drugs. *Nat Cell Biol* **13**: 2. doi:10.1038/ncb0111-2

Jones PA, Taderera JV, Hawtrey AO. 1972. Transformation of hamster cells in vitro by 1-β-D arabinofuranosylcytosine, 5-fluorodeoxyuridine and hydroxyurea. *Eur J Cancer* **8**: 595–599. doi:10.1016/0014-2964(72)90138-7

Jones PA, Ohtani H, Chakravarthy A, De Carvalho DD. 2019. Epigenetic therapy in immune-oncology. *Nat Rev Cancer* **19**: 151–161. doi:10.1038/s41568-019-0109-9

Josse J, Kaiser AD, Kornberg A. 1961. Enzymatic synthesis of deoxyribonucleic acid. VIII. Frequencies of nearest neighbor base sequences in deoxyribonucleic acid. *J Biol Chem* **236**: 864–875.

Jost KL, Bertulat B, Cardoso MC. 2012. Heterochromatin and gene positioning: inside, outside, any side? *Chromosoma* **121**: 555–563. doi:10.1007/s00412-012-0389-2

Joubert BR, Felix JF, Yousefi P, Bakulski KM, Just AC, Breton C, Reese SE, Markunas CA, Richmond RC, Xu C-J, et al. 2016. DNA methylation in newborns and maternal smoking in pregnancy: genome-wide consortium meta-analysis. *Am J Hum Genet* **98**: 680–696. doi:10.1016/j.ajhg.2016.02.019

Kaelin CB, McGowan KA, Trotman JC, Koroma DC, David VA, Menotti-Raymond M, Graff EC, Schmidt-Küntzel A, Oancea E, Barsh GS. 2024. Molecular and genetic characterization of sex-linked orange coat color in the domestic cat. bioRxiv doi:10.1101/2024.11.21.624608

Kang GH, Lee S, Kim WH, Lee HW, Kim JC, Rhyu M-G, Ro JY. 2002. Epstein–Barr virus–positive gastric carcinoma demonstrates frequent aberrant methylation of multiple genes and constitutes CpG island methylator phenotype-positive gastric carcinoma. *Am J Pathol* **160**: 787–794. doi:10.1016/S0002-9440(10)64901-2

Kaniskan HÜ, Konze KD, Jin J. 2015. Selective inhibitors of protein methyltransferases. *J Med Chem* **58**: 1596–1629. doi:10.1021/jm501234a

Kant I. 2006. *The critique of judgement. Part one. The critique of aesthetic judgement*, p. 124. Digireads.com, Overland Park, KS.

Kaprio J. 2012. Twins and the mystery of missing heritability: the contribution of gene–environment interactions. *J Intern Med* **272**: 440–448. doi:10.1111/j.1365-2796.2012.02587.x

Karczewski KJ, Tatonetti NP, Landt SG, Yang X, Slifer T, Altman RB, Snyder M. 2011. Cooperative transcription factor associations discovered using regulatory variation. *Proc Natl Acad Sci* **108**: 13353–13358. doi:10.1073/pnas.1103105108

Kataoka K, Mochizuki K. 2011. Programmed DNA elimination in *Tetrahymena*: a small RNA-mediated genome surveillance mechanism. *Adv Exp Med Biol* **722**: 156–173. doi:10.1007/978-1-4614-0332-6_10

Katti A, Diaz BJ, Caragine CM, Sanjana NE, Dow LE. 2022. CRISPR in cancer biology and therapy. *Nat Rev Cancer* **22**: 259–279. doi:10.1038/s41568-022-00441-w

Kaya-Okur HS, Wu SJ, Codomo CA, Pledger ES, Bryson TD, Henikoff JG, Ahmad K, Henikoff S. 2019. CUT&Tag for efficient epigenomic profiling of small samples and single cells. *Nat Commun* **10**: 1930. doi:10.1038/s41467-019-09982-5

Kazachenka A, Bertozzi TM, Sjoberg-Herrera MK, Walker N, Gardner J, Gunning R, Pahita E, Adams S, Adams D, Ferguson-Smith AC. 2018. Identification, characterization, and heritability of murine metastable epialleles: implications for non-genetic inheritance. *Cell* **175**: 1259–1271.e13. doi:10.1016/j.cell.2018.09.043

Keele GR, Quach BC, Israel JW, Chappell GA, Lewis L, Safi A, Simon JM, Cotney P, Crawford GE, Valdar W, et al. 2020. Integrative QTL analysis of gene expression and chromatin accessibility identifies multi-tissue patterns of genetic regulation. *PLoS Genet* **16**: e1008537. doi:10.1371/journal.pgen.1008537

Kellum R, Schedl P. 1992. A group of scs elements function as domain boundaries in an enhancer-blocking assay. *Mol Cell Biol* **12**: 2424–2431. doi:10.1128/mcb.12.5.2424-2431.1992

Kerkel K, Spadola A, Yuan E, Kosek J, Jiang L, Hod E, Li K, Murty VV, Schupf N, Vilain E, et al. 2008. Genomic surveys by methylation-sensitive SNP analysis identify sequence-dependent allele-specific DNA methylation. *Nat Genet* **40**: 904–908. doi:10.1038/ng.174

Keshet I, Schlesinger Y, Farkash S, Rand E, Hecht M, Segal E, Pikarski E, Young RA, Niveleau A, Cedar H, Simon I. 2006. Evidence for an instructive mechanism of de novo methylation in cancer cells. *Nat Genet* **38**: 149–153. doi:10.1038/ng1719

Khil PP, Smagulova F, Brick KM, Camerini-Otero RD, Petukhova GV. 2012. Sensitive mapping of recombination hotspots using sequencing-based detection of ssDNA. *Genome Res* **22**: 957–965. doi:10.1101/gr.130583.111

Khulan B, Thompson RF, Ye K, Fazzari MJ, Suzuki M, Stasiek E, Figueroa ME, Glass JL, Chen Q, Montagna C, et al. 2006. Comparative isoschizomer profiling of cytosine methylation: the HELP assay. *Genome Res* 16: 1046–1055. doi:10.1101/gr.5273806

Kim S-K, Jung I, Lee H, Kang K, Kim M, Jeong K, Kwon CS, Han Y-M, Kim YS, Kim D, Lee D. 2012. Human histone H3K79 methyltransferase DOT1L protein [corrected] binds actively transcribing RNA polymerase II to regulate gene expression. *J Biol Chem* 287: 39698–39709. doi:10.1074/jbc.M112.384057

Kim S, Becker J, Bechheim M, Kaiser V, Noursadeghi M, Fricker N, Beier E, Klaschik S, Boor P, Hess T, et al. 2014. Characterizing the genetic basis of innate immune response in TLR4-activated human monocytes. *Nat Commun* 5: 5236. doi:10.1038/ncomms6236

Klein CA. 2011. Framework models of tumor dormancy from patient-derived observations. *Curr Opin Genet Dev* 21: 42–49. doi:10.1016/j.gde.2010.10.011

Knittle JL, Hirsch J. 1968. Effect of early nutrition on the development of rat epididymal fat pads: cellularity and metabolism. *J Clin Invest* 47: 2091–2098. doi:10.1172/JCI105894

Knopik VS. 2009. Maternal smoking during pregnancy and child outcomes: real or spurious effect? *Dev Neuropsychol* 34: 1–36. doi:10.1080/87565640802564366

Knutson SK, Wigle TJ, Warholic NM, Sneeringer CJ, Allain CJ, Klaus CR, Sacks JD, Raimondi A, Majer CR, Song J, et al. 2012. A selective inhibitor of EZH2 blocks H3K27 methylation and kills mutant lymphoma cells. *Nat Chem Biol* 8: 890–896. doi:10.1038/nchembio.1084

Koestler DC, Avissar-Whiting M, Houseman EA, Karagas MR, Marsit CJ. 2013. Differential DNA methylation in umbilical cord blood of infants exposed to low levels of arsenic in utero. *Environ Health Perspect* 121: 971–977. doi:10.1289/ehp.1205925

Koestler DC, Usset J, Christensen BC, Marsit CJ, Karagas MR, Kelsey KT, Wiencke JK. 2017. DNA methylation–derived neutrophil-to-lymphocyte ratio: an epigenetic tool to explore cancer inflammation and outcomes. *Cancer Epidemiol Biomarkers Prev* 26: 328–338. doi:10.1158/1055-9965.EPI-16-0461

Kofoid CA. 1927. Das Krebsproblem, Rückblicke und Ausblicke, Grund- und Scheinprobleme der Krebsforschung, -Behandlung und -Verhütung. *Am J Public Health* 17: 502. doi:10.2105/AJPH.17.5.502-a

Kolchinsky EI, Kutschera U, Hossfeld U, Levit GS. 2017. Russia's new Lysenkoism. *Curr Biol* 27: R1042–R1047. doi:10.1016/j.cub.2017.07.045

Kong Y, Rastogi D, Seoighe C, Greally JM, Suzuki M. 2019a. Insights from deconvolution of cell subtype proportions enhance the interpretation of functional genomic data. *PLoS One* 14: e0215987. doi:10.1371/journal.pone.0215987

Kong Y, Rose CM, Cass AA, Williams AG, Darwish M, Lianoglou S, Haverty PM, Tong A-J, Blanchette C, Albert ML, et al. 2019b. Transposable element expression in tumors is associated with immune infiltration and increased antigenicity. *Nat Commun* 10: 5228. doi:10.1038/s41467-019-13035-2

Koren A, Polak P, Nemesh J, Michaelson JJ, Sebat J, Sunyaev SR, McCarroll SA. 2012. Differential relationship of DNA replication timing to different forms of human mutation and variation. *Am J Hum Genet* 91: 1033–1040. doi:10.1016/j.ajhg.2012.10.018

Kottek SS. 1981. Embryology in Talmudic and Midrashic literature. *J Hist Biol* **14**: 299–315.

Krawetz SA. 2005. Paternal contribution: new insights and future challenges. *Nat Rev Genet* **6**: 633–642. doi:10.1038/nrg1654

Krawetz SA, Kruger A, Lalancette C, Tagett R, Anton E, Draghici S, Diamond MP. 2011. A survey of small RNAs in human sperm. *Hum Reprod* **26**: 3401–3412. doi:10.1093/humrep/der329

Kroker AJ, Bruning JB. 2015. Review of the structural and dynamic mechanisms of PPARγ partial agonism. *PPAR Res* **2015**: 816856. doi:10.1155/2015/816856

Kropinski AM, Turner D, Nash JHE, Ackermann H-W, Lingohr EJ, Warren RA, Ehrlich KC, Ehrlich M. 2018. The sequence of two bacteriophages with hypermodified bases reveals novel phage–host interactions. *Viruses* **10**: 217. doi:10.3390/v10050217

Krump NA, You J. 2018. Molecular mechanisms of viral oncogenesis in humans. *Nat Rev Microbiol* **16**: 684–698. doi:10.1038/s41579-018-0064-6

Kuo TT, Huang TC, Wu RY, Chen CP. 1968. Phage Xp12 of *Xanthomonas oryzae* (Uyeda et Ishiyama) Dowson. *Can J Microbiol* **14**: 1139–1142. doi:10.1139/m68-190

Lambert R. 2009. *A Jackson Laboratory resource manual: breeding strategies for maintaining colonies of laboratory mice*. Jackson Laboratory, Farmington, CT. https://research.uci.edu/wp-content/uploads/JAX-breeding-strategies.pdf

Lambert SA, Jolma A, Campitelli LF, Das PK, Yin Y, Albu M, Chen X, Taipale J, Hughes TR, Weirauch MT. 2018. The human transcription factors. *Cell* **172**: 650–665. doi:10.1016/j.cell.2018.01.029

Lämke J, Brzezinka K, Altmann S, Bäurle I. 2016. A hit-and-run heat shock factor governs sustained histone methylation and transcriptional stress memory. *EMBO J* **35**: 162–175. doi:10.15252/embj.201592593

Lapeyre JN, Becker FF. 1979. 5-Methylcytosine content of nuclear DNA during chemical hepatocarcinogenesis and in carcinomas which result. *Biochem Biophys Res Commun* **87**: 698–705. doi:10.1016/0006-291x(79)92015-1

Laplace PS. 2009. *Essai philosophique sur les probabilités*. Cambridge University Press, Cambridge, UK. doi:10.1017/CBO9780511693182

Lappalainen T, Greally JM. 2017. Associating cellular epigenetic models with human phenotypes. *Nat Rev Genet* **18**: 441–451. doi:10.1038/nrg.2017.32

Lee E, Iskow R, Yang L, Gokcumen O, Haseley P, Luquette LJ, Lohr JG, Harris CC, Ding L, Wilson RK, et al. 2012. Landscape of somatic retrotransposition in human cancers. *Science* **337**: 967–971. doi:10.1126/science.1222077

Lee MN, Ye C, Villani A-C, Raj T, Li W, Eisenhaure TM, Imboywa SH, Chipendo PI, Ran FA, Slowikowski K, et al. 2014. Common genetic variants modulate pathogen-sensing responses in human dendritic cells. *Science* **343**: 1246980. doi:10.1126/science.1246980

Lee C-J, Ahn H, Jeong D, Pak M, Moon JH, Kim S. 2020. Impact of mutations in DNA methylation modification genes on genome-wide methylation landscapes and downstream gene activations in pan-cancer. *BMC Med Genom* **13**(Suppl 3): 27. doi:10.1186/s12920-020-0659-4

Lee R, Feinbaum R, Ambros V. 2004. A short history of a short RNA. *Cell* **116**(2 Suppl): S89–S92, 1 p following S96. doi:10.1016/s0092-8674(04)00035-2

Leek JT, Johnson WE, Parker HS, Jaffe AE, Storey JD. 2012. The sva package for removing batch effects and other unwanted variation in high-throughput experiments. *Bioinformatics* **28**: 882–883. doi:10.1093/bioinformatics/bts034

Levine AJ, Berger SL. 2017. The interplay between epigenetic changes and the p53 protein in stem cells. *Genes Dev* **31**: 1195–1201. doi:10.1101/gad.298984.117

Lewis CM, Vassos E. 2020. Polygenic risk scores: from research tools to clinical instruments. *Genome Med* **12**: 44. doi:10.1186/s13073-020-00742-5

Liang G, Gonzalgo ML, Salem C, Jones PA. 2002. Identification of DNA methylation differences during tumorigenesis by methylation-sensitive arbitrarily primed polymerase chain reaction. *Methods* **27**: 150–155. doi:10.1016/s1046-2023(02)00068-3

Lichinchi G, Gao S, Saletore Y, Gonzalez GM, Bansal V, Wang Y, Mason CE, Rana TM. 2016. Dynamics of the human and viral m(6)A RNA methylomes during HIV-1 infection of T cells. *Nat Microbiol* **1**: 16011. doi:0.1038/nmicrobiol.2016.11

Lichter P, Cremer T, Borden J, Manuelidis L, Ward DC. 1988. Delineation of individual human chromosomes in metaphase and interphase cells by in situ suppression hybridization using recombinant DNA libraries. *Hum Genet* **80**: 224–234. doi:10.1007/BF01790090

Li X, Fu X-D. 2019. Chromatin-associated RNAs as facilitators of functional genomic interactions. *Nat Rev Genet* **20**: 503–519. doi:10.1038/s41576-019-0135-1

Li C, Lumey LH. 2017. Exposure to the Chinese famine of 1959–61 in early life and long-term health conditions: a systematic review and meta-analysis. *Int J Epidemiol* **46**: 1157–1170. doi:10.1093/ije/dyx013

Li E, Bestor TH, Jaenisch R. 1992. Targeted mutation of the DNA methyltransferase gene results in embryonic lethality. *Cell* **69**: 915–926. doi:10.1016/0092-8674(92)90611-f

Li R, Grimm SA, Chrysovergis K, Kosak J, Wang X, Du Y, Burkholder A, Janardhan K, Mav D, Shah R, et al. 2014. Obesity, rather than diet, drives epigenomic alterations in colonic epithelium resembling cancer progression. *Cell Metab* **19**: 702–711. doi:10.1016/j.cmet.2014.03.012

Li Q, Suzuki M, Wendt J, Patterson N, Eichten SR, Hermanson PJ, Green D, Jeddeloh J, Richmond T, Rosenbaum H, et al. 2015. Post-conversion targeted capture of modified cytosines in mammalian and plant genomes. *Nucl Acids Res* **43**: e81. doi:10.1093/nar/gkv244

Li X, Xiong X, Zhang M, Wang K, Chen Y, Zhou J, Mao Y, Lv J, Yi D, Chen X-W, et al. 2017. Base-resolution mapping reveals distinct m^1A methylome in nuclear- and mitochondrial-encoded transcripts. *Mol Cell* **68**: 993–1005.e9. doi:10.1016/j.molcel.2017.10.019

Lin SY, Riggs AD. 1971. Lac repressor binding to operator analogues: comparison of poly(d(A-T)), poly(d(A-BrU)), and poly(d(A-U)). *Biochem Biophys Res Commun* **45**: 1542–1547. doi:10.1016/0006-291x(71)90195-1

Lin TC, Hou HA, Chou WC, Ou DL, Yu SL, Tien HF, Lin LI. 2011. CEBPA methylation as a prognostic biomarker in patients with de novo acute myeloid leukemia. *Leukemia* **25**: 32–40. doi:10.1038/leu.2010.222

Lin B, Srikanth P, Castle AC, Nigwekar S, Malhotra R, Galloway JL, Sykes DB, Rajagopal J. 2018. Modulating cell fate as a therapeutic strategy. *Cell Stem Cell* **23**: 329–341. doi:10.1016/j.stem.2018.05.009

Lind MI, Spagopoulou F. 2018. Evolutionary consequences of epigenetic inheritance. *Heredity* **121**: 205–209. doi:10.1038/s41437-018-0113-y

Link VM, Duttke SH, Chun HB, Holtman IR, Westin E, Hoeksema MA, Abe Y, Skola D, Romanoski CE, Tao J, et al. 2018. Analysis of genetically diverse macrophages reveals local and domain-wide mechanisms that control transcription factor binding and function. *Cell* **173**: 1796–1809.e17. doi:10.1016/j.cell.2018.04.018

Lister R, Pelizzola M, Dowen RH, Hawkins RD, Hon G, Tonti-Filippini J, Nery JR, Lee L, Ye Z, Ngo Q-M, et al. 2009. Human DNA methylomes at base resolution show widespread epigenomic differences. *Nature* **462**: 315–322. doi:10.1038/nature08514

Lister R, Pelizzola M, Kid YS, Hawkins RD, Nery JR, Hon G, Antosiewicz-Bourget J, O'Malley R, Castanon R, Klugman S, et al. 2011. Hotspots of aberrant epigenomic reprogramming in human induced pluripotent stem cells. *Nature* **471**: 68–73. doi:10.1038/nature09798

Liu J, Perumal NB, Oldfield CJ, Su EW, Uversky VN, Dunker AK. 2006. Intrinsic disorder in transcription factors. *Biochemistry* **45**: 6873–6888. doi:10.1021/bi0602718

Liu Y, Li C, Shen S, Chen X, Szlachta K, Edmonson MN, Shao Y, Ma X, Hyle J, Wright S, et al. 2020. Discovery of regulatory noncoding variants in individual cancer genomes by using *cis*-X. *Nat Genet* **52**: 811–818. doi:10.1038/s41588-020-0659-5

Llanos C, Friedman DM, Saxena A, Izmirly PM, Tseng C-E, Dische R, Abellar RG, Halushka M, Clancy RM, Buyon JP. 2012. Anatomical and pathological findings in hearts from fetuses and infants with cardiac manifestations of neonatal lupus. *Rheumatology (Oxford, Engl)* **51**: 1086–1092. doi:10.1093/rheumatology/ker515

Lo P-W, Shie J-J, Chen C-H, Wu C-Y, Hsu T-L, Wong C-H. 2018. *O*-GlcNAcylation regulates the stability and enzymatic activity of the histone methyltransferase EZH2. *Proc Natl Acad Sci* **115**: 7302–7307. doi:10.1073/pnas.1801850115

Logsdon GA, Rozanski AN, Ryabov F, Potapova T, Shepelev VA, Catacchio CR, Porubsky D, Mao Y, Yoo D, Rautiainen M, et al. 2024. The variation and evolution of complete human centromeres. *Nature* **629**: 136–145. doi:10.1038/s41586-024-07278-3

Long HK, Sims D, Heger A, Blackledge NP, Kutter C, Wright ML, Grützner F, Odom DT, Patient R, Ponting CP, Klose RJ. 2013. Epigenetic conservation at gene regulatory elements revealed by non-methylated DNA profiling in seven vertebrates. *eLife* **2**: e00348. doi:0.7554/eLife.00348

López-Moyado IF, Tsagaratou A, Yuita H, Seo H, Delatte B, Heinz S, Benner C, Rao A. 2019. Paradoxical association of TET loss of function with genome-wide DNA hypomethylation. *Proc Natl Acad Sci* **116**: 16933–16942. doi:10.1073/pnas.1903059116

Lu Z, Carter AC, Chang HY. 2017. Mechanistic insights in X-chromosome inactivation. *Philos Trans R Soc Lond B Biol Sci* **372**: 20160356. doi:10.1098/rstb.2016.0356

Lucifero D, Mann MRW, Bartolomei MS, Trasler JM. 2004. Gene-specific timing and epigenetic memory in oocyte imprinting. *Hum Mol Genet* **13**: 839–849. doi:10.1093/hmg/ddh104

Lumey LH. 2001. Glucose tolerance in adults after prenatal exposure to famine. *Lancet* **357**: 472–473. doi:0.1016/s0140-6736(00)04003-4

Luo G-Z, Blanco MA, Greer EL, He C, Shi Y. 2015. DNA N(6)-methyladenine: a new epigenetic mark in eukaryotes? *Nat Rev Mol Cell Biol* **16**: 705–710. doi:10.1038/nrm4076

Luo S, Valencia CA, Zhang J, Lee N-C, Slone J, Gui B, Wang X, Li Z, Dell S, Brown J, et al. 2018. Biparental inheritance of mitochondrial DNA in humans. *Proc Natl Acad Sci* **115**: 13039–13044. doi:10.1073/pnas.1810946115

Luria SE, Human ML. 1952. A nonhereditary, host-induced variation of bacterial viruses. *J Bacteriol* **64**: 557–569.

Luyckx VA, Brenner BM. 2005. Low birth weight, nephron number, and kidney disease. *Kidney Int Suppl* **97**: S68–S77. doi:10.1111/j.1523-1755.2005.09712.x

Lyon MF. 1992. Some milestones in the history of X-chromosome inactivation. *Ann Rev Genet* **26**: 16–28. doi:10.1146/annurev.ge.26.120192.000313

Madsen JGS, Rauch A, Van Hauwaert EL, Schmidt SF, Winnefeld M, Mandrup S. 2018. Integrated analysis of motif activity and gene expression changes of transcription factors. *Genome Res* **28**: 243–255. doi:10.1101/gr.227231.117

Makova KD, Hardison RC. 2015. The effects of chromatin organization on variation in mutation rates in the genome. *Nat Rev Genet* **16**: 213–223. doi:10.1038/nrg3890

Malmström A, Łysiak M, Kristensen BW, Hovey E, Henriksson R, Söderkvist P. 2020. Do we really know who has an *MGMT* methylated glioma? Results of an international survey regarding use of *MGMT* analyses for glioma. *Neuro-Oncol Pract* **7**: 68–76. doi:10.1093/nop/npz039

Mansour MR, Abraham BJ, Anders L, Berezovskaya A, Gutierrez A, Durbin AD, Etchin J, Lawton L, Sallan SE, Silverman LB, et al. 2014. Oncogene regulation. An oncogenic super-enhancer formed through somatic mutation of a noncoding intergenic element. *Science* **346**: 1373–1377. doi:10.1126/science.1259037

Manuelidis L. 1985. Individual interphase chromosome domains revealed by in situ hybridization. *Hum Genet* **71**: 288–293. doi:10.1007/BF00388453

Marchese FP, Raimondi I, Huarte M. 2017. The multidimensional mechanisms of long noncoding RNA function. *Genome Biol* **18**: 206. doi:10.1186/s13059-017-1348-2

Marks PA, Breslow R. 2007. Dimethyl sulfoxide to vorinostat: development of this histone deacetylase inhibitor as an anticancer drug. *Nat Biotechnol* **25**: 84–90. doi:10.1038/nbt1272

Marks EI, Brown VS, Dizon DS. 2020. Genomic and molecular abnormalities in gynecologic clear cell carcinoma. *Am J Clin Oncol* **43**: 139–145. doi:10.1097/COC.0000000000000641

Marmur J, Brandon C, Neubort S, Ehrlich M, Mandel M, Konvicka J. 1972. Unique properties of nucleic acid from *Bacillus subtilis* phage SP-15. *Nat New Biol* **239**: 68–70. doi:10.1038/newbio239068a0

Marré J, Trave EC, Jose AM. 2016. Extracellular RNA is transported from one generation to the next in *Caenorhabditis elegans. Proc Natl Acad Sci* **113**: 12496–12501. doi:10.1073/pnas.1608959113

Martienssen R, Moazed D. 2015. RNAi and heterochromatin assembly. *Cold Spring Harb Perspect Biol* **7**: a019323. doi:10.1101/cshperspect.a019323

Martin AR, Kanai M, Kamatani Y, Okada Y, Neale BM, Daly MJ. 2019. Clinical use of current polygenic risk scores may exacerbate health disparities. *Nat Genet* **51:** 584–591. doi:10.1038/s41588-019-0379-x

Martin BJE, Brind'Amour J, Kuzmin A, Jensen KN, Liu ZC, Lorincz M, Howe LJ. 2021. Transcription shapes genome-wide histone acetylation patterns. *Nat Commun* **12:** 210. doi:10.1038/s41467-020-20543-z

Marx V. 2021. Method of the year: spatially resolved transcriptomics. *Nat Meth* **18:** 9–14. doi:10.1038/s41592-020-01033-y

Maslov AY, Lee M, Gundry M, Gravina S, Strogonova N, Tazearslan C, Bendebury A, Suh Y, Vijg J. 2012. 5-Aza-2′-deoxycytidine-induced genome rearrangements are mediated by DNMT1. *Oncogene* **31:** 5172–5179. doi:10.1038/onc.2012.9

Mateos-Aparicio P, Rodríguez-Moreno A. 2019. The impact of studying brain plasticity. *Front Cell Neurosci* **13:** 66. doi:10.3389/fncel.2019.00066

Matthews BJ, Waxman DJ. 2018. Computational prediction of CTCF/cohesin-based intra-TAD loops that insulate chromatin contacts and gene expression in mouse liver. *eLife* **7:** 34077. doi:10.7554/eLife.34077

Maunakea AK, Nagarajan RP, Bilenky M, Ballinger TJ, D'Souza C, Fouse SD, Johnson BE, Hong C, Nielsen C, Zhao Y, et al. 2010. Conserved role of intragenic DNA methylation in regulating alternative promoters. *Nature* **466:** 253–257. doi:10.1038/nature09165

McCart AE, Vickaryous NK, Silver A. 2008. Apc mice: models, modifiers and mutants. *Pathol Res Pract* **204:** 479–490. doi:10.1016/j.prp.2008.03.004

McCarthy S, Das S, Kretzschmar W, Delaneau O, Wood AR, Teumer A, Kang HM, Fuchsberger C, Danecek P, Sharp K, et al. 2016. A reference panel of 64,976 haplotypes for genotype imputation. *Nat Genet* **48:** 1279–1283. doi:10.1038/ng.3643

McDaniell R, Lee B-K, Song L, Liu Z, Boyle AP, Erdos MR, Scott LJ, Morken MA, Kucera KS, Battenhouse A, et al. 2010. Heritable individual-specific and allele-specific chromatin signatures in humans. *Science* **328:** 235–239. doi:10.1126/science.1184655

McGhee JD, Ginder GD. 1979. Specific DNA methylation sites in the vicinity of the chicken β-globin genes. *Nature* **280:** 419–420. doi:10.1038/280419a0

McGrath J, Solter D. 1984a. Maternal T^{hp} lethality in the mouse is a nuclear, not cytoplasmic, defect. *Nature* **308:** 550–551. doi:10.1038/308550a0

McGrath J, Solter D. 1984b. Completion of mouse embryogenesis requires both the maternal and paternal genomes. *Cell* **37:** 179–183. doi:10.1016/0092-8674(84)90313-1

McLaughlin-Drubin ME, Munger K. 2008. Viruses associated with human cancer. *Biochim Biophys Acta* **1782:** 127–150. doi:10.1016/j.bbadis.2007.12.005

McLaughlin K, Flyamer IM, Thomson JP, Mjoseng HK, Shukla R, Williamson I, Grimes GR, Illingworth RS, Adams IR, Pennings S, et al. 2019. DNA methylation directs Polycomb-dependent 3D genome re-organization in naive pluripotency. *Cell Rep* **29:** 1974–1985.e6. doi:10.1016/j.celrep.2019.10.031

Medawar PB, Medawar JS. 1983. *Aristotle to zoos: a philosophical dictionary of biology*, New Ed, p. 320. Harvard University Press, Boston.

Meissner A, Gnirke A, Bell GW, Ramsahoye B, Lander ES, Jaenisch R. 2005. Reduced representation bisulfite sequencing for comparative high-resolution DNA methylation analysis. *Nucl Acids Res* **33**: 5868–5877. doi:10.1093/nar/gki901

Melton C, Reuter JA, Spacek DV, Snyder M. 2015. Recurrent somatic mutations in regulatory regions of human cancer genomes. *Nat Genet* **47**: 710–716. doi:10.1038/ng.3332

Mendel G. 1866. Versuche über Pflanzen-Hybriden. *Verhandlungen Des Naturforschenden Vereines in Brünn* **4**: 3–47.

Mentc SJ, Locasale JW. 2016. One-carbon metabolism and epigenetics: understanding the specificity. *Ann NY Acad Sci* **1363**: 91–98. doi:10.1111/nyas.12956

Mews P, Egervari G, Nativio R, Sidoli S, Donahue G, Lombroso SI, Alexander DC, Riesche SL, Heller EA, Nestler EJ, et al. 2019. Alcohol metabolism contributes to brain histone acetylation. *Nature* **574**: 717–721. doi:10.1038/s41586-019-1700-7

Meyer KD, Saletore Y, Zumbo P, Elemento O, Mason CE, Jaffrey SR. 2012. Comprehensive analysis of mRNA methylation reveals enrichment in 3′ UTRs and near stop codons. *Cell* **149**: 1635–1646. doi:10.1016/j.cell.2012.05.003

Michelotti EF, Sanford S, Levens D. 1997. Marking of active genes on mitotic chromosomes. *Nature* **388**: 895–899. doi:10.1038/42282

Millán-Zambrano G, Burton A, Bannister AJ, Schneider R. 2022. Histone post-translational modifications—cause and consequence of genome function. *Nat Revi Genet* **23**: 563–580. doi:10.1038/s41576-022-00468-7

Mills JL. 1993. Data torturing. *New Engl J Med* **329**: 1196–1199. doi:10.1056/NEJM199310143291613

Miraldi ER, Pokrovskii M, Watters A, Castro DM, De Veaux N, Hall JA, Lee J-Y, Ciofani M, Madar A, Carriero N, et al. 2019. Leveraging chromatin accessibility for transcriptional regulatory network inference in T helper 17 cells. *Genome Res* **29**: 449–463. doi:0.1101/gr.238253.118

Mita MM, Mita AC. 2020. Bromodomain inhibitors a decade later: a promise unfulfilled? *Br J Cancer* **123**: 1713–1714. doi:10.1038/s41416-020-01079-x

Mitchell KJ. 2018. *Innate: how the wiring of our brains shapes who we are.* Princeton University Press, Princeton, NJ.

Mohandas T, Sparkes RS, Shapiro LJ. 1981. Reactivation of an inactive human X chromosome: evidence for X inactivation by DNA methylation. *Science* **211**: 393–396. doi:10.1126/science.6164095

Molinie B, Giallourakis CC. 2017. Genome-wide location analyses of N6-methyladenosine modifications (m⁶A-Seq). *Meth Mol Biol* **1562**: 45–53. doi:10.1007/978-1-4939-6807-7_4

Monick MM, Beach SRH, Plume J, Sears R, Gerrard M, Brody GH, Philibert RA. 2012. Coordinated changes in AHRR methylation in lymphoblasts and pulmonary macrophages from smokers. *Am J Med Genet B Neuropsychiatr Genet* **159B**: 141–151. doi:10.1002/ajmg.b.32021

Monk M. 1990. Changes in DNA methylation during mouse embryonic development in relation to X-chromosome activity and imprinting. *Philos Trans R Soc Lond B Biol Sci* **326**: 299–312. doi:10.1098/rstb.1990.0013

Monk M, Boubelik M, Lehnert S. 1987. Temporal and regional changes in DNA methylation in the embryonic, extraembryonic and germ cell lineages during mouse embryo development. *Development* **99:** 371–382. doi:10.1242/dev.99.3.371

Montano C, Taub MA, Jaffe A, Briem E, Feinberg JI, Trygvadottir R, Idrizi A, Runarsson A, Berndsen B, Gur RC, et al. 2016. Association of DNA methylation differences with schizophrenia in an epigenome-wide association study. *JAMA Psychiatry* **73:** 506–514. doi:10.1001/jamapsychiatry.2016.0144

Morgan TH. 1910. Chromosomes and heredity. *Am Nat* **44:** 449–496.

Morgan HD, Sutherland HG, Martin DI, Whitelaw E. 1999. Epigenetic inheritance at the agouti locus in the mouse. *Nat Genet* **23:** 314–318. doi:10.1038/15490

Morgan HD, Jin XL, Li A, Whitelaw E, O'Neill C. 2008. The culture of zygotes to the blastocyst stage changes the postnatal expression of an epigenically labile allele, agouti viable yellow, in mice. *Biol Reprod* **79:** 618–623. doi:10.1095/biolreprod.108.068213

Morley M, Molony CM, Weber TM, Devlin JL, Ewens KG, Spielman RS, Cheung VG. 2004. Genetic analysis of genome-wide variation in human gene expression. *Nature* **430:** 743–747. doi:10.1038/nature02797

Morris JA, Caragine C, Daniloski Z, Domingo J, Barry T, Lu L, Davis K, Ziosi M, Glinos DA, Hao S, et al. 2023. Discovery of target genes and pathways at GWAS loci by pooled single-cell CRISPR screens. *Science* **380:** eadh7699. doi:10.1126/science.adh7699

Moser AR, Pitot HC, Dove WF. 1990. A dominant mutation that predisposes to multiple intestinal neoplasia in the mouse. *Science* **247:** 322–324. doi:10.1126/science.2296722

Muller HJ. 1930. Types of visible variations induced by X-rays in *Drosophila*. *J Genet* **22:** 299–334. doi:10.1007/BF02984195

Murphy PJ, Cipriany BR, Wallin CB, Ju CY, Szeto K, Hagarman JA, Benitez JJ, Craighead HG, Soloway PD. 2013a. Single-molecule analysis of combinatorial epigenomic states in normal and tumor cells. *Proc Natl Acad Sci* **110:** 7772–7777. doi:0.1073/pnas.1218495110

Murphy SK, Yang H, Moylan CA, Pang H, Dellinger A, Abdelmalek MF, Garrett ME, Ashley-Koch A, Suzuki A, Tillmann HL, et al. 2013b. Relationship between methylome and transcriptome in patients with nonalcoholic fatty liver disease. *Gastroenterology* **145:** 1076–1087. doi:10.1053/j.gastro.2013.07.047

Murray K. 1964. The occurrence of ε-N-methyl lysine in histones. *Biochemistry* **3:** 10–15.

Nadeau JH. 2017. Do gametes woo? Evidence for their nonrandom union at fertilization. *Genetics* **207:** 369–387. doi:10.1534/genetics.117.300109

Nadel J, Athanasiadou R, Lemetre C, Wijetunga NA, Ó Broin P, Sato H, Zhang Z, Jeddeloh J, Montagna C, Golden A, et al. 2015. RNA:DNA hybrids in the human genome have distinctive nucleotide characteristics, chromatin composition, and transcriptional relationships. *Epigenet Chromatin* **8:** 46. doi:10.1186/s13072-015-0040-6

Nagamori I, Kobayashi H, Shiromoto Y, Nishimura T, Kuramochi-Miyagawa S, Kono T, Nakano T. 2015. Comprehensive DNA methylation analysis of retrotransposons in male germ cells. *Cell Rep* **12:** 1541–1547. doi:10.1016/j.celrep.2015.07.060

Nanney DL. 1958. Epigenetic control systems. *Proc Natl Acad Sci* **44:** 712–717. doi:10.1073/pnas.44.7.712

Nanney DL. 2004. *Candide in Academe*. Nanney autobiographic essays. http://www.life.illinois.edu/nanney/autobiography/candide.html

National Academy of Sciences. 2001. *Biographical memoirs*, Vol. 80. National Academies, Washington. doi:10.17226/10269

Nédélec Y, Sanz J, Baharian G, Szpiech ZA, Pacis A, Dumaine A, Grenier J-C, Freiman A, Sams AJ, Hebert S, et al. 2016. Genetic ancestry and natural selection drive population differences in immune responses to pathogens. *Cell* 167: 657–669.e21. doi:10.1016/j.cell.2016.09.025

Needham J. 1931. *Chemical embryology*, Vol. 3. The MacMillan Company, New York.

Needham J. 1934. *A history of embryology*. Cambridge University Press, Cambridge.

Nègre N, Brown CD, Shah PK, Kheradpour P, Morrison CA, Henikoff JG, Feng X, Ahmad K, Russell S, White RAH, et al. 2010. A comprehensive map of insulator elements for the *Drosophila* genome. *PLoS Genet* 6: e1000814. doi:10.1371/journal.pgen.1000814

Nehme R, Barrett LE. 2020. Using human pluripotent stem cell models to study autism in the era of big data. *Mol Autism* 11: 21. doi:10.1186/s13229-020-00322-9

Nergadze SG, Piras FM, Gamba R, Corbo M, Cerutti F, McCarter JGW, Cappelletti E, Gozzo F, Harman RM, Antczak DF, et al. 2018. Birth, evolution, and transmission of satellite-free mammalian centromeric domains. *Genome Res* 28: 789–799. doi:10.1101/gr.231159.117

Nestor CE, Ottaviano R, Reddington J, Sproul D, Reinhardt D, Dunican D, Katz E, Dixon JM, Harrison DJ, Meehan RR. 2012. Tissue type is a major modifier of the 5-hydroxymethylcytosine content of human genes. *Genome Res* 22: 467–477. doi:10.1101/gr.126417.111

Neugebauer R, Paneth N. 1992. Epidemiology and the wider world: celebrating Zena Stein and Mervyn Susser. *Paed Perinatal Epidemiol* 6: 122–132. doi:10.1111/j.1365-3016.1992.tb00753.x

Newman AM, Liu CL, Green MR, Gentles AJ, Feng W, Xu Y, Hoang CD, Diehn M, Alizadeh AA. 2015. Robust enumeration of cell subsets from tissue expression profiles. *Nat Meth* 12: 453–457. doi:10.1038/nmeth.3337

Newman AM, Steen CB, Liu CL, Gentles AJ, Chaudhuri AA, Scherer F, Khodadoust MS, Esfahani MS, Luca BA, Steiner D, et al. 2019. Determining cell type abundance and expression from bulk tissues with digital cytometry. *Nat Biotechnol* 37: 773–782. doi:0.1038/s41587-019-0114-2

Newmark HL, Lipkin M, Maheshwari N. 1990. Colonic hyperplasia and hyperproliferation induced by a nutritional stress diet with four components of Western-style diet. *J Natl Cancer Inst* 82: 491–496. doi:10.1093/jnci/82.6.491

Newmark HL, Yang K, Lipkin M, Kopelovich L, Liu Y, Fan K, Shinozaki H. 2001. A Western-style diet induces benign and malignant neoplasms in the colon of normal C57Bl/6 mice. *Carcinogenesis* 22: 1871–1875. doi:10.1093/carcin/22.11.1871

Nguyen HT, Duong H-Q. 2018. The molecular characteristics of colorectal cancer: implications for diagnosis and therapy. *Oncol Lett* 16: 9–18. doi:10.3892/ol.2018.8679

Nielsen SCA, Roskin KM, Jackson KJL, Joshi SA, Nejad P, Lee J-Y, Wagar LE, Pham TD, Hoh RA, Nguyen KD, et al. 2019. Shaping of infant B cell receptor repertoires by environmental factors and infectious disease. *Sci Transl Med* **11**: eaat2004. doi:10.1126/scitranslmed.aat2004

Nigon VM, Félix M-A. 2017. History of research on *C. elegans* and other free-living nematodes as model organisms. *Wormbook: the online review of C. elegans biology*, **2017**, pp. 1–84. doi:10.1895/wormbook.1.181.1

Nosek BA, Ebersole CR, DeHaven AC, Mellor DT. 2018. The preregistration revolution. *Proc Natl Acad Sci* **115**: 2600–2606. doi:10.1073/pnas.1708274114

Noto T, Kataoka K, Suhren JH, Hayashi A, Woolcock KJ, Gorovsky MA, Mochizuki K. 2015. Small-RNA-mediated genome-wide *trans*-recognition network in *Tetrahymena* DNA elimination. *Mol Cell* **59**: 229–242. doi:10.1016/j.molcel.2015.05.024

Ntumngia FB, Thomson-Luque R, Pires CV, Adams JH. 2016. The role of the human Duffy antigen receptor for chemokines in malaria susceptibility: current opinions and future treatment prospects. *J Receptor Ligand Channel Res* **9**: 1–11. doi:10.2147/JRLCR.S99725

O'Brown ZK, Boulias K, Wang J, Wang SY, O'Brown NM, Hao Z, Shibuya H, Fady P-E, Shi Y, He C, et al. 2019. Sources of artifact in measurements of 6mA and 4mC abundance in eukaryotic genomic DNA. *BMC Genom* **20**: 445. doi:10.1186/s12864-019-5754-6

Oetjen KA, Lindblad KE, Goswami M, Gui G, Dagur PK, Lai C, Dillon LW, McCoy JP, Hourigan CS. 2018. Human bone marrow assessment by single-cell RNA sequencing, mass cytometry, and flow cytometry. *JCI Insight* **6**: e124928. doi:10.1172/jci.insight.124928

Ohno S, Kaplan WD, Kinosita R. 1959. Formation of the sex chromatin by a single X-chromosome in liver cells of *Rattus norvegicus*. *Exp Cell Res* **18**: 415–418. doi:10.1016/0014-4827(59)90031-x

Ohno S. 1972. So much "junk" DNA in our genome. *Brookhaven Symp Biol* **23**: 366–370.

Okada Y, Hirota T, Kamatani Y, Takahashi A, Ohmiya H, Kumasaka N, Higasa K, Yamaguchi-Kabata Y, Hosono N, Nalls MA, et al. 2011. Identification of nine novel loci associated with white blood cell subtypes in a Japanese population. *PLoS Genet* **7**: e1002067. doi:10.1371/journal.pgen.1002067

Ono R, Taki T, Taketani T, Taniwaki M, Kobayashi H, Hayashi Y. 2002. LCX, leukemia-associated protein with a CXXC domain, is fused to MLL in acute myeloid leukemia with trilineage dysplasia having t(10;11)(q22;q23). *Cancer Res* **62**: 4075–4080.

Onuchic V, Lurie E, Carrero I, Pawliczek P, Patel RY, Rozowsky J, Galeev T, Huang Z, Altshuler RC, Zhang Z, et al. 2018. Allele-specific epigenome maps reveal sequence-dependent stochastic switching at regulatory loci. *Science* **361**: eaar3146. doi:10.1126/science.aar3146

Orrù V, Steri M, Sole G, Sidore C, Virdis F, Dei M, Lai S, Zoledziewska M, Busonero F, Mulas A, et al. 2013. Genetic variants regulating immune cell levels in health and disease. *Cell* **155**: 242–256. doi:10.1016/j.cell.2013.08.041

Otani J, Arita K, Kato T, Kinoshita M, Kimura H, Suetake I, Tajima S, Ariyoshi M, Shirakawa M. 2013. Structural basis of the versatile DNA recognition ability of the methyl-CpG binding domain of methyl-CpG binding domain protein 4. *J Biol Chem* **288**: 6351–6362. doi:10.1074/jbc.M112.431098

Oyoshi K, Katano K, Yunose M, Suzuki N. 2020. Memory of 5-min heat stress in *Arabidopsis thaliana. Plant Signaling Behav* **15**: 1778919. doi:10.1080/15592324.2020.17 78919

Pandey NB, Chodchoy N, Liu TJ, Marzluff WF. 1990. Introns in histone genes alter the distribution of 3′ ends. *Nucl Acids Res* **18**: 3161–3170. doi:0.1093/nar/18.11.3161

Paneth N, Susser M. 1995. Early origin of coronary heart disease (the "Barker hypothesis"). *BMJ Clin Res Ed* **310**: 411–412. doi:10.1136/bmj.310.6977.411

Park J-M, Jo S-H, Kim M-Y, Kim T-H, Ahn Y-H. 2015. Role of transcription factor acetylation in the regulation of metabolic homeostasis. *Protein Cell* **6**: 804–813. doi:10.1007/s13238-015-0204-y

Parker GA. 1970. Sperm competition and its evolutionary consequences in the insects. *Biol Rev* **45**: 525–567. doi:10.1111/j.1469-185X.1970.tb01176.x

Paul IJ, Duerksen JD. 1975. Chromatin-associated RNA content of heterochromatin and euchromatin. *Mol Cell Biochem* **9**: 9–16. doi:10.1007/BF01731728

Pearson H. 2016. *The life project: the extraordinary story of 70,000 ordinary lives*, p. 256. Soft Skull, New York.

Pelikan RC, Kelly JA, Fu Y, Lareau CA, Tessneer KL, Wiley GB, Wiley MM, Glenn SB, Harley JB, Guthridge JM, et al. 2018. Enhancer histone-QTLs are enriched on autoimmune risk haplotypes and influence gene expression within chromatin networks. *Nat Commun* **9**: 2905. doi:10.1038/s41467-018-05328-9

Pérez A, Castellazzi CL, Battistini F, Collinet K, Flores O, Deniz O, Ruiz ML, Torrents D, Eritja R, Soler-López M, Orozco M. 2012. Impact of methylation on the physical properties of DNA. *Biophys J* **102**: 2140–2148. doi:10.1016/j.bpj.2012.03.056

Pertea M, Shumate A, Pertea G, Varabyou A, Breitwieser FP, Chang Y-C, Madugundu AK, Pandey A, Salzberg SL. 2018. CHESS: a new human gene catalog curated from thousands of large-scale RNA sequencing experiments reveals extensive transcriptional noise. *Genome Biol* **19**: 208. doi:10.1186/s13059-018-1590-2

Phillips-Cremins JE, Corces VG. 2013. Chromatin insulators: linking genome organization to cellular function. *Mol Cell* **50**: 461–474. doi:10.1016/j.molcel.2013.04.018

Pintacuda G, Young AN, Cerase A. 2017. Function by structure: spotlights on Xist long non-coding RNA. *Front Mol Biosci* **4**: 90. doi:10.3389/fmolb.2017.00090

Pinzón N, Bertrand S, Subirana L, Busseau I, Escrivá H, Seitz H. 2019. Functional lability of RNA-dependent RNA polymerases in animals. *PLoS Genet* **15**: e1007915. doi:10.1371/journal.pgen.1007915

Poetsch AR, Boulton SJ, Luscombe NM. 2018. Genomic landscape of oxidative DNA damage and repair reveals regioselective protection from mutagenesis. *Genome Biol* **19**: 215. doi:10.1186/s13059-018-1582-2

Popay TM, Dixon JR. 2022. Coming full circle: on the origin and evolution of the looping model for enhancer-promoter communication. *J Biol Chem* **298**: 102117. doi:10.1016/j.jbc.2022.102117

Prasad A. 2012. *Like a virgin: how science is redesigning the rules of sex*. Oneworld, London.

Preissl S, Gaulton KJ, Ren B. 2023. Characterizing *cis*-regulatory elements using single-cell epigenomics. *Nat Rev Genet* **24**: 21–43. doi:10.1038/s41576-022-00509-1

Prescott SL, Logan AC. 2016. Transforming life: a broad view of the developmental origins of health and disease concept from an ecological justice perspective. *Int J Environ Res Public Health* **13**: 1075. doi:10.3390/ijerph13111075

Ptashne M. 2007. On the use of the word "epigenetic". *Curr Biol* **17**: R233-236. doi:10.1016/j.cub.2007.02.030

Quach H, Rotival M, Pothlichet J, Loh Y-HE, Dannemann M, Zidane N, Laval G, Patin E, Harmant C, Lopez M, et al. 2016. Genetic adaptation and Neandertal admixture shaped the immune system of human populations. *Cell* **167**: 643–656.e17. doi:10.1016/j.cell.2016.09.024

Rabl C. 1885. Über Zelltheilung. *Morpholgisches Jahrbuch* **10**: 214–330.

Radaeva M, Ton A-T, Hsing M, Ban F, Cherkasov A. 2021. Drugging the "undruggable". Therapeutic targeting of protein–DNA interactions with the use of computer-aided drug discovery methods. *Drug Discov Today*, **26**: 2660–2679. doi:10.1016/j.drudis.2021.07.018

Raddatz G, Gao Q, Bender S, Jaenisch R, Lyko F. 2012. Dnmt3a protects active chromosome domains against cancer-associated hypomethylation. *PLoS Genet* **8**: e1003146. doi:10.1371/journal.pgen.1003146

Radford EJ, Ito M, Shi H, Corish JA, Yamazawa K, Isganaitis E, Seisenberger S, Hore TA, Reik W, Erkek S, et al. 2014. In utero effects. In utero undernourishment perturbs the adult sperm methylome and intergenerational metabolism. *Science* **345**: 1255903. doi:10.1126/science.1255903

Rahmani E, Shenhav L, Schweiger R, Yousefi P, Huen K, Eskenazi B, Eng C, Huntsman S, Hu D, Galanter J, et al. 2017. Genome-wide methylation data mirror ancestry information. *Epigenet Chromatin* **10**: 1. doi:10.1186/s13072-016-0108-y

Rakyan VK, Chong S, Champ ME, Cuthbert PC, Morgan HD, Luu KVK, Whitelaw E. 2003. Transgenerational inheritance of epigenetic states at the murine Axin(Fu) allele occurs after maternal and paternal transmission. *Proc Natl Acad Sci* **100**: 2538–2543. doi:10.1073/pnas.0436776100

Rando OJ. 2012. Combinatorial complexity in chromatin structure and function: revisiting the histone code. *Curr Opin Genet Dev* **22**: 148–155. doi:10.1016/j.gde.2012.02.013

Rao S, Chiu T-P, Kribelbauer JF, Mann RS, Bussemaker HJ, Rohs R. 2018. Systematic prediction of DNA shape changes due to CpG methylation explains epigenetic effects on protein–DNA binding. *Epigenet Chromatin* **11**: 6. doi:10.1186/s13072-018-0174-4

Rathke C, Baarends WM, Awe S, Renkawitz-Pohl R. 2014. Chromatin dynamics during spermiogenesis. *Biochim Biophys Acta* **1839**: 155–168. doi:10.1016/j. bbagrm.2013.08.004

Rauch T, Pfeifer GP. 2005. Methylated-CpG island recovery assay: a new technique for the rapid detection of methylated-CpG islands in cancer. *Lab Invest* **85**: 1172–1180. doi:10.1038/labinvest.3700311

Ravelli GP, Stein ZA, Susser MW. 1976. Obesity in young men after famine exposure in utero and early infancy. *New Engl J Med* **295**: 349–353. doi:10.1056/ NEJM197608122950701

Razin A, Riggs AD. 1980. DNA methylation and gene function. *Science* **210**: 604–610. doi:10.1126/science.6254144

Rea S, Eisenhaber F, O'Carroll D, Strahl BD, Sun ZW, Schmid M, Opravil S, Mechtler K, Ponting CP, Allis CD, Jenuwein T. 2000. Regulation of chromatin structure by site-specific histone H3 methyltransferases. *Nature* **406**: 593–599. doi:10.1038/35020506

Reed CE, Fenton SE. 2013. Exposure to diethylstilbestrol during sensitive life stages: a legacy of heritable health effects. *Birth Defects Res C Embryo Today Rev* **99**: 134–146. doi:10.1002/bdrc.21035

Reich D, Nalls MA, Kao WHL, Akylbekova EL, Tandon A, Patterson N, Mullikin J, Hsueh W-C, Cheng C-Y, Coresh J, et al. 2009. Reduced neutrophil count in people of African descent is due to a regulatory variant in the Duffy antigen receptor for chemokines gene. *PLoS Genet* **5**: e1000360. doi:0.1371/journal.pgen.1000360

Reinhardt C, Travis AS. 2000. *Heinrich Caro and the creation of modern chemical industry.* Springer, Dordrecht, Netherlands. doi:10.1007/978-94-015-9353-3

Reiss D, Zhang Y, Rouhi A, Reuter M, Mager DL. 2010. Variable DNA methylation of transposable elements: the case study of mouse early transposons. *Epigenetics* **5**: 68–79. doi:10.4161/epi.5.1.10631

Ren C, Liu F, Ouyang Z, An G, Zhao C, Shuai J, Cai S, Bo X, Shu W. 2017. Functional annotation of structural ncRNAs within enhancer RNAs in the human genome: implications for human disease. *Sci Rep* **7**: 15518. doi:0.1038/s41598-017-15822-7

Ren B, Robert F, Wyrick JJ, Aparicio O, Jennings EG, Simon I, Zeitlinger J, Schreiber J, Hannett N, Kanin E, et al. 2000. Genome-wide location and function of DNA binding proteins. *Science* **290**: 2306–2309. doi:10.1126/science.290.5500.2306

Rentzsch P, Witten D, Cooper GM, Shendure J, Kircher M. 2019. CADD: predicting the deleteriousness of variants throughout the human genome. *Nucl Acids Res* **47**: D886–D894. doi:10.1093/nar/gky1016

Richmond RC, Simpkin AJ, Woodward G, Gaunt TR, Lyttleton O, McArdle WL, Ring SM, Smith ADAC, Timpson NJ, Tilling K, et al. 2015. Prenatal exposure to maternal smoking and offspring DNA methylation across the lifecourse: findings from the Avon Longitudinal Study of Parents and Children (ALSPAC). *Hum Mol Genet* **24**: 2201–2217. doi:10.1093/hmg/ddu739

Richmond RC, Sharp GC, Ward ME, Fraser A, Lyttleton O, McArdle WL, Ring SM, Gauny TR, Lawlor DA, Davey Smith G, Relton CL. 2016a. DNA methylation and

BMI: investigating identified methylation sites at *HIF3A* in a causal framework. *Diabetes* **65**: 1231–1244. doi:10.2337/db15-0996

Richmond RC, Hemani G, Tilling K, Davey Smith G, Relton CL. 2016b. Challenges and novel approaches for investigating molecular mediation. *Hum Mol Genet* **25**: R149–R156. doi:10.1093/hmg/ddw197

Riggs AD. 1975. X inactivation, differentiation, and DNA methylation. *Cytogenet Cell Genet* **14**: 9–25.

Riggs AD. 2002. X chromosome inactivation, differentiation, and DNA methylation revisited, with a tribute to Susumu Ohno. *Cytogenet Genome Res* **99**: 17–24. doi:10.1159/000071569

Riggs AD, Bourgeois S. 1968. On the assay, isolation and characterization of the *lac* repressor. *J Mol Biol* **34**: 361–364. doi:10.1016/0022-2836(68)90260-X

Riggs AD, Bourgeois S, Newby RF, Cohn M. 1968. DNA binding of the *lac* repressor. *J Mol Biol* **34**: 365–368.

Riising E M, Comet I, Leblanc B, Wu X, Johansen J V, Helin K. 2014. Gene silencing triggers polycomb repressive complex 2 recruitment to CpG islands genome wide. *Mol Cell* **55**: 347–360. doi:10.1016/j.molcel.2014.06.005

Rinaldi L, Datta D, Serrat J, Morey L, Solanas G, Avgustinova A, Blanco E, Pons JI, Matallanas D, Von Kriegsheim A, et al. 2016. Dnmt3a and Dnmt3b associate with enhancers to regulate human epidermal stem cell homeostasis. *Cell Stem Cell* **19**: 491–501. doi:10.1016/j.stem.2016.06.020

Rius R, Cowley MJ, Riley L, Puttick C, Thorburn DR, Christodoulou J. 2019. Biparental inheritance of mitochondrial DNA in humans is not a common phenomenon. *Genet Med* **21**: 2823–2826. doi:10.1038/s41436-019-0568-0

Robertson A. 1977. Conrad Hal Waddington, 8 November 1905–26 September 1975. *Biogr Mem Fellows R Soc* **23**: 575–622. doi:10.1098/rsbm.1977.0022

Robinson WP, Price EM. 2015. The human placental methylome. *Cold Spring Harb Perspect Med* **5**: a023044. doi:10.1101/cshperspect.a023044

Rodgers AB, Morgan CP, Bronson SL, Revello S, Bale TL. 2013. Paternal stress exposure alters sperm microRNA content and reprograms offspring HPA stress axis regulation. *J Neurosci* **33**: 9003–9012. doi:10.1523/JNEUROSCI.0914-13.2013

Rodriguez J, Frigola J, Vendrell E, Risques R-A, Fraga MF, Morales C, Moreno V, Esteller M, Capellà G, Ribas M, Peinado MA. 2006. Chromosomal instability correlates with genome-wide DNA demethylation in human primary colorectal cancers. *Cancer Res* **66**: 8462–9468. doi:10.1158/0008-5472.CAN-06-0293

Roederer M, Quaye L, Mangino M, Beddall MH, Mahnke Y, Chattopadhyay P, Tosi I, Napolitano L, Terranova Barberio M, Menni C, et al. 2015. The genetic architecture of the human immune system: a bioresource for autoimmunity and disease pathogenesis. *Cell* **161**: 387–403. doi:10.1016/j.cell.2015.02.046

Romanelli V, Meneses HNM, Fernández L, Martínez-Glez V, Gracia-Bouthelier R, Fraga M, Guillén E, Nevado J, Gean E, Martorell L, et al. 2011. Beckwith–Wiedemann syndrome and uniparental disomy 11p: fine mapping of the recombination breakpoints and evaluation of several techniques. *Eur J Human Genet* **19**: 416–421. doi:10.1038/ejhg.2010.236

Romanov GA, Vanyushin BF. 1981. Methylation of reiterated sequences in mammalian DNAs: effects of the tissue type, age, malignancy and hormonal induction. *Biochim Biophys Acta Nucl Acids Protein Synth* **653**: 204–218. doi:0.1016/0005-2787(81)90156-8

Rossi MJ, Lai WKM, Pugh BF. 2018. Simplified ChIP-exo assays. *Nat Commun* **9**: 2842. doi:10.1038/s41467-018-05265-7

Roundtree IA, Evans ME, Pan T, He C. 2017. Dynamic RNA modifications in gene expression regulation. *Cell* **169**: 1187–1200. doi:10.1016/j.cell.2017.05.045

Russell LB, Bangham JW. 1961. Variegated-type position effects in the mouse. *Genetics* **46**: 509–525.

Russo VEA, Martienssen RA, Riggs AD (eds). 1996. *Epigenetic mechanisms of gene regulation* (illustrated ed.). Cold Spring Harbor Laboratory Press, Cold Spring Harbor, NY.

Rustom A, Saffrich R, Markovic I, Walther P, Gerdes H-H. 2004. Nanotubular highways for intercellular organelle transport. *Science* **303**: 1007–1010. doi:10.1126/science.1093133

Ruvinsky AO, Agulnik AI. 1990. Gametic imprinting and the manifestation of the fused gene in the house mouse. *Dev Genet* **11**: 263–269. doi:0.1002/dvg.1020110404

Sadikovic B, Levy MA, Kerkhof J, Aref-Eshghi E, Schenkel L, Stuart A, McConkey H, Henneman P, Venema A, Schwartz CE, et al. 2021. Clinical epigenomics: genome-wide DNA methylation analysis for the diagnosis of Mendelian disorders. *Genet Med* **23**: 1065–1074. doi:10.1038/s41436-020-01096-4

Sager R. 1986. Genetic suppression of tumor formation: a new frontier in cancer research. *Cancer Res* **46**(4 Pt 1): 1573–1580.

Sager R, Sheng S, Pemberton P, Hendrix MJ. 1996. Maspin: a tumor suppressing serpin. *Curr Topics Microbiol Immunol* **213**(Pt 1): 51–64. doi:10.1007/978-3-642-61107-0_4

Saletore Y, Meyer K, Korlach J, Vilfan ID, Jaffrey S, Mason CE. 2012. The birth of the epitranscriptome: deciphering the function of RNA modifications. *Genome Biol* **13**: 175. doi:10.1186/gb-2012-13-10-175

Sapp J. 1987. *Beyond the gene: cytoplasmic inheritance and the struggle for authority in genetics*. Oxford University Press, Oxford, UK.

Sardina JL, Collombet S, Tian TV, Gómez A, Di Stefano B, Berenguer C, Brumbaugh J, Stadhouders R, Segura-Morales C, Gut M, et al. 2018. Transcription factors drive Tet2-mediated enhancer demethylation to reprogram cell fate. *Cell Stem Cell* **23**: 727–741.e9. doi:10.1016/j.stem.2018.08.016

Sasse SK, Gerber AN. 2015. Feed-forward transcriptional programming by nuclear receptors: regulatory principles and therapeutic implications. *Pharmacol Therapeut* **145**: 85–91. doi:10.1016/j.pharmthera.2014.08.004

Sato H, Wu B, Delahaye F, Singer RH, Greally J. 2018. Re-targeting of macroH2A following mitosis to cytogenetic-scale heterochromatic domains. bioRxiv doi:10.1101/333468

Sato H, Wu B, Delahaye F, Singer RH, Greally J. M. 2019. Retargeting of macroH2A following mitosis to cytogenetic-scale heterochromatic domains. *J Cell Biol* **218**: 1810–1823. doi:10.1083/jcb.201811109

Sawan C, Herceg Z. 2010. Histone modifications and cancer. *Adv Genet* **70:** 57–85. doi:10.1016/B978-0-12-380866-0.60003-4

Scarano E. 1971. The control of gene function in cell differentiation and in embryogenesis. *Adv Cytopharmacol* **1:** 13–24.

Schadt EE, Monks SA, Drake TA, Lusis AJ, Che N, Colinayo V, Ruff TG, Milligan SB, Lamb JR, Cavet G, et al. 2003. Genetics of gene expression surveyed in maize, mouse and man. *Nature* **422:** 297–302. doi:10.1038/nature01434

Schaedler R. 2006. René Dubos, friend of the good earth: microbiologist, medical scientist, environmentalist. *Emerg Infect Dis* **12:** 876–877. doi:0.3201/eid1205.060354

Schardin M, Cremer T, Hager HD, Lang M. 1985. Specific staining of human chromosomes in Chinese hamster x man hybrid cell lines demonstrates interphase chromosome territories. *Human Genet* **71:** 281–287. doi:10.1007/BF00388452

Schenk T, Chen WC, Göllner S, Howel L, Jin L, Hebestreit K, Klein H-U, Popescu AC, Burnett A, Mills K, et al. 2012. Inhibition of the LSD1 (KDM1A) demethylase reactivates the all-*trans*-retinoic acid differentiation pathway in acute myeloid leukemia. *Nat Med* **18:** 605–611. doi:10.1038/nm.2661

Scheuerlein H, Henschke F, Köckerling F. 2017. Wilhelm von Waldeyer-Hartz—a great forefather: his contributions to anatomy with particular attention to "his" fascia. *Front Surg* **4:** 74. doi:10.3389/fsurg.2017.00074

Schmid M, Durussel T, Laemmli UK. 2004. ChIC and ChEC; genomic mapping of chromatin proteins. *Mol Cell* **16:** 147–157. doi:10.1016/j.molcel.2004.09.007

Schmidt-Küntzel A, Nelson G, David VA, Schäffer AA, Eizirik E, Roelke ME, Kehler JS, Hannah SS, O'Brien SJ, Menotti-Raymond M. 2009. A domestic cat X chromosome linkage map and the sex-linked orange locus: mapping of orange, multiple origins and epistasis over nonagouti. *Genetics* **181:** 1415–1425. doi:10.1534/genetics.108.095240

Schneider S, Kaufmann W, Buesen R, van Ravenzwaay B. 2008. Vinclozolin—the lack of a transgenerational effect after oral maternal exposure during organogenesis. *Reproduct Toxicol* **25:** 352–360. doi:10.1016/j.reprotox.2008.04.001

Schultz MD, He Y, Whitaker JW, Hariharan M, Mukamel EA, Leung D, Rajagopal N, Nery JR, Urich MA, Chen H, et al. 2015. Human body epigenome maps reveal noncanonical DNA methylation variation. *Nature* **523:** 212–216. doi:10.1038/nature14465

Schumacher A, Kapranov P, Kaminsky Z, Flanagan J, Assadzadeh A, Yau P, Virtanen C, Winegard N, Cheng J, Gingeras T, Petronis A. 2006. Microarray-based DNA methylation profiling: technology and applications. *Nucl Acids Res* **34:** 528–542. doi:10.1093/nar/gkj461

Schutsky EK, Nabel CS, Davis AKF, DeNizio JE, Kohli RM. 2017. APOBEC3A efficiently deaminates methylated, but not TET-oxidized, cytosine bases in DNA. *Nucl Acids Res* **45:** 7655–7665. doi:10.1093/nar/gkx345

Schwartz S, Meshorer E, Ast G. 2009. Chromatin organization marks exon-intron structure. *Nat Struct Mol Biol* **16:** 990–995. doi:10.1038/nsmb.1659

Scott-Browne JP, López-Moyado IF, Trifari S, Wong V, Chavez L, Rao A, Pereira RM. 2016. Dynamic changes in chromatin accessibility occur in CD8+ T cells responding to viral infection. *Immunity* **45:** 1327–1340. doi:10.1016/j.immuni.2016.10.028

Sdelci S, Rendeiro AF, Rathert P, You W, Lin J-MG, Ringler A, Hofstätter G, Moll HP, Gürtl B, Farlik M, et al. 2019. MTHFD1 interaction with BRD4 links folate metabolism to transcriptional regulation. *Nat Genet* **51**: 990–998. doi:10.1038/s41588-019-0413-z

Sendžikaitė G, Kelsey G. 2019. The role and mechanisms of DNA methylation in the oocyte. *Essays Biochem* **63**: 691–705. doi:10.1042/EBC20190043

Sentürk Cetin N, Kuo C-C, Ribarska T, Li R, Costa IG, Grummt I. 2019. Isolation and genome-wide characterization of cellular DNA:RNA triplex structures. *Nucl Acids Res* **47**: 2306–2321. doi:10.1093/nar/gky1305

Serag A, Ion-Margineanu A, Qureshi H, McMillan R, Saint Martin M-J, Diamond J, O'Reilly P, Hamilton P. 2019. Translational AI and deep learning in diagnostic pathology. *Front Med* **6**: 185. doi:10.3389/fmed.2019.00185

Serra RW, Fang M, Park SM, Hutchinson L, Green MR. 2014. A KRAS-directed transcriptional silencing pathway that mediates the CpG island methylator phenotype. *eLife* **3**: e02313. doi:10.7554/eLife.02313

Sever R, Glass CK. 2013. Signaling by nuclear receptors. *Cold Spring Harb Perspect Biol* **5**: a016709. doi:10.1101/cshperspect.a016709

Shariati SA, Dominguez A, Xie S, Wernig M, Qi LS, Skotheim JM. 2019. Reversible disruption of specific transcription factor–DNA interactions using CRISPR/Cas9. *Mol Cell* **74**: 622–633.e4. doi:10.1016/j.molcel.2019.04.011

Sharma AB, Dimitrov S, Hamiche A, Van Dyck E. 2019. Centromeric and ectopic assembly of CENP-A chromatin in health and cancer: old marks and new tracks. *Nucl Acids Res* **47**: 1051–1069. doi:10.1093/nar/gky1298

Sharp GC, Lawlor DA, Richardson SS. 2018. It's the mother! How assumptions about the causal primacy of maternal effects influence research on the developmental origins of health and disease. *Soc Sci Med* **213**: 20–27. doi:10.1016/j.socscimed.2018.07.035

Sharp GC, Schellhas L, Richardson SS, Lawlor DA. 2019. Time to cut the cord: recognizing and addressing the imbalance of DOHaD research towards the study of maternal pregnancy exposures. *J Devl Origins Health Dis* **10**: 509–512. doi:10.1017/S2040174419000072

Shen H, Laird PW. 2013. Interplay between the cancer genome and epigenome. *Cell* **153**: 38–55. doi:10.1016/j.cell.2013.03.008

Shen SY, Singhania R, Fehringer G, Chakravarthy A, Roehrl MHA, Chadwick D, Zuzarte PC, Borgida A, Wang TT, Li T, et al. 2018. Sensitive tumour detection and classification using plasma cell-free DNA methylomes. *Nature* **563**: 579–583. doi:10.1038/s41586-018-0703-0

Shen N, Wang T, Li D, Zhu Y, Xie H, Lu Y. 2019. Hypermethylation of the *SEPT9* gene suggests significantly poor prognosis in cancer patients: a systematic review and meta-analysis. *Front Genet* **10**: 887. doi:10.3389/fgene.2019.00887

Shenker NS, Polidoro S, van Veldhoven K, Sacerdote C, Ricceri F, Birrell MA, Belvis MG, Brown R, Vineis P, Flanagan JM. 2013. Epigenome-wide association study in the European Prospective Investigation into Cancer and Nutrition (EPIC-Turin) identifies novel genetic loci associated with smoking. *Human Mol Genet* **22**: 843–851. doi:10.1093/hmg/dds488

Shen-Orr SS, Tibshirani R, Khatri P, Bodian DL, Staedtler F, Perry NM, Hastie T, Sarwal MM, Davis MM, Butte AJ. 2010. Cell type–specific gene expression differences in complex tissues. *Nat Meth* 7: 287–289. doi:10.1038/nmeth.1439

Sheppard PAS, Choleris E, Galea LAM. 2019. Structural plasticity of the hippocampus in response to estrogens in female rodents. *Mol Brain* 12: 22. doi:0.1186/s13041-019-0442-7

Sherwood RI, Hashimoto T, O'Donnell CW, Lewis S, Barkal AA, van Hoff JP, Karun V, Jaakkola T, Gifford DK. 2014. Discovery of directional and nondirectional pioneer transcription factors by modeling DNase profile magnitude and shape. *Nat Biotechnol* 32: 171–178. doi:10.1038/nbt.2798

Shi J, Teschendorff AE, Chen W, Chen L, Li T. 2020. Quantifying Waddington's epigenetic landscape: a comparison of single-cell potency measures. *Brief Bioinformat* 21: 248–261. doi:10.1093/bib/bby093

Shim E-H, Liv CB, Rakheja D, Tan J, Benson D, Parekh V, Kho E-Y, Ghosh AP, Kirkman R, Velu S, et al. 2014. L-2-Hydroxyglutarate: an epigenetic modifier and putative oncometabolite in renal cancer. *Cancer Discov* 4: 1290–1298. doi:10.1158/2159-8290.CD-13-0696

Shin T, Kraemer D, Pryor J, Liu L, Rugila J, Howe L, Buck S, Murphy K, Lyons L, Westhusin M. 2002. A cat cloned by nuclear transplantation. *Nature* 415: 859. doi:10.1038/nature723

Shorstova T, Foulkes WD, Witcher M. 2021. Achieving clinical success with BET inhibitors as anti-cancer agents. *Br J Cancer* 124: 1478–1490. doi:10.1038/s41416-021-01321-0

Sibbritt T, Patel HR, Preiss T. 2013. Mapping and significance of the mRNA methylome. *Wiley Interdisc Rev* 4: 397–422. doi:10.1002/wrna.1166

Simmons RA, Templeton LJ, Gertz SJ. 2001. Intrauterine growth retardation leads to the development of type 2 diabetes in the rat. *Diabetes* 50: 2279–2286. doi:10.2337/diabetes.50.10.2279

Skene PJ, Henikoff S. 2017. An efficient targeted nuclease strategy for high-resolution mapping of DNA binding sites. *eLife* 6: 21856. doi:10.7554/eLife.21856

Skinner MK, Anway MD, Savenkova MI, Gore AC, Crews D. 2008. Transgenerational epigenetic programming of the brain transcriptome and anxiety behavior. *PLoS One* 3: e3745. doi:10.1371/journal.pone.0003745

Slaughter DP, Southwick HW, Smejkal W. 1953. Field cancerization in oral stratified squamous epithelium; clinical implications of multicentric origin. *Cancer* 6: 963–968.

Smemo S, Tena JJ, Kim K-H, Gamazon ER, Sakabe NJ, Gómez-Marín C, Aneas I, Credidio FL, Sobreira DR, Wasserman NF, et al. 2014. Obesity-associated variants within in *FTO* form long-range functional connections with *IRX3*. *Nature* 507: 371–375. doi:10.1038/nature13138

Smith CA. 1947a. Effects of maternal under nutrition upon the newborn infant in Holland (1944–1945). *J Pediatr* 30: 229–243. doi:10.1016/s0022-3476(47)80158-1

Smith CA. 1947b. The effect of wartime starvation in Holland upon pregnancy and its product. *Am J Obstetr Gynecol* 53: 599–608. doi:10.1016/0002-9378(47)90277-9

Smith CA. 1976. Famine and human development: The Dutch Hunger Winter 1944–1945. *Arch Pediatr Adolesc Med* **130**: 222. doi:10.1001/archpedi.1976.02120030112028

Smith JM. 1987. When learning guides evolution. *Nature* **329**: 761–762. doi:10.1038/329761a0

Smith GD, Lawlor DA, Harbord R, Timpson N, Day I, Ebrahim S. 2007. Clustered environments and randomized genes: a fundamental distinction between conventional and genetic epidemiology. *PLoS Med* **4**: e352. doi:10.1371/journal.pmed.0040352

Snell GD. 1946. An analysis of translocations in the mouse. *Genetics* **31**: 157–180.

Snyder MW, Kircher M, Hill AJ, Daza RM, Shendure J. 2016. Cell-free DNA comprises an in vivo nucleosome footprint that informs its tissues-of-origin. *Cell* **164**: 57–68. doi:10.1016/j.cell.2015.11.050

Solomon MJ, Varshavsky A. 1985. Formaldehyde-mediated DNA–protein crosslinking: a probe for in vivo chromatin structures. *Proc Natl Acad Sci* **82**: 6470–6474. doi:10.1073/pnas.82.19.6470

Solter D. 2008. In Memoriam: Salome Glucksohn-Waelsch (1907–2007). *Dev Cell* **14**: 22–24. doi:10.1016/j.devcel.2007.12.018

Solter D. 2015. A conversation with Davor Solter. *Cold Spring Harb Symp Quant Biol* **80**: 346–347. doi:10.1101/sqb.2015.80.030130

Sosa MS, Avivar-Valderas A, Bragado P, Wen H-C, Aguirre-Ghiso JA. 2011. ERK1/2 and p38α/β signaling in tumor cell quiescence: opportunities to control dormant residual disease. *Clin Cancer Res* **17**: 5850–5857. doi:10.1158/1078-0432.CCR-10-2574

Sosa MS, Parikh F, Maia AG, Estrada Y, Bosch A, Bragado P, Ekpin E, George A, Zheng Y, Lam H-M, et al. 2015. NR2F1 controls tumour cell dormancy via SOX9- and RARβ-driven quiescence programmes. *Nat Commun* **6**: 6170. doi:10.1038/ncomms7170

Spemann H, Mangold H. 1924. Über Induktion von Embryonalanlagen Durch Implantation artfrem der Organisatoren. *Archiv Mikros Anat Entwicklungsmechanik* **100**: 599–638. doi:10.1007/BF02108133

Stark R, Grzelak M, Hadfield J. 2019. RNA sequencing: the teenage years. *Nat Rev Genet* **20**: 631–656. doi:10.1038/s41576-019-0150-2

Stäubli A, Peters AH. 2021. Mechanisms of maternal intergenerational epigenetic inheritance. *Curr Opin Genet Dev* **67**: 151–162. doi:10.1016/j.gde.2021.01.008

Stefansson OA, Sigurpalsdottir BD, Rognvaldsson S, Halldorsson GH, Juliusson K, Sveinbjornsson G, Gunnarsson B, Beyter D, Jonsson H, Gudjonsson SA, et al. 2024. The correlation between CpG methylation and gene expression is driven by sequence variants. *Nat Genet* **56**: 1624–1631. doi:10.1038/s41588-024-01851-2

Stein Z, Susser M, Saenger G, Marolla F. 1972. Nutrition and mental performance. *Science* **178**: 708–713. doi:10.1126/science.178.4062.708

Stein ZA, Susser MW, Saenger G, Marolla F. 1975. *Famine and human development: the Dutch Hunger Winter of 1944–1945*. Oxford University Press, Oxford, UK.

Steiner FA, Henikoff S. 2015. Diversity in the organization of centromeric chromatin. *Curr Opin Genet Dev* **31**: 28–35. doi:10.1016/j.gde.2015.03.010

Stephenson W, Razaghi R, Busan S, Weeks KM, Timp W, Smibert P. 2022. Direct detection of RNA modifications and structure using single-molecule nanopore sequencing. *Cell Genomics* 2: doi:10.1016/j.xgen.2022.100097

Stergachis AB, Debo BM, Haugen E, Churchman LS, Stamatoyannopoulos JA. 2020. Single-molecule regulatory architectures captured by chromatin fiber sequencing. *Science* 368: 1449–1454. doi:0.1126/science.aaz1646

Stern CD. 2000. Conrad H. Waddington's contributions to avian and mammalian development, 1930–1940. *Int J Dev Biol* 44: 15–22.

Stoeckius M, Hafemeister C, Stephenson W, Houck-Loomis B, Chattopadhyay PK, Swerdlow H, Satija R, Smibert P. 2017. Simultaneous epitope and transcriptome measurement in single cells. *Nat Meth* 14: 865–868. doi:10.1038/nmeth.4380

Strahl BD, Allis CD. 2000. The language of covalent histone modifications. *Nature* 403: 41–45. doi:10.1038/47412

Struhl K, Segal E. 2013. Determinants of nucleosome positioning. *Nat Struct Mol Biol* 20: 267–273. doi:10.1038/nsmb.2506

Suetake I, Shinozaki F, Miyagawa J, Takeshima H, Tajima S. 2004. DNMT3L stimulates the DNA methylation activity of Dnmt3a and Dnmt3b through a direct interaction. *J Biol Chem* 279: 27816–27823. doi:10.1074/jbc.M400181200

Sun Y, Han J, Wang Z, Li X, Sun Y, Hu Z. 2020. Safety and efficacy of bromodomain and extra-terminal inhibitors for the treatment of hematological malignancies and solid tumors: a systematic study of clinical trials. *Front Pharmacol* 11: 621093. doi:10.3389/fphar.2020.621093

Supernat A, Vidarsson OV, Steen VM, Stokowy T. 2018. Comparison of three variant callers for human whole genome sequencing. *Sci Rep* 8: 17851. doi:10.1038/s41598-018-36177-7

Surani MA, Barton SC, Norris ML. 1984. Development of reconstituted mouse eggs suggests imprinting of the genome during gametogenesis. *Nature* 308: 548–550. doi:10.1038/308548a0

Susser E, St Clair D. 2013. Prenatal famine and adult mental illness: interpreting concordant and discordant results from the Dutch and Chinese Famines. *Soc Sci Med* 97: 325–330. doi:10.1016/j.socscimed.2013.02.049

Suzuki M, Oda M, Ramos M-P, Pascual M, Lau K, Stasiek E, Agyiri F, Thompson RF, Glass JL, Jing Q, et al. 2011. Late-replicating heterochromatin is characterized by decreased cytosine methylation in the human genome. *Genome Res* 21: 1833–1840. doi:10.1101/gr.116509.110

Szankasi P, Ho AK, Bahler DW, Efimova O, Kelley TW. 2011. Combined testing for CCAAT/enhancer-binding protein alpha (CEBPA) mutations and promoter methylation in acute myeloid leukemia demonstrates shared phenotypic features. *Leukemia Res* 35: 200–207. doi:10.1016/j.leukres.2010.09.018

Tahiliani M, Koh KP, Shen Y, Pastor WA, Bandukwala H, Brudno Y, Agarwal S, Iyer LM, Liu DR, Aravind L, Rao A. 2009. Conversion of 5-methylcytosine to 5-hydroxymethylcytosine in mammalian DNA by MLL partner TET1. *Science* 324: 930–935. doi:0.1126/science.1170116

Takahashi N, Coluccio A, Thorball CW, Planet E, Shi H, Offner S, Turelli P, Imbeault M, Ferguson-Smith AC, Trono D. 2019. ZNF445 is a primary regulator of genomic imprinting. *Genes Dev* **33**: 49–54. doi:10.1101/gad.320069.118

Takayama S, Dhahbi J, Roberts A, Mao G, Heo S-J, Pachter L, Martin DIK, Boffelli D. 2014. Genome methylation in *D. melanogaster* is found at specific short motifs and is independent of DNMT2 activity. *Genome Res* **24**: 821–830. doi:10.1101/gr.162412.113

Talbert PB, Henikoff S. 2021. Histone variants at a glance. *J Cell Sci* **134**: jcs244749. doi:10.1242/jcs.244749

Talbert PB, Ahmad K, Almouzni G, Ausió J, Berger F, Bhalla PL, Bonner WM, Cande WZ, Chadwick BP, Chan SWL, et al. 2012. A unified phylogeny-based nomenclature for histone variants. *Epigenet Chromatin* **5**: 7. doi:10.1186/1756-8935-5-7

Tan C, Takada S. 2020. Nucleosome allostery in pioneer transcription factor binding. *Proc Natl Acad of Sci* **117**: 20586–20596. doi:10.1073/pnas.2005500117

Tang F, Barbacioru C, Wang Y, Nordman E, Lee C, Xu N, Wang X, Bodeau J, Tuch BB, Siddiqui A, et al. 2009. mRNA-Seq whole-transcriptome analysis of a single cell. *Nat Meth* **6**: 377–382. doi:10.1038/nmeth.1315

Tang WWC, Dietmann S, Irie N, Leitch HG, Floros VI, Bradshaw CR, Hackett JA, Chinnery PF, Surani MA. 2015. A unique gene regulatory network resets the human germline epigenome for development. *Cell* **161**: 1453–1467. doi:10.1016/j.cell.2015.04.053

Tao Y, Kang B, Petkovich DA, Bhandari YR, In J, Stein-O'Brien G, Kong X, Xie W, Zachos N, Maegawa S, et al. 2019. Aging-like spontaneous epigenetic silencing facilitates Wnt activation, stemness, and $Braf^{V600E}$-induced tumorigenesis. *Cancer Cell* **35**: 315–328. e6. doi:10.1016/j.ccell.2019.01.005

Tehranchi AK, Myrthil M, Martin T, Hie BL, Golan D, Fraser HB. 2016. Pooled ChIP-Seq links variation in transcription factor binding to complex disease risk. *Cell* **165**: 730–741. doi:10.1016/j.cell.2016.03.041

Thomas ML, Marcato P. 2018. Epigenetic modifications as biomarkers of tumor development, therapy response, and recurrence across the cancer care continuum. *Cancers* **10**: 101. doi:10.3390/cancers10040101

Thompson WR. 1957. Influence of prenatal maternal anxiety on emotionality in young rats. *Science* **125**: 698–699. doi:10.1126/science.125.3250.698

Thompson RF, Fazzari MJ, Niu H, Barzilai N, Simmons RA, Greally JM. 2010. Experimental intrauterine growth restriction induces alterations in DNA methylation and gene expression in pancreatic islets of rats. *J Biol Chem* **285**: 15111–15118. doi:10.1074/jbc.M109.095133

Thompson M, Hill BL, Rakocz N, Chiang JN, Sankararaman S, Hofer I, Cannesson M, Zaitlen N, Halperin E. 2022. Methylation risk scores are associated with a collection of phenotypes within electronic health record systems. medRxiv doi:10.1101/2022.02.07.22270047

Tijsseling D, Wijnberger LDE, Derks JB, van Velthoven CTJ, de Vries WB, van Bel F, Nikkels PGJ, Visser GHA. 2012. Effects of antenatal glucocorticoid therapy on hippocampal histology of preterm infants. *PLoS One* **7**: e33369. doi:10.1371/journal.pone.0033369

Tilghman SM. 2014. Twists and turns: a scientific journey. *Ann Rev Cell Dev Biol* **30**: 1–21. doi:10.1146/annurev-cellbio-100913-013512

Timp W, Bravo HC, McDonald OG, Goggins M, Umbricht C, Zeiger M, Feinberg AP, Irizarry RA. 2014. Large hypomethylated blocks as a universal defining epigenetic alteration in human solid tumors. *Genome Med* **6**: 61. doi:10.1186/s13073-014-0061-y

Tobi EW, Goeman JJ, Monajemi R, Gu H, Putter H, Zhang Y, Slieker RC, Stok AP, Thijssen PE, Müller F, et al. 2014. DNA methylation signatures link prenatal famine exposure to growth and metabolism. *Nat Commun* **5**: 5592. doi:10.1038/ncomms6592

Toh H, Au Y, WK, Unoki M, Matsumoto Y, Miki Y, Matsumura Y, Baba Y, Sado T, Nakamura Y, Matsuda M, Sasaki H. 2024. A deletion at the X-linked *ARHGAP36* gene locus is associated with the orange coloration of tortoiseshell and calico cats. bioRxiv doi:10.1101/2024.11.19.624036

Tootle TL, Rebay I. 2005. Post-translational modifications influence transcription factor activity: a view from the ETS superfamily. *Bioessays News Rev Mol Cell Dev Biol* **27**: 285–298. doi:10.1002/bies.20198

Toyota M, Ahuja N, Ohe-Toyota M, Herman JG, Baylin SB, Issa JP. 1999a. CpG island methylator phenotype in colorectal cancer. *Proc Natl Acad Sci* **96**: 8681–8686. doi:10.1073/pnas.96.15.8681

Toyota M, Ho C, Ahuja N, Jair KW, Li Q, Ohe-Toyota M, Baylin SB, Issa JP. 1999b. Identification of differentially methylated sequences in colorectal cancer by methylated CpG island amplification. *Cancer Res* **59**: 2307–2312.

Trojer P. 2022. Targeting BET bromodomains in cancer. *Annual Rev Cancer Biol* **6**: 313–336. doi:10.1146/annurev-cancerbio-070120-103531

Tsankov AM, Gu H, Akopian V, Ziller MJ, Donaghey J, Amit I, Gnirke A, Meissner A. 2015. Transcription factor binding dynamics during human ES cell differentiation. *Nature* **518**: 344–349. doi:10.1038/nature14233

Tsiarli MA, Rudine A, Kendall N, Pratt MO, Krall R, Thiels E, DeFranco DB, Monaghan AP. 2017. Antenatal dexamethasone exposure differentially affects distinct cortical neural progenitor cells and triggers long-term changes in murine cerebral architecture and behavior. *Transl Psychiatry* **7**: e1153. doi:10.1038/tp.2017.65

Tsuchida T, Friedman SL. 2017. Mechanisms of hepatic stellate cell activation. *Nat Rev Gastroenterol Hepatol* **14**: 397–411. doi:10.1038/nrgastro.2017.38

Turcan S, Rohle D, Goenka A, Walsh LA, Fang F, Yilmaz E, Campos C, Fabius AWM, Lu C, Ward PS, et al. 2012. *IDH1* mutation is sufficient to establish the glioma hypermethylator phenotype. *Nature* **483**: 479–483. doi:10.1038/nature10866

Turing AM. 1952. The chemical basis of morphogenesis. *Philos Trans R Soc Lond B Biol Sci* **237**: 37–72. doi:10.1098/rstb.1952.0012

Udvardy A, Maine E, Schedl P. 1985. The 87A7 chromomere. Identification of novel chromatin structures flanking the heat shock locus that may define the boundaries of higher order domains. *J Mol Biol* **185**: 341–358. doi:10.1016/0022-2836(85)90408-5

Ulahannan N, Greally JM. 2015. Genome-wide assays that identify and quantify modified cytosines in human disease studies. *Epigenet Chromatin* **8**: 5. doi:10.1186/1756-8935-8-5

Unruh D, Zewde M, Buss A, Drumm MR, Tran AN, Scholtens DM, Horbinski C. 2019. Methylation and transcription patterns are distinct in *IDH* mutant gliomas compared to other *IDH* mutant cancers. *Sci Rep* **9**: 8946. doi:10.1038/s41598-019-45346-1

Vanyushin BF, Mazin AL, Vasilyev VK, Belozersky AN. 1973. The content of 5-methylcytosine in animal DNA: the species and tissue specificity. *Biochim Biophys Acta* **299**: 397–403. doi:10.1016/0005-2787(73)90264-5

van den Tweel JG, Taylor CR. 2010. A brief history of pathology: preface to a forthcoming series that highlights milestones in the evolution of pathology as a discipline. *Virch Arch Int J Pathol* **457**: 3–10. doi:10.1007/s00428-010-0934-4

van der Meulen J. 2001. Glucose tolerance in adults after prenatal exposure to famine. *Lance* **357**: 1797–1798. doi:10.1016/S0140-6736(00)04907-2

van Steensel B, Belmont AS. 2017. Lamina-associated domains: links with chromosome architecture, heterochromatin, and gene repression. *Cell* **169**: 780–791. doi:10.1016/j.cell.2017.04.022

van Steensel B, Henikoff S. 2003. Epigenomic profiling using microarrays. *Biotechniques* **35**: 346–350, 352. doi:10.2144/03352rv01

Vargon S. 2001. The date of composition of the Book of Job in the context of S.D. Luzzatto's attitude to biblical criticism. *Jewish Q Rev* **91**: 377. doi:10.2307/1455552

Vasicek TJ, Zeng L, Guan XJ, Zhang T, Costantini F, Tilghman SM. 1997. Two dominant mutations in the mouse fused gene are the result of transposon insertions. *Genetics* **147**: 777–786. doi:10.1093/genetics/147.2.777

Vastenhouw NL, Brunschwig K, Okihara KL, Müller F, Tijsterman M, Plasterk RHA. 2006. Gene expression: long-term gene silencing by RNAi. *Nature* **442**: 882. doi:10.1038/442882a

Vasu K, Nagaraja V. 2013. Diverse functions of restriction-modification systems in addition to cellular defense. *Microbiol Mol Biol Rev* **77**: 53–72. doi:10.1128/MMBR.00044-12

Vicente-Dueñas C, Hauer J, Cobaleda C, Borkhardt A, Sánchez-García I. 2018. Epigenetic priming in cancer initiation. *Trends Cancer* **4**: 408–417. doi:10.1016/j.trecan.2018.04.007

Viegas J. 2016. Profile of Peter A. Jones. *Proc Natl Acad Sci* **113**: 13546–13548. doi:10.1073/pnas.1617318113

Vierstra J, Lazar J, Sandstrom R, Halow J, Lee K, Bates D, Diegel M, Dunn D, Ner F, Haugen E, et al. 2020. Global reference mapping and dynamics of human transcription factor footprints. bioRxiv doi:10.1101/2020.01.31.927798

Virgin HW, Wherry EJ, Ahmed R. 2009. Redefining chronic viral infection. *Cell* **138**: 30–50. doi:10.1016/j.cell.2009.06.036

Vogelstein B, Fearon ER, Hamilton SR, Kern SE, Preisinger AC, Leppert M, Nakamura Y, White R, Smits AM, Bos JL. 1988. Genetic alterations during colorectal-tumor development. *New Engl J Med* **319**: 525–532. doi:10.1056/NEJM198809013190901

Von Hoff DD, Slavik M, Muggia FM. 1976. 5-Azacytidine. A new anticancer drug with effectiveness in acute myelogenous leukemia. *Ann Intern Med* **85**: 237–245. doi:10.7326/0003-4819-85-2-237

Waddington CH. 1939. *An introduction to modern genetics*. The MacMillan Company, New York.

Waddington CH. 1940. *Organisers and genes*. Cambridge University Press, Cambridge, UK.

Waddington CH. 1942. Canalization of development and the inheritance of acquired characters. *Nature* **150**: 563–565. doi:10.1038/150563a0

Waddington CH. 1953. Genetic assimilation of an acquired character. *Evolution* **7**: 118–126. doi:10.1111/j.1558-5646.1953.tb00070.x

Waddington CH. 1956a. *Principles of embryology*. Macmillan, New York. doi:0.5962/bhl.title.7217

Waddington CH. 1956b. Genetic assimilation of the *bithorax* phenotype. *Evolution* **10**: 1–13. doi:10.1111/j.1558-5646.1956.tb02824.x

Waddington CH. 1957. *The strategy of the genes: a discussion of some aspects of theoretical biology*. Allen and Unwin, London.

Waddington CH. 1959. Canalization of development and genetic assimilation of acquired characters. *Nature* **183**: 1654–1655. doi:10.1038/1831654a0

Waddington CH. 1962. *New patterns in genetics & development*, Vol. 21. Columbia University Press, New York.

Waddington CH. 1968. *Towards a theoretical biology: an IUBS symposium*. Edinburgh UP, Edinburgh.

Wadhwa PD, Buss C, Entringer S, Swanson JM. 2009. Developmental origins of health and disease: brief history of the approach and current focus on epigenetic mechanisms. *Sem Reproduct Med* **27**: 358–368. doi:10.1055/s-0029-1237424

Wahl S, Drong A, Lehne B, Loh M, Scott WR, Kunze S, Tsai P-C, Ried JS, Zhang W, Yang Y, et al. 2017. Epigenome-wide association study of body mass index, and the adverse outcomes of adiposity. *Nature* **541**: 81–86. doi:10.1038/nature20784

Walker MS, Becker FF. 1981. DNA methylase activity of normal liver, regenerating liver, and a transplantable hepatocellular carcinoma. *Cancer Biochem Biophys* **5**: 169–173.

Waller J. 2004. *Leaps in the dark: the making of scientific reputations*. Oxford University Press, Oxford, UK.

Walter S, Mejía-Guevara I, Estrad K, Liu SY, Glymour MM. 2016. Association of a genetic risk score with body mass index across different birth cohorts. *J Am Med Assoc* **316**: 63–69. doi:10.1001/jama.2016.8729

Wang Z-Y, Chen Z. 2008. Acute promyelocytic leukemia: from highly fatal to highly curable. *Blood* **111**: 2505–2515. doi:10.1182/blood-2007-07-102798

Wang X, Yamaguchi N. 2024. Cause or effect: probing the roles of epigenetics in plant development and environmental responses. *Curr Opin Plant Biol* **81**: 102569. doi:10.1016/j.pbi.2024.102569

Wang Y, Li Y, Toth JI, Petroski MD, Zhang Z, Zhao JC. 2014. N^6-methyladenosine modification destabilizes developmental regulators in embryonic stem cells. *Nat Cell Biol* **16**: 191–198. doi:10.1038/ncb2902

Warrington NM, Beaumont RN, Horikoshi M, Day FR, Helgeland Ø, Laurin C, Bacelis J, Peng S, Hao K, Feenstra B, et al. 2019. Maternal and fetal genetic effects on birth weight and their relevance to cardio-metabolic risk factors. *Nat Genet* **51**: 804–814. doi:10.1038/s41588-019-0403-1

Wartlick O, Kicheva A, González-Gaitán M. 2009. Morphogen gradient formation. *Cold Spring Harb Perspect Biol* **1**: a001255. doi:10.1101/cshperspect.a001255

Wassenegger M, Heimes S, Riedel L, Sänger HL. 1994. RNA-directed de novo methylation of genomic sequences in plants. *Cel* **76**: 567–576. doi:10.1016/0092-8674(94)90119-8

Watanabe T, Tomizawa S, Mitsuya K, Totoki Y, Yamamoto Y, Kuramochi-Miyagawa S, Iida N, Hoki Y, Murphy PJ, Toyoda A, et al. 2011. Role for piRNAs and noncoding RNA in de novo DNA methylation of the imprinted mouse *Rasgrf1* locus. *Science* **332**: 848–852. doi:10.1126/science.1203919

Waterland RA, Dolinoy DC, Lin J-R, Smith CA, Shi X, Tahiliani KG. 2006. Maternal methyl supplements increase offspring DNA methylation at Axin Fused. *Genesis* **44**: 401–406. doi:10.1002/dvg.20230

Wattacheril JJ, Raj S, Knowles DA, Greally JM. 2023. Using epigenomics to understand cellular responses to environmental influences in diseases. *PLoS Genet* **19**: e1010567. doi:10.1371/journal.pgen.1010567

Weaver ICG, Cervoni N, Champagne FA, D'Alessio AC, Sharma S, Seckl JR, Dymov S, Szyf M, Meaney MJ. 2004. Epigenetic programming by maternal behavior. *Nat Neurosci* **7**: 847–854. doi:10.1038/nn1276

Weber M, Davies JJ, Wittig D, Oakeley EJ, Haase M, Lam WL, Schübeler D. 2005. Chromosome-wide and promoter-specific analyses identify sites of differential DNA methylation in normal and transformed human cells. *Nat Genet* **37**: 853–862. doi:10.1038/ng1598

Webster G, Berul CI. 2013. An update on channelopathies: from mechanisms to management. *Circulation* **127**: 126–140. doi:10.1161/CIRCULATIONAHA.111.060343

Weidemüller P, Kholmatov M, Petsalaki E, Zaugg JB. 2021. Transcription factors: bridge between cell signaling and gene regulation. *Proteomics* **21**: e2000034. doi:10.1002/pmic.202000034

Weindling P. 2012. Julian Huxley and the continuity of eugenics in twentieth-century Britain. *J Modern Eur Hist* **10**: 480–499. doi:10.17104/1611-8944_2012_4

Weintraub H, Flint SJ, Leffa IM, Groudine M, Grainger RM. 1978. The generation and propagation of variegated chromosome structures. *Cold Spring Harb Symp Quant Biol* **42**: 401–407. doi:10.1101/SQB.1978.042.01.042

Weisenberger DJ. 2014. Characterizing DNA methylation alterations from The Cancer Genome Atlas. *J Clin Invest* **124**: 17–23. doi:10.1172/JCI69740

Weisenberger DJ, Siegmund KD, Campan M, Young J, Long T, Faasse MA, Kang GH, Widschwendter M, Weener D, Buchanan D, et al. 2006. CpG island methylator phenotype underlies sporadic microsatellite instability and is tightly associated with *BRAF* mutation in colorectal cancer. *Nat Genet* **38**: 787–793. doi:10.1038/ng1834

White MK, Pagano JS, Khalili K. 2014. Viruses and human cancers: a long road of discovery of molecular paradigms. *Clin Microbiol Rev* **27**: 463–481. doi:10.1128/CMR.00124-13

Whitelaw NC, Chong S, Morgan DK, Nestor C, Bruxner TJ, Ashe A, Lambley E, Meehan R, Whitelaw E. 2010. Reduced levels of two modifiers of epigenetic gene silencing, Dnmt3a and Trim28, cause increased phenotypic noise. *Genome Biol* **11**; R111. doi:10.1186/gb-2010-11-11-r111

Whitney WD, Smith BE (eds). 1914. *The Century dictionary: an encyclopedic lexicon of the English language*, 17th ed., Vols. 1–10. The Century Company, New York.

Widdowson EM, McCance RA. 1960. Some effects of accelerating growth. I. General somatic development. *Proc R Soc London B Biol Sci* **152**: 188–206. doi:10.1098/rspb.1960.0032

Wijetunga NA, Delahaye F, Zhao YM, Golden A, Mar JC, Einstein FH, Greally JM. 2014. The meta-epigenomic structure of purified human stem cell populations is defined at *cis*-regulatory sequences. *Nat Commun* **5**: 5195. doi:10.1038/ncomms6195

Wijetunga NA, Pascual M, Tozour J, Delahaye F, Alani M, Adeyeye M, Wolkoff AW, Verma A, Greally JM. 2017. A pre-neoplastic epigenetic field defect in HCV-infected liver at transcription factor binding sites and Polycomb targets. *Oncogene* **36**: 2030–2044. doi:10.1038/onc.2016.340

Wilcox AJ. 2003. A conversation with Zena Stein. *Epidemiology* **14**: 498–501. doi:10.1097/01.ede.0000071471.35756.96

Willard HF. 2010. 2009 William Allan Award address: Life in the sandbox: unfinished business. *Am J Human Genet* **86**: 318–327. doi:10.1016/j.ajhg.2010.01.037

Williams CAC, Soufi A, Pollard SM. 2020. Post-translational modification of SOX family proteins: key biochemical targets in cancer? *Sem Cancer Biol* **67**: 30–38. doi:10.1016/j.semcancer.2019.09.009

Williamson I, Berlivet S, Eskeland R, Boyle S, Illingworth RS, Paquette D, Dostie J, Bickmore WA. 2014. Spatial genome organization: contrasting views from chromosome conformation capture and fluorescence in situ hybridization. *Genes Dev* **28**: 2778–2791. doi:10.1101/gad.251694.114

Willis RA. 1934. *The spread of tumours in the human body*. J & A Churchill, London.

Wolff GL. 1965. Body composition and coat color correlation in different phenotypes of "viable yellow" mice. *Science* **147**: 1145–1147.

Wolff GL, Kodell RL, Moore SR, Cooney CA. 1998. Maternal epigenetics and methyl supplements affect agouti gene expression in Avy/a mice. *FASEB J* **12**: 949–957.

Wu J, Zhang R, Shen F, Yang R, Zhou D, Cao H, Chen G, Pan Q, Fan J. 2018. Altered DNA methylation sites in peripheral blood leukocytes from patients with simple steatosis and nonalcoholic steatohepatitis (NASH). *Med Sci Monitor* **24**: 6946–6967. doi:10.12659/MSM.909747

Wutz A, Rasmussen TP, Jaenisch R. 2002. Chromosomal silencing and localization are mediated by different domains of *Xist* RNA. *Nat Genet* **30**: 167–174. doi:10.1038/ng820

Wynder EL. 1988. Tobacco and health: a review of the history and suggestions for public health policy. *Publ Health Rep* **103**: 8–18.

Xia B, de Belle JS. 2018. Non-genetic transgenerational inheritance of acquired traits in *Drosophila*. In *Drosophila melanogaster—model for recent advances in genetics and therapeutics* (Perveen FK. ed). InTech, London. doi:10.5772/intechopen.71643

Xiao C-L, Zhu S, He M, Chen D, Zhang Q, Chen Y, Yu G, Liu J, Xie S-Q, Luo F, et al. 2018. N^6-methyladenine DNA modification in the human genome. *Mol Cell* 71: 306–318. e7. doi:10.1016/j.molcel.2018.06.015

Xu W, Yang H, Liu Y, Yang Y, Wang P, Kim S.-H, Ito S, Yang C, Wang P, Xiao M-T, et al. 2011. Oncometabolite 2-hydroxyglutarate is a competitive inhibitor of α-ketoglutarate-dependent dioxygenases. *Cancer Cell* 19: 17–30. doi:10.1016/j.ccr.2010.12.014

Yamamoto T, Kyo M, Kamiya T, Tanaka T, Engel JD, Motohashi H, Yamamoto M. 2006. Predictive base substitution rules that determine the binding and transcriptional specificity of Maf recognition elements. *Genes Cells* 11: 575–591. doi:10.1111/j.1365-2443.2006.00965.x

Yang H, Ye D, Guan K-L, Xiong Y. 2012. *IDH1* and *IDH2* mutations in tumorigenesis: mechanistic insights and clinical perspectives. *Clin Cancer Res* 18: 5562–5571. doi:10.1158/1078-0432.CCR-12-1773

Yang S-M, Kim BJ, Norwood Toro L, Skoultchi AI. 2013. H1 linker histone promotes epigenetic silencing by regulating both DNA methylation and histone H3 methylation. *Proc Natl Acad Sci* 110: 1708–1713. doi:10.1073/pnas.1213266110

Yang HW, Chung M, Kudo T, Meyer T. 2017. Competing memories of mitogen and p53 signalling control cell-cycle entry. *Nature* 549: 404–408. doi:10.1038/nature23880

Yehuda R, Daskalakis NP, Bierer LM, Bader HN, Klengel T, Holsboer F, Binder EB. 2016. Holocaust exposure induced intergenerational effects on FKBP5 methylation. *Biol Psychiatry* 80: 372–380. doi:10.1016/j.biopsych.2015.08.005

Yoder JA, Walsh CP, Bestor TH. 1997. Cytosine methylation and the ecology of intragenomic parasites. *Trends Genet* 13: 335–340. doi:10.1016/s0168-9525(97)01181-5

Yoshida M, Kijima M, Akita M, Beppu T. 1990. Potent and specific inhibition of mammalian histone deacetylase both in vivo and in vitro by trichostatin A. *J Biol Chem* 265: 17174–17179. doi:10.1016/S0021-9258(17)44885-X

Yoshida K, Muratani M, Araki H, Miura F, Suzuki T, Dohmae N, Katou Y, Shirahige K, Ito T, Ishii S. 2018. Mapping of histone-binding sites in histone replacement-completed spermatozoa. *Nat Commun* 9: 3885. doi:10.1038/s41467-018-06243-9

You JS, Jones PA. 2012. Cancer genetics and epigenetics: two sides of the same coin? *Cancer Cell* 22: 9–20. doi:10.1016/j.ccr.2012.06.008

Yuan V, Price EM, Del Gobbo G, Mostafavi S, Cox B, Binder AM, Michels KB, Marsit C, Robinson WP. 2019. Accurate ethnicity prediction from placental DNA methylation data. *Epigenet Chromatin* 12: 51. doi:10.1186/s13072-019-0296-3

Yue Y, Liu J, He C. 2015. RNA N^6-methyladenosine methylation in post-transcriptional gene expression regulation. *Genes Dev* 29: 1343–1355. doi:10.1101/gad.262766.115

Zeineldin M, Neufeld KL. 2013. Understanding phenotypic variation in rodent models with germline *Apc* mutations. *Cancer Res* 73: 2389–2399. doi:10.1158/0008-5472.CAN-12-4607

Zeng H. 2022. What is a cell type and how to define it? *Cell* **185**: 2739–2755. doi:10.1016/j. cell.2022.06.031

Zernicka-Goetz M, Huang S. 2010. Stochasticity versus determinism in development: a false dichotomy? *Nat Rev Genet* **11**: 743–744. doi:10.1038/nrg2886

Zhang T, Cooper S, Brockdorff N. 2015. The interplay of histone modifications—writers that read. *EMBO Rep* **16**: 1467–1481. doi:10.15252/embr.201540945

Zhang R-N, Pan Q, Zheng R-D, Mi Y-Q, Shen F, Zhou D, Chen G-Y, Zhu C-Y, Fan J-G. 2018. Genome-wide analysis of DNA methylation in human peripheral leukocytes identifies potential biomarkers of nonalcoholic fatty liver disease. *Int J Mol Med* **42**: 443–452. doi:10.3892/ijmm.2018.3583

Zhang D, Tang Z, Huang H, Zhou G, Cui C, Weng Y, Liu W, Kim S, Lee S, Perez-Neut M, et al. 2019. Metabolic regulation of gene expression by histone lactylation. *Nature* **574**: 575–580. doi:0.1038/s41586-019-1678-1

Zhang Z, Chen N, Yin N, Liu R, He Y, Li D, Tong M, Gao A, Lu P, Zhao Y, et al. 2023. The rs1421085 variant within *FTO* promotes brown fat thermogenesis. *Nat Metab* **5**: 1337–1351. doi:10.1038/s42255-023-00847-2

Zhao Y, Garcia BA. 2015. Comprehensive catalog of currently documented histone modifications. *Cold Spring Harb Perspect Biol* **7**: a025064. doi:10.1101/cshperspect. a025064

Zhou W, Dinh HQ, Ramjan Z, Weisenberger DJ, Nicolet CM, Shen H, Laird PW, Berman BP. 2018. DNA methylation loss in late-replicating domains is linked to mitotic cell division. *Nat Genet* **50**: 591–602. doi:10.1038/s41588-018-0073-4

Zoghbi HY, Beaudet AL. 2016. Epigenetics and human disease. *Cold Spring Harb Perspect Biol* **8**: a019497. doi:10.1101/cshperspect.a019497

Index

Page references followed by f denote a figure on the corresponding page.